DRUG AND ALCOHOL CONSUMPTION
AS FUNCTIONS OF SOCIAL STRUCTURES

This book has been awarded
The Adèle Mellen Prize
for its distinguished contribution to scholarship.

DRUG AND ALCOHOL CONSUMPTION AS FUNCTIONS OF SOCIAL STRUCTURES

A Cross-Cultural Sociology

James E. Hawdon

Mellen Studies in Sociology
Volume 47

The Edwin Mellen Press
Lewiston•Queenston•Lampeter

Library of Congress Cataloging-in-Publication Data

Hawdon, James.
  Drug and alcohol consumption as functions of social structures : a cross-cultural
sociology / James E. Hawdon.
      p. cm. -- (Mellen studies in sociology ; v. 47)
  Includes bibliographical references and index.
  ISBN 0-7734-6187-6
    1. Drug abuse. 2. Drug abuse--Social aspects. 3. Narcotics, Control of. I. Title. II.
Series.

HV5801.H38 2005
306'.1--dc22

                                                                                      2004065635

This is volume 47 in the continuing series
Mellen Studies in Sociology
Volume 47 ISBN 0-7734-6187-6
MSS Series ISBN 0-88946-630-0

A CIP catalog record for this book is available from the British Library

The Edwin Mellen Press   The Edwin Mellen Press
        Box 450   Box 67
   Lewiston, New York   Queenston, Ontario
    USA 14092-0450   CANADA L0S 1L0

The Edwin Mellen Press, Ltd.
Lampeter, Ceredigion, Wales
UNITED KINGDOM SA48 8LT

Printed in the United States of America

This book is dedicated with much love to my wife, Donna Ann Sedgwick, and our daughter, Gwendolyn Jude Hawdon. May your worlds always be enchanted.

Table of Contents

Foreword   i

Preface by Randolph Atkins, Jr.   xi

Acknowledgments   xvii

Chapter 1

On the Sociology of Drugs   1

    Why do individuals use drugs?   3

    Why are some drugs illegal and why do we care?   5

    On The Sociology of Drugs   7

        Drugs Defined   7

        How Drugs Vary   10

    A Note on Methodology   16

Chapter 2

The Empirical Patterns of Legal Drug Use   21

    Caffeine   22

    Tobacco   28

    Alcohol   34

    Pharmaceutical Drugs   40

    Summary   46

Chapter 3

The Empirical Patterns of Illegal Drug Use   49

    Global Patterns of Illicit Drug Use   50

    Regional Patterns   52

        Africa   53

        Asia and Middle East   62

        Europe   76

        North America   88

        South America, Central America and the Caribbean 96

        Oceania   107

    Summary   111

Chapter 4

Theoretic Framework for a Sociology of Drug Use   115

    The Sacred and the Profane   122

    Rationalization and Modernization   136

        Early Pre-Modern Societies: Culture   137

        Early Pre-Modern Societies: Structure   139

        Pre-Modern Societies: Culture   142

        Pre-Modern Societies: Structure   143

        Advanced Pre-Modern: Culture   144

        Advanced Pre-Modern: Structure   146

        Modern Societies: Culture   147

        Modern Societies: Structure   148

    Some Consequences of Rationalization and Modernization   152

    Rationalization, Modernization and Drug Consumption   158

Summary 163

Chapter 5

The "Why" of Drug Use 165

    The Social Functions of Drug Use 165

        Drugs and Religion 166

        Drugs as a Source of Magic 168

        Drugs to Mark a Rite of Passage 168

        Drugs to Mark Social Boundaries 169

        Drugs to Promote Solidarity 171

        Drugs to Consummate a Social Contract 173

        Drugs to Promote Economic Productivity **174**

        Drugs to Promote Efficient Warfare 176

        Drugs as a Social Lubricant 177

        Drugs for Recreation 179

        Drugs as Medicine 179

    Rationalization and Drug Use 181

    The Religious Sub-functions of Drug Use **185**

    The Irrationalization of Religious Drug Use 192

    Summary 201

Chapter 6

Empirical Tests of a Sociological Theory of Drug Use 203

    What is Used 203

        Variety 203

        Drugs in Combination 207

    How Much is Used 211

        Modernization and Drug Use in the Contemporary World 212

        Modernization and Drug Use Rates over Time 218

        Modernization and Increased Formal Restrictions over Use 224

Illegal Use   231

Who Uses   232

    Ascribed versus Achieved Status and Drug Use   238

Summary: The Contradictions of Modernity   246

Chapter 7

Fluctuations in Rates of Use Over Time   249

    Marginalization of Users and the Rate of Drug Use   249

    Social Mobility and Rates of Drug Use   259

      Empirical Evidence   263

      Social Mobility and Drug Use in the United States   263

      Social Mobility and Drug Use in Britain   269

        Social Mobility: Wave One   271

        Drug Use: Wave Two   275

        Social Mobility: Wave Two   277

      Social Mobility and Drug Use in Japan   285

        A Brief History of Drug Use in Japan   286

        Social Mobility in Japan since World War II   289

        Israel: Social Mobility, Drug Use and the Six-Day War   299

        Drugs and Social Mobility in the Soviet Republics   303

    Conclusion   308

Chapter 8

Applying the Theory   311

    Theories Explaining Drug Use by Individuals   312

      Attitudinal Theories of Drug Use   312

        Self-degradation/Self-esteem Theory   312

        Self-control Theory   313

        Problem-Behavior Proneness Theory   314

      Behavioral Theories of Drug Use   315

       Differential Association/ Social Learning Theory   316

       Summarizing What We Know   318

       Science and Attitudinal Theories of Drug Use   320

       Individualism and Attitudinal Theories of Drug Use   324

Rationalization, Disenchantment and the Drug Subculture   329

       The Subculture as Religion   334

       The Subculture as a Search for Re-enchantment   339

The Evolution of the Subculture: The Rationalization of Religion   342

Summary   357

Postscript
Considering Policy   359

Bibliography   365

Index   395

List of Tables

Table 2.1	Alcohol Consumption by Region	35
Table 2.2	Global Pharmaceutical Sales by Region	43
Table 2.3	Per Capita Sales of Pharmaceuticals	44
Table 3.1	Types of Illegal Drug Used	52
Table 3.2	Annual Use by High School Seniors	95
Table 3.3	Percent Adult Population Using in Australia and New Zealand	110
Table 3.4	Illegal Drug Use by Region	113
Table 4.1	Relative Variations in Use Patterns between Sacred and Profane Drug Use	135
Table 6.1	Percent Reporting Nations Reporting the Use of Specific Drugs by World-System Status	206
Table 6.2	Correlations between Drug Use and Modernization	214
Table 6.3	Correlations between Per Capita GDP and Drugs	215
Table 6.4	Mean Level of Use by World-System Status	217

Foreword

Since the "psychoactive revolution" (Courtwright 2001), people have found a seemingly never-ending supply of substances with which to alter their consciousness. They have found new drugs, new ways to prepare old drugs, new combinations of drugs, and new techniques of ingesting drugs. They have searched the world for cures, tonics and elixirs. They have applied technology to make more potent and efficient drugs. They have used business principles to cultivate, produce and distribute drugs more efficiently. They have created organizations to protect and promote their interests in drugs. Yes, we, as a race of beings, have pursued the ever-better and ever-lasting buzz.

With this interest in drugs, it is not surprising that, each year, hundreds of books, articles, technical reports, newspaper articles and magazine stories are printed that discuss drugs in some way, shape or form. What follows is yet another book about drugs. I hope, however, that it is different than most. The book does not discuss the chemical properties of drugs. The book does not attempt to explain the biology of drug use. The book does not address the aspects of criminal justice that are influenced by the use of drugs. There is no discussion of the drugs–crime connection or the drugs–creativity connection. Drugs used to enhance athletic performance occurs but is ignored in this work. This work is not a guide for the "just say no" campaign, nor is its purpose to advocate the use of drugs by anyone for any reason. To use or not to use is a question of personal choice and values. It cannot be answered empirically. I therefore ignore this and other issues of morality. The book is neither pharmacological nor psychological. It is not policy analysis, advocacy, or moral philosophy. The book is, I hope, theoretical, empirical, and unabashedly sociological.

The purpose of this work is to apply *sociological theory* to the study of drugs. Instead of trying to explain why people use drugs with such variables as self-esteem, self-control, deviant peer relations, or any of the host of usual suspects, I offer an explanation that relates drug use to levels of development and modernization, institutional differentiation, and social stratification. I try to explain variations in rates of use with such concepts as social mobility and heterogeneity. I account for attitudes toward drugs and drug users with sociological concepts such as social and cultural distance. Thus, I intend to offer a sociological theory of drugs.

Saying this is not meant to imply that the tremendous body of work that uses a different perspective or disciplinary slant is not valuable. To do so would be ludicrous. What we have learned about the chemical properties of drugs, the biological effects they have, the psychology of those who use them, and the social relations that promote or deter use is astonishing and extremely valuable. Likewise, the collection of anthropological and historical work that has opened our eyes to the different worlds of drugs, taught us the many ways cultures use drugs, and demonstrated how drugs are woven into a culture's social fabric is not only fascinating but extremely useful. And, from a humanistic perspective, what psychiatrists, psychologists, sociologists and social workers have learned about assisting those who use drugs realize their desire to quit is important and undoubtedly socially beneficial. All of this work has its place and utility.

I do believe that the existing body of work, despite its value and utility, is incomplete. While there are excellent accounts of drug use in different cultures, countries and times, there is not, to the best of my knowledge, any work that attempts to explain these patterns using aggregate-level, sociological variables. That is, there is no "general theory" of drugs that tries to explain the observed cross-cultural and historical patterns of use. Bits and pieces of such a theory are there, to be sure, but no one has tried to tie these together. I do not pretend that this work does this totally or even that I achieved partial success. My aspirations are more modest. I hope to show that sociology has much more to say about drugs than it has and to stimulate greater interest in the line of inquiry I pursue.

The book's first chapter discusses the study of drugs from a sociological perspective. I argue that sociology has a lot to say about drug use; however, to date, it has not been fully utilized to explain patterns of drug consumption. Sociology can explain variations in the extent of use over time and place, reasons for use, methods of use, what drugs are used, and social attitudes toward use and users. Chapter One offers my working definition of drugs and describes how drugs vary. How sociology can address each of the ways in which drugs varies is also described. Finally, I discuss the methodology used in the text. Although cross-cultural and historical data on drug use is often vague and incomplete, we can use the best available data to test our assertions and theories, at least tentatively. I triangulate data from the United Nations, various national governments, intergovernmental organizations, anthropological studies and case studies to build a reasonably valid dataset.

Chapters Two and Three are devoted to the empirical patterns of drug use. Chapter Two discusses the use of legal drugs, including caffeine, alcohol, tobacco and pharmaceutical drugs. In Chapter Three, contemporary patterns of drug use in the modern world are presented. The discussion is based on United Nations and other governmental data, anthropological accounts and ethnographic case studies. The best data available on the use of illicit drugs such as marijuana, cocaine, opiate drugs and amphetamine-type stimulants (ATS). Regional and sub-regional variations in rates of use are discussed in both of these chapters to provide the reader with a sense of the variation in consumption found in the contemporary world. Although it is undoubtedly interesting, the history of these drugs is largely ignored. I make passing note of the history of some drugs simply to emphasize the antiquity of their use. The interested reader should consult any one of the several outstanding texts that discuss the history of drugs. Although there are dozens of these books, one recent and highly readable one is David Courtwright's *Forces of Habit: Drugs and the Making of the Modern World* (Harvard University Press, 2001).

Chapter Four offers the basic theoretic framework that will be used in an attempt to explain contemporary patterns of drug consumption. I note the importance of geography and trade and then discuss why these explanations are inadequate. While the structure of the drug trade influences what drugs are popular in certain regions of the world, the drug trade routes developed where they did for a reason. Namely, drug trade routes typically lead from developing to developed nations because the demand for these products is highest in developed nations. Therefore, arguing that drug consumption is solely a function of the production and distribution system is begging the question. Such arguments fail to recognize that the structure of the international drug distribution system was a function of the process of drug use. We therefore need a more dynamic explanation.

I then present the basis of a general sociological theory of drug consumption. I outline the differences between sacred and profane drug use. By definition, sacred objects are treated with more reverence and are more tightly controlled than profane objects. As the process of rationalization alters the definition of the world so that more objects are defined as profane, drugs lose their sacred status. The use of drugs is therefore less tightly controlled. To further the theory, I synthesize the work of Weber, Habermas, Parsons and Durkheim and summarize the general processes of rationalization and modernization. The discussion highlights the changes in culture and structure these processes induce. Finally, I relate these changes to patterns of drug consumption.

In Chapter Five, the theory is applied to explain changes in the social functions of drug use. I discuss the social functions that drug use can fulfill ranging from inducing spirit possession to the purely symbolic. I argue that these functions become more concrete and less abstract as modernization advances. Moreover, the functions of use become more individualized and less communally-oriented as rationalization advances. Next, I outline the religious sub-functions of drugs. I explain the relatively widespread use of drugs in shamanistic religions, the symbolic use of drugs in the "higher religions" and the lack of religious drug use in the "this-worldly religions" such as Buddhism and Confucianism. From this discussion, we see that religious drug use varies inversely with the number of deities associated with the religion. Thus, over time, drug use within a religious context becomes "irrationalized" as the functions of use become more abstract and more communal. Finally, I discuss the differences between religious rites and secular laws in controlling drug use. Examples from the Judaic-Christian Bible and the Qu'ran support the argument.

Chapter Six presents general propositions that explain what drugs are used. I then present data supporting the argument that drug use increases with modernization. Then, I refine the theory by noting modernization's influence on law and definitions of deviance. While modernization varies directly with rates of drug use if viewed cross-culturally at any given point in history, there is a curvilinear relationship between modernization and rates of use over time. At least in the most modernized nations, rates of use increase as modernization increases until modernization alters the nature of law and definitions of deviance. After modernization leads to greater amounts of law, drug use, or more accurately, some drug use is defined negatively. Once drug use becomes a means of earning a deviant status, rates of use typically decrease. Next, I discuss the declining significance of ascribed statuses and the increasing significance of achieved statuses for patterning drug consumption. Ascribed statuses, especially gender and age, determine who uses drugs in less-developed nations. In more modernized nations, drug consumption is no longer strongly correlated with ascribed statuses. Instead, achieved statuses correlate with drug consumption.

Chapter Seven adds a more dynamic dimension to the theory by considering how social mobility influences rates of drug use over time. Social mobility, by increasing inter-group contacts, associations, heterogeneity and individuation alters definitions of "right" and "wrong." The "deviance structure" expands to tolerate once forbidden behaviors. Drug use becomes more acceptable and, therefore, rates of use increase. Conversely, when social mobility decreases, society's moral boundaries contract and define previously acceptable behaviors as deviant. Consequently, rates of drug use decrease. I provide tentative tests of this theoretical extension by considering rates of use in American, British, Japanese and Israeli history. To test the theory, I present data on drug use and social mobility from the mid-19th century until 2000 in the cases of the United States and Great Britain. In Japan, I discuss drug use and social mobility beginning in the post-war period. In the Israeli case, the period immediately following the Six Day War is considered. In all of these cases, rates of use increase as social mobility increase and decrease as social mobility decreases. Then, I consider the recent influence of social mobility on rates of use in the former eastern bloc and Soviet Republics. Again, available data support the theory.

I briefly summarize existing theories of drug use that explain why certain individuals are more likely to use drugs. I then show how the current theory can account for similar correlates of use at the individual level. In short, many of the social-psychological correlates of use at the individual level such as "flexible thinking," "broad intellectual interests," "low religiosity," and a lack of respect for tradition and authority are reflective of a scientific world-view. Other social-psychological correlates of use such as "low self control" and "risk taking" reflect the rampant individualism associated with modernization. Social correlates of use, such as associating with drug using peers and involvement in the drug subculture, can be understood as an attempt to find enchantment in a disenchanted world. I then apply the theory to the evolution of the American drug subculture since the 1950s. I argue that the subculture emerged as an alternative to the disenchantment that is associated with the rationalization process. However, the same forces of modernization and rationalization that created the subculture changed it. The process has led to a dialectic relationship between the dominant culture and drug subculture. The subculture itself, once justified by its members as a path toward spiritual enlightenment, has been rationalized. Drug use is now "normal" and defended by users as simply an alternative means of recreation. I conclude by considering policy implications.

Throughout the book, I try to maintain a "value-free perspective." I do not mean to pass moral judgment on those who use drugs or those who judge them. These issues cannot be addressed sociologically, at least if sociology is to be "scientific." These issues are a matter of morality and personal values. As I discuss throughout the book, science is devoid of morality. Thus, while we can scientifically discuss why certain drugs are considered "good" and others are "bad," we cannot discuss whether drugs are "good" or "bad." We can say that approximately four percent of the world's population uses illegal drugs. We cannot say if this is a "good" or "bad" thing. Therefore, I try to remain objective and to bracket my personal feelings and values about drugs. I therefore try to avoid the terms "abuse" and "misuse" since these are, by their vary nature, value-ladened terms. I also try to avoid implying that drugs are "the key to reality" or "a source of enlightenment" or any other magical cure for our problems or means to achieve our dreams. If I do imply these values, I apologize. Please consider any such implications as accidents of poor word choice.

I take this "scientific" approach because I believe the field is currently too moralistic. There are few texts or research manuscripts that do not explicitly denounce drug use. Those that do not, typically praise it far too much. While it is true that drugs can ruin lives, cause misery and kill, it is also true they can improve one's quality of life, cause joy and save lives. Yet, reading the literature, we often are led down a specific value-oriented path without knowing even the most fundamental facts about the drug in question such as "how many people use this drug" or "is this drug only used for recreation." I hopefully provide an objective description about the empirical facts of drug consumption and then an explanation for these patterns that is devoid of "blame" or "praise."

With that said, it would be disingenuous of me to not make my personal value positions concerning drugs explicit. First, with respect to the use of drugs, I honestly believe that "it depends." Whether or not the use of a specific drug, from caffeine to heroin, is "good" or "bad" depends on the setting in which it is used. That is, drugs can be used for either good or evil. Explicating all the possible settings in which drug use can occur is nearly impossible. Thus, in my opinion, I cannot really say if drug use is "good" or "bad." All I can say is "it depends." Sometimes it is "good," sometimes it is "bad."

With respect to our policies toward drug use, however, I do have a clear value position. I firmly believe that our policies are misdirected. While I do not advocate unequivocal legalization or even decriminalization, because these strategies are hampered with problems too, it is clear that the path we have taken has not achieved the stated goal. While I do not know what we should do, I am convinced we should not keep doing what we are doing. A major problem with our current policies, I believe, is the tendency for us to treat drugs as something special. When we consider current drug policies, it seems that drugs are considered enchanted elements. They are somehow "magical" and allegedly do not behave in a similar manner as other objects. For example, let us consider the Reagan administrations approach to the "war on drugs." To curtail the "cocaine epidemic" of the mid-1980s, the Reagan administration used the Drug Enforcement Administration (DEA) and Colombian officials to break-up the Medellin cartel. This strategy may appear logical at first glance: stop, or at least hinder, those who are distributing drugs. However, if we recall Reagan's approach to economics in general, this strategy does not make sense. Reagan's "supply side" economics is based on the belief that economic growth occurs when competition is stimulated. Competition, the logic goes, generates better products at lower prices which, in turn, stimulates demand. Thus, the Reagan administration actively pursued economic policies that deregulated industries. Somehow, however, the administration failed to apply their own economic logic to the drug trade. By breaking up the Medellin cartel, which is believed to have controlled nearly 90% of the world's cocaine at one time, the Reagan administration increased competition. As a result, the purity of cocaine increased, the availability increased and the price decreased. It was supply side economics at its best!

So why did the administration not apply its own economic logic to the cocaine distribution system? We are left with two possible answers. First, as some have argued, it was a conspiracy. The Reagan administration wanted the inner cities to be flooded with crack cocaine. They wanted poor minorities to use the drug so (1) they could aggressively target them with law enforcement and effectively imprison a large segment of an "undesirable population,"(2) they could further destabilize inner-cities and use this instability in a war against the welfare state (welfare mothers became crack whores and their sons became "super predator" crack heads), and (3) they could divert the nation's attention from other issues. While some have made convincing arguments along these lines, I am not a huge fan of conspiracy theories. For one, conspiracies require intelligence and secrecy beyond the capabilities of most people.

The second alternative is far less sinister but just as damning. The Reagan administration believed that cocaine was "magical." They believed that cocaine did not follow the same laws of supply and demand that legal commodities are prone to follow. They failed to realize, if their general economic logic was indeed correct, that by breaking up the Medellin cartel, they would open the door for the Cali cartel and others. If their general economic logic was correct, why not buy all the cocaine and heroin from the major foreign producers and destroy it. That is, let the U.S. government become the illicit drug monopoly. Then, they could control how much of the drug was on the street. After all, the U.S. produces very little cocaine or heroin. They therefore would not need to concern themselves with the thousands of street-level "pharmaceutical distributors" who peddled the drugs to "children and other victims." They could have gone to the top. To the Medellin cartel, the Shan United Army and the relatively short list of other major suppliers of cocaine and heroin.

Could it be the Reagan administration (and those that preceded and followed it) was blinded by the "magic" of cocaine? To me, that is the more plausible answer. That is why moral entrepreneurs believe we can "rid society of the scourge." They too often get caught up in the morality of the issue without paying attention to empirical reality. While one's moral position need not be based on empirical reality, it is helpful if policies are.

So, with this said, my goal is to offer a summary of the empirical patterns of drug consumption in the contemporary world, to place these patterns in historical perspective and to offer an explanation of these patterns that, to my knowledge, has not been offered previously. I believe this perspective can help demystify some aspects of drug use. If we really do want to "do something about drug use," why do we empower drugs more than they need to be? Hopefully, by showing how drug use is a function of the rationalization process and the disenchantment that this process engenders, we can disenchant drugs!

Preface

It is with great pleasure that I introduce this important book on drug use. While books on the subject abound, it is always refreshing to find a scholarly text on drug use that offers a new vantage point on this complicated and ever present social phenomenon. This is such a book. James Hawdon has skillfully synthesized classic sociological thought to craft a general theory of drugs that provides us with significant insights into human drug use. He has also painstakingly gathered the existing data on drug use throughout the world to put his new theory to the test. The result is a broad macro-sociological theory of drug use, firmly grounded in a wealth of empirical evidence, which has much to offer both academics and policy makers alike.

Recognizing the often-problematic nature of defining what is considered a drug and what is not, the book provides a working definition of drugs that includes both the psychoactive aspects of substances and the political reality that goes into defining what substances society recognizes as drugs. Drugs have become extremely politicized. Whether it is moral entrepreneurs concerned with saving souls, political entrepreneurs concerned with constituencies and elections, or some other interested parties, drugs have come to be defined as "magical" substances that are somehow different from other things. Hawdon demonstrates that this special status that drugs have acquired is largely unfounded. While drugs can be very powerful substances, treating drugs as totally different from all other commodities has led many to approach issues related to drug use in a manner that is often misguided or even counterproductive. It is important to remember that drugs, both legal and illegal, are basically just commodities. The same economic forces of supply and demand that influence the consumption patterns of other commodities impact the consumption of drugs. What is more, larger social forces, including modernization and rationalization, also shape these consumption patterns. And demonizing these substances tends to obscure the social reality of drugs and drug use.

The nature of drug use is largely predicated on the context in which the drug use takes place. Hawdon points out that whether or not a drug has been socially defined as sacred by a social group plays an essential role in how a drug is used and the extent to which it is abused by members of that group. There is nothing inherently sacred about any given drug. A drug becomes sacred only when the collectivity defines it as such and maintains beliefs and rites that support the drug's sacred status. Moreover, social forces such as modernization and scientific rationality have increasingly impacted religious practices and, in turn, changed the nature of sacred drug use. This influence is especially evident in the patterns of drug use in more modernized western societies.

Hawdon notes that the differences in social control over sacred versus profane drug using behaviors are important. While defining drugs as sacred restricts drug using behaviors, it prescribes certain drug using behaviors as well. In contrast, restrictions on drugs defined as profane are basically negative in nature, either restricting or prohibiting drug use, but not requiring drug use. The difference has significant ramifications. Sacred drug use requires the use of the sacred drugs by certain people at specific times and in a specific manner. At the same time, generally, the proscriptions of sacred drug use tend to make abuse of these drugs much less likely and the rituals related to sacred use also serve an integrative function for the people within this belief system. Conversely, the use of profane drugs is not so influenced, thus drugs defined as profane are prone to greater variations in who, when, and how they are used. Profane drugs are also more likely to be abused and to be socially disintegrative with regard to the larger society, fostering the development of distinct subgroups. And while groups within a society may disagree on what is sacred drug use and what is not, these insights can have important policy implications.

Hawdon's theory maintains that modernization and rationalization have changed the nature of sacred and profane drug use. Pre-modern societies saw a world filled with the supernatural in which sacred drug use could literally transform people, facilitate spiritual journeys to other worlds, and manipulate the gods. In modern societies, however, the growing influence of modernization, science and rational thought has led to a demystification of the world, which has reduced the emphasis on religion and dealing directly with the supernatural. As the predominant worldview has grown more secular, drug use has become more profane and less subject to the sacred proscriptions of earlier times. Sacred drug use has become more abstract, symbolic, and otherworldly in focus with less direct control on drug use. Meanwhile, an increased emphasis on rational thought and science has produced a stronger emphasis on individual instrumental action, resulting in an increase in recreational drug use. Secular society is a society based largely on laws but, unlike the absolute nature of religious beliefs, laws are more relative and change much more rapidly. Modern religions tend to focus on more general moral teachings in which control of drug use is more derivative than direct. Thus, modern western societies that glorify individualism and the freedom to make personal choices by their very nature reduce the influence of communal restraints and increase the likelihood of greater variation in who uses drugs, what drugs they use, and how they use them. Subcultures may develop in reaction to the disenchantment of the world and use their own sacred drugs to reintroduce the mystical, but the rationalization process eventually changes even these groups.

Hawdon's work, supported by numerous examples and global data, show that rates of drug use are higher in nations or in regions that are more developed. The rise of synthetic drugs and the continuous growth and spread of pharmaceutical knowledge makes many new drugs readily available. Modern factories produce drugs faster. Drugs become cheaper and easier to obtain. Thus, the process of modernization increases the variety of drugs available and the variety of drugs used for all segments of society. Modernization also affects the structure of social control mechanisms related to drug use.

This book demonstrates, through numerous historical examples, the general pattern of drug use in modernizing societies throughout the 19th and 20th centuries. As industrialization rapidly modernizes various aspects of a given society, drug use expands rapidly, and then slowly stabilizes. This is followed by a dramatic decrease in drug use. This curvilinear pattern is related to changes in social control mechanisms. Traditional sources of informal social control are weakened by the processes of modernization and eventually replaced by formal social control in the form of anti-drug laws. The changing nature of work and the growing interdependence of social institutions, both nationally and internationally, contribute to a new emphasis on sobriety. This has been coupled with a shifting emphasis on the importance of achieved over ascribed status in modern societies. The result is an increasing correlation of drug use patterns with achieved social status in contrast to less modernized societies where ascribed status plays a much greater role in determining drug use patterns. So, generally, there are fewer distinctions between groups with regard to drug use as societies become more modern and more egalitarian.

Hawdon provides ample evidence to demonstrate how cyclical patterns of drug use found within societies are closely related to the status of those who are using the drugs and the perceived dangers of the drugs being used. Typically, new drugs come along or old drugs are rediscovered by societal elites. Over time, the use of these drugs spreads to other segments of society and eventually to people in the lower segments of society. Then the use of these drugs falls out of favor in elite circles, perhaps due to the arrival of another new drug or the increased social costs of being associated with a drug that is now identified with low social status. It is at this point in the cycle that anti-drug laws tend to appear which target these drugs that are now primarily used by people with lower social status. Not coincidentally, these lower status users have fewer resources to influence the law making process or to conceal their drug use. This shift is reinforced by an increased societal perception of the harm of these drugs; a perception often built on a circular logic, i.e. people at the bottom of the social ladder use these drugs, so they must be at the bottom because they use these drugs. These groups then get demonized and their drug use blamed for all sorts of social ills, while ignoring the larger pattern, the previous elite use of these drugs, and the policy alternatives the larger pattern may indicate. Unfortunately, this type of scapegoating behavior is often heartily embraced by political and moral entrepreneurs looking for someone to blame for social problems and the countless economic entrepreneurs looking to cash in by providing their "solutions" to these problems. It is a lot easier to sell scapegoats than the need for larger structural change.

This book shows how the timing of new anti-drug campaigns and anti-drug legislation is also closely tied to another social process – social mobility. Building on his earlier work, Hawdon's discussion here of the impact of the frequent fluctuations of social mobility on the cyclical nature of drug use patterns and social control efforts is extremely informative. He shows how changes in social mobility affect deviance structures, which set the behavioral boundaries of what societies consider acceptable and what is defined as deviant. When social mobility is relatively high, there is greater opportunity for individuals to interact with a greater number of social networks. This exposes individuals to more diverse behaviors, beliefs and norms. It also reduces the costs of exiting a group. This increases the emphasis on individualism and reduces the ability of any particular group to exercise social control over an individual. As social mobility increases, deviance structures expand to allow more behaviors including increased and more varied drug using behaviors. Conversely, when mobility decreases, social networks are more limited and deviance structures contract. This leads to a reduction in drug use. In these periods of contraction, not only are opportunities for use reduced, but the range of behaviors deemed acceptable is also narrowed. Anti-drug crusades often follow, after the level of drug use has already significantly decreased, as part of the narrowing definitions of deviant behavior.

Hawdon also offers a useful summary of theories of drug use that focus on micro-level individual causes. Hawdon notes the utility of these explanations of drug use. He also provides a good case for considering how macro-level forces, such as modernization and a scientific worldview, influence their individual social-psychological correlates. This connection is often overlooked. It is something that deserves much greater attention in policy-making circles.

This book makes significant contributions to the literature on drug use. It is a rare find; a refreshing "value free" approach to the study of drugs and society. By standing on the shoulders of sociological giants, James Hawdon has succeeded in producing a grand theoretical explanation of drug use with a solid empirical foundation that allows all of us to see much farther than we previously had seen. The discussion of sacred and profane drug use and its implications is enlightening. The book also demonstrates how macro-level forces, such as modernization, rationalization, and social mobility, can account for patterns of drug use. His focus on these big-picture variables provides new insights and new challenges for policy makers who must come to terms with the need to address larger issues in their attempts to resolve the issue of drug use. As Hawdon notes, people often empower drugs inappropriately. Drugs become these socially constructed demons, rather than commodities subject to the same social forces as other commodities. Many policy makers and moral entrepreneurs remain mystified by drug use, but this book provides us with a rational lens to better understand it. If we truly want to change things for the better, we first need to develop a rational understanding of the big picture. This book offers a major contribution to that understanding.

Randolph Atkins, Jr., Ph.D.
The Walsh Group, P.A.
Bethesda, Maryland

Acknowledgments

The number of people who deserve recognition for helping me on this manuscript is far too numerous to list. I have been working on this text for over 20 years in one way or another. I will undoubtedly forget someone, and I apologize to them. First, I would like to thank my longtime friends Joel Leeman, Vince Pajerski, Michael Roland and Janet Johnstone. All of these people have influenced me in too many ways to count, and their friendship has been critical to my overall mental health. In addition, I would like to thank the literally thousand of students at the University of Virginia and Clemson University that sat through lectures and provided feedback to the earliest versions and latter refinements of the ideas presented in this text. I also need to thank my good friends and colleagues at the University of Virginia and Clemson University. I extend a special thanks to my good friend and colleague John Ryan. He was a tremendous "soundboard" for many of my ideas and provided numerous suggestions that undoubtedly improved this project. Several former graduate students also assisted on this project by finding literature, discussing ideas or reading drafts of the manuscript. These students and friends include Jessica Cooper, Stacey Willocks and Megan Linz. Megan's assistance with editing the manuscript was invaluable and I am extremely grateful for the time and care she dedicated to it. I truly appreciate the time and effort Dr. Randolph Atkins dedicated to writing the preface of this book.

Obviously, this work and all the work I have accomplished to date would not have been possible without the support and love of my family. My father George, who passed away while I was writing this manuscript, was not only a great father, he was the best teacher I ever had. I miss him terribly. My mother

Dori has always been, and continues to be, a source of tremendous support. Her strength of will is amazing and inspiring. The unconditional love both my parents gave me allowed me to achieve what I have. They are both greatly appreciated and loved very much. I also thank my brothers, George and John, for being good brothers and for all the wonderful times we spent together while growing up.

There are no words to express what my wife Donna has meant to me. I probably do not deserve the love, encouragement, support, understanding, companionship and patience she has provided me. Deserving or not, I am certainly grateful. In addition to being a wonderful wife and mother to our daughter, Donna is also a tremendous sociologist in her own right. Her ability to see flaws in my thinking and willingness to help me correct them has helped me refine my arguments. Finally, I would like to thank my daughter Gwen. Her innocence of youth and zeal for life has shown me once again that the world can be enchanting. To all of these people, "thank you."

CHAPTER ONE
ON THE SOCIOLOGY OF DRUGS

Drugs are amazing substances. What other substance or set of substances has an entire discipline devoted to it and still is discussed by several other scientific fields? Although pharmacology lays an obvious claim to the study of drugs, drugs are also studied and discussed by biochemists, chemists, psychologists, economists, historians, anthropologists and sociologists. These substances attract the attention of political scientists, policy analysts, criminal justice scholars, community development specialists, and social workers. They are discussed and debated in political bodies ranging from city councils to the United Nations. Entire occupational groupings, including the jobs of pharmacists, DEA agents, tobacco farmers, brewers, bartenders and drug rehabilitation workers, exist solely because of drugs. Occupations such as physicians, nurses, psychiatrists and veterinarians are highly dependent on drugs -- so to speak.

Drugs are indeed amazing substances. What other substance or set of substances play central roles in more than one major social institution? Drugs are obviously central substances to the health-care institution; however, they also assume prominent roles in religious institutions, the economy, and the state. While health-care professionals dispense drugs as prophylactics, analgesics and cures, other health-care professionals spend countless hours trying to cure or rehabilitate those who use drugs. While religious leaders use drugs to symbolize, communicate with, or appease their God or gods, other religious leaders devote their professional careers combating the evils of drugs. While billions of dollars are made -- legally and illegally -- developing, cultivating, producing, advertising and distributing drugs each year, billions of additional dollars are spent -- legally and illegally -- by those wishing to consume them. While governments raise billions of narco-dollars in taxes each year by issuing licensing fees and collecting taxes on their sale, these same governments spend billions of dollars each year trying to regulate or prevent the use of some drugs and punishing the use of others.

Drugs are truly amazing substances. What other substance or set of substances has symbolized so many divergent and contradictory things? If we think about what drugs symbolize or have symbolized it becomes apparent how inherently social these substances are and how embedded in the social fabric drugs have become. Depending on the reference group, drugs have symbolized progress or regression, hope or despair, freedom or enslavement, health or illness, holiness or sinfulness, modernization or tradition, maturity or childishness, conformity or deviance, power or weakness, eliteness or poverty, purity or depravity, a key to or escape from reality, goodness or evil. Given the powerful symbolizing elements of drugs, it is of little wonder why these substances are referred to in holy scriptures, medical texts and criminal codes. It is of little wonder why millions of written pages are devoted to drugs. It is understandable why drugs find a voice in fiction, pseudo-non-fiction, and non-fiction literature.

Given the divergence of disciplines interested in drugs, the amount of energy spent studying various aspects of the "drug reality," and the competing interests that profess knowledge in the subject, any discussion or study of drugs is bound to be incomplete. Indeed, any discussion is doomed to be limited by personal and professional choices, personal and disciplinary ignorance, and simple incompetence. Something will always be left out, over-looked or ignored. While this is undoubtedly true for any discussion of any topic, it may be especially true for a study of drugs. This study will suffer that same fate, as have all that preceded it and all that will follow. Given the volumes written on the topic and the amazing amount of knowledge we have about the subject, it is hard to imagine that there is still more to know. Yet there is so much we do not know about drugs. This study, despite its many flaws and limitations, will hopefully shed light on the topic in a new way and provide a strategy for answering some of the many unanswered questions we still have.

What follows in these pages is a *sociological study of drugs*. Although sociologists have studied drugs for several decades and have contributed priceless insights to our collective understanding of the subject, I believe that we have only begun to tap the power of sociological analysis in the sub-field of the sociology of drugs. In my opinion, we have focused on an overly limited range of questions. I hope to expand that range. This book's goal is to account for drug patterns by considering social processes such as modernization, rationalization, stratification, mobility and heterogeneity. To accomplish this task, let us first consider the traditional sociological framing of the drug issue. In so doing, I hope to highlight the utility of my approach.

Why do individuals use drugs?

By far, the question most frequently asked by sociologists is "why do individuals use drugs?" To answer this question we have typically turned to quantitative studies of individuals, usually youth. We have occasionally conducted field research and intensive interviews; yet, we typically frame the issue in similar terms, conceptualize it at the same level of analysis, and, ultimately, ask the same questions. From these efforts, we have learned about the types of individuals likely to use drugs, at least illegal ones. We now know that drug users tend to suffer from low self-esteem (see Kaplan and Fukurai 1992; Kaplan and Johnson 1991; Kaplan, Johnson and Baily 1986; Kaplan, Johnson and Baily 1987; Miller et al 2000; Vega and Gil 1998; contrast Jang and Thornberry. 1998), have low self-control (Gottfredson and Hirschi 1990; Leeman and Wapner 2001), and have "a concern with autonomy, a lack of interest in the goals of conventional institutions, like church and school, a jaundiced view of the larger society, and a more tolerant view of transgression" (Jessor and Jessor 1980: 109; also see Jessor and Jessor 1977; Donovan, Jessor and Costa 1991; 1993). We have also learned that drug users -- or more accurately illegal drug users -- engage in specific types of leisure activities (see Hawdon 1996a; 1999; 2003; 2004; Osgood et al. 1996) and adopt a well-defined social role (Stephens 1991). We know that illegal drug users associate with drug-using peers (see, for example, Akers et. al. 1979; Ary et al 1993; Aseltine 1995; Bailey and Hubbard 1991;Becker 1963; Brook et al 1990; Elliot, Huizinga, and Ageton 1985; Hawdon 1996a; Hawdon 1999; Hawdon 2003; Hawdon 2004; Kandel 1980; Kandel and Yamaguchi 1993; Kandel and Yamaguchi 1999; Kandel and Yamaguchi 2002; LaGrange and White 1985; Marcos, Bahr and Johnson 1986; Pope 1971; Thornberry 1987; Thornberry 1996; Thornberry et al 1991; Thornberry et al, 1994; Krohn et al 1996) or deviant parents (Adler and Adler 1978).[1]

As valuable as this information about who uses drugs is, it answers only part of the question. In my opinion, this individualistic perspective, although valuable in its own right, has led to a reductionary tendency and an over-reliance on social psychology. From historical accounts we know that *rates of drug use* have varied dramatically over time (see, for example, Ashley 1975; Hawdon 1996b; Inciardi 1992; Morgan 1974; Morgan 1981; Musto 2002; Rublowsky 1974; Terry and Pellens [1928] 1970). Assuming the social-psychological type of individual who uses drugs is relatively stable over time, we are left with the question of why a greater percentage of the U.S. population was risk-taking, jaundiced, low-self-esteemed, leisure-loving, and deviant-peer-associating in the 1970s than in the 1950s? Did the American people of the mid-1980s have more self-control than those of the late-1970s despite the fact that many of them were the same people? If they did have differing levels of self-control, why did they? Could there be larger social processes occurring that determine how individual's

----

[1] This list of potential causal variables is in no way complete. I provide those listed only as examples of some of the variables typically used to explain why people use drugs. I emphasize again that this literature almost exclusively addresses why people use illegal drugs or legal drugs illegally. The majority of the literature on alcohol and tobacco use focuses on the use of these substances by youth. Thus, much of this use is also illegal.

social-psychological attributes develop? These questions remain unanswered by even the most elaborate, well-designed longitudinal studies of drug use conducted at an individual level of analysis.

We have rarely asked the question "why do rates of drug use vary over time?" This question, while still ultimately concerned with why individuals use drugs, begins to conceptualize the issue at a more social level. Instead of focusing on the attributes, behaviors and associations of individuals, this question raises the level of analysis from the individual to that of social processes. The few studies and theoretical explanations that address rates of use focus on social variables such as the strength and cohesion of the drug subculture (e.g. Goode 2003), rates of social mobility and integration (Hawdon 1996b), or the global spread of drugs fueled by European colonialism and capitalistic expansion (Courtwright 2001). Although knowledge of how social processes help pattern the nature and extent of drug use is scientifically valuable and has obvious policy implications, few studies at this level of analysis have been performed and traditional sponsors of drug-related research would fund even fewer. Moreover, variations in macro-level variables could conceivably influence the social-psychological variables associated with individual drug use. Researchers have not asked such questions.

Why are some drugs illegal and why do we care?

Another line of inquiry that sociologists have pursued -- albeit to some lesser extent -- is to uncover what determines the legal status of drugs. Why, for example, is marijuana illegal while the physically more dangerous drugs of alcohol and tobacco remain legal? Why were various drug laws passed when they were? We have learned from such studies that drug laws have little to do with the objective harm associated with their use (see, for example, Goode 1997). Instead, drug legislation has been used to promote national interests in the international policy arena (see Duster [1970] 1989; Reasons 1974), as a means of cultural imperialism (Szasz 1974; Szasz 1985) and economic stimulation (see Morgan 1981), to advance an ideological position and hegemonic strategy (Beckett and

Sasson 2000), or to control and repress various ethnic groups (Helmer 1975).

Related to the studies of drug legality and acceptability is a relatively large body of literature on drugs and moral panics (see, especially, Cohen 1986; Ben-Yehuda 1986; Goode and Ben-Yehuda 1994; Goode and Ben-Yehuda 1994b). Studies have demonstrated how public and political concern about drugs is not always proportionate to the threat they pose (see, for example, Cohen 1980; Goode 1990; Goode and Ben-Yehuda 1994; Hawdon 2001; Jensen, Gerber, and Babcock 1991, Levine and Reinarman 1988, Orcutt and Turner 1993). We have learned through this research about the important role the media plays in creating our perceptions of the "drug problem" (see, for example, Beckett 1995; Merriam 1989; Reinarman 1996; Reinarman and Levine 1989; Shoemaker, Wanata, and Leggett 1989). This literature also substantiates the importance of the state in creating moral panics about drugs and crime (see Beckett 1994; Hawdon 2001). It has also documented how the drug issue can come to symbolize status politics and how the rhetoric used to discuss the "drug problem" can influence public attitudes toward drugs (Elwood 1994; Gusfield 1963; Gusfield 1981; Gusfield 1996). Moreover, why the common response to such "drug scares" is so often punitive and results in challenges to civil liberties has been explained (Cohen 1980; Goode and Ben-Yehuda 1994).

The studies that address the questions of drug legality and social attitudes toward drugs ask social questions, focus on sociological variables, and offer tremendous insight. Yet additional sociological questions remain unanswered by these efforts. First, do the various legal strategies correlate with other macro-level phenomena such as economic, political or demographic trends? Why do these strategies appear to be more popular in the United States and the United Kingdom than they are in other western powers or less-developed countries (for a possible explanation, see Savelsberg 1994). Second, these studies are constrained by a legalistic definition of drugs. Although this is a critical dimension of the drug reality, social definitions of acceptable and unacceptable drug use vary not only between legal and illegal drugs but also within legal categories. Why, for example, do Americans consider it to be more "sophisticated" to drink wine than beer? Why has cigar smoking become "chic" and "in" while cigarette smoking is positively "out?" Who is more despised: the cigarette smoker or the alcoholic? Why was cocaine "fashionable" in the early 1980s and loathed in the late 1980s? Why have attitudes concerning the legalization of marijuana fluctuated wildly since the 1970s? Why were opiate users considered to be among the most respected members of society in the nineteenth century (see Terry and Pellens [1928] 1970; Musto 1987)? Are our attitudes about drugs and drug users and the attempts to manipulate those attitudes related to processes of modernization, social stratification and mobility? These are questions that sociology could, but has not, answer adequately.

On The Sociology of Drugs

It is hopefully clear that sociology has barely scratched the surface of the study of intoxicants. What I hope to offer is a baseline study that applies basic sociological theory to the study of drugs. I hope to illustrate how fundamental social processes pattern the setting of drug use. The "who," "how much," and "why" of drug use can be explained sociologically by relating patterns of use to social forces such as modernization, rationalization, social control, mobility and stratification. Before attempting this, however, we must first define what drugs are, how drugs vary, and the nature of the data used to demonstrate the basic propositions developed throughout this work.

Drugs defined

First, we must specify what a drug is. As we will see, most definitions of the term are rather vague and either exclude too many substances that we, in a common-sense way, want to consider a drug or include too many substances that basic logic would tell us are not drugs. For example, Lyman and Potter (1998: 19) define a drug as, "any substance that causes or creates significant psychological and/or physiological changes in the body." This definition would certainly classify heroin, marijuana, alcohol, amphetamines, and penicillin as drugs. Unfortunately, it would also incorporate almost anything that humans ingest. Refined sugar causes physiological and psychological changes in the body. A bullet can also cause "significant" physiological changes in the body. However, we tend not to define sugar or bullets as drugs. Similarly, Levinthal (2002: 3) defines a drug as, "a chemical substance that, when taken into the body, alters the structure or functioning of the body in some way, excluding those nutrients considered to be related to normal functioning." Again, this definition includes a considerable range of materials. After all, every tangible thing is, at its core, a chemical substance. Thus, a bullet would still be a drug by this definition. And, is refined sugar a nutrient? It certainly is unneeded for "normal functioning" since many humans lived their entire lives without consuming it. Others define drugs based on the psychoactivity of the substance. For example, Goode (1997: 11) defines drugs as, "any and all substances that influence or alter the workings of the human mind." This definition also includes too many substances that we typically do not consider to be drugs. Again, the bullet comes to mind! In reality, there is no objective property that all substances we wish to classify as drugs share that other substances we typically do not consider to be drugs lack. As Goode (1999: 58) states, "there is no effect that is common to all substances that are referred to as drugs, that is, at the same time, not by definition a property of those things no one would call drugs."

Other definitions, such as those adopted by the U.S. and other governments (see Goode 1997), rely on legalistic definitions. According to this perspective, law defines drugs. A substance may be a drug if it is outlawed. If it is legal to consume, it cannot be a drug. Thus, tobacco, alcohol and prescribed pharmaceutical drugs, provided they are used legally, are not drugs according to those adopting a strict legalistic definition of drugs. This definition, in my opinion, is not overly useful. It ignores the most widely used substances, such as caffeine, and raises problems of consistency. For example, someone adopting a strict legalistic definition may not consider morphine a drug if administered by a physician while the user was recovering from a painful surgery, but it would be a drug if it were self-administered while recovering from a hangover. As Goode (1997: 11) says,

> If the currently illegal drugs were legalized, would that mean that, overnight -- according to this (legalistic) definition -- they would magically *cease to be* drugs? This is what the legalization definition would be forced to say (emphasis in the original).

Similarly, by this definition, alcohol became a drug in the United States when the 18th Amendment outlawed its manufacturing, transportation and distribution. However, alcohol was no longer a drug once congress repealed the 18th Amendment. These problematic inconsistencies render the definition useless for scientific investigation.

Constructionists offer another definition of drugs. Drugs, according to this perspective, are "something that has been defined by certain segments of the society as a drug" (Goode 1999: 58). Goode (1999: 58) continues,

> Although all substances called drugs do not share certain pharmacological traits that set them apart from other, nondrug substances, they do share the trait of being labeled drugs by some members of a society. The classification is partly an artificial one; it resides in the mind, not solely in the substances themselves. But it is no less real because it is in part artificial and arbitrary. Society defines what a drug is, and this social definition shapes our attitudes toward the class of substances so labeled.

Indeed, this definition is somewhat, although not entirely, arbitrary. Then again, so are social definitions of drugs. Drugs are, at least in part, cultural artifacts. Yet a constructionist definition of drugs is not without its problems. The social definition of alcohol has changed over time and, consequently, so would its classification as a drug or non-drug by constructionists. Prior to the 1950s, few Americans defined alcohol as a drug. Once the AMA declared alcoholism a disease, however, Americans began to change their collective position on the status of alcohol, although not totally. Today, however, most Americans consider alcohol a drug, even if begrudgingly (see, for example, Fort 1973; Quindlen 2000). Definitions of what are and are not drugs also vary cross-culturally. Islamic cultures have considered alcohol a drug since the rise of Islam.

What, then, are we to do? How can drugs be defined? A complete definition would likely include the psychoactive and social dimensions of drugs. The physiological dimension is not needed. Substances such as penicillin and the host of drugs used to control ailments like hypertension are undoubtedly drugs but do not cause psychoactive effects. However, most people would define these as drugs and therefore these would be included due to their meeting the social dimension of the definition. The social dimension is needed so that substances such as sugar and bullets are not usually included. Thus, the working definition of a drug that I will use is *any substance that society labels a drug, causes psychoactive effects, or both.* Drugs can be legal or illegal, mildly or powerfully psychoactive, used for utilitarian or non-utilitarian purposes, natural or synthetic. They are all, however, defined as drugs by some large or powerful segment of society.

How Drugs Vary

The dependent variable defines a sub-discipline. Thus, criminology is the study of crime, the sociology of the family is the study of family life and familial arrangements, the sociology of science is the study of scientific inquiry and discovery, and the sociology of law studies the "behavior of law" (see Black 1976; Black 2000). The focus for research within the given sub-discipline then becomes what "causes" variation in that variable. Thus, in criminology, we seek to understand what influences or causes the observed patterns of crime. Although discovering and documenting the causes of anything social is a daunting task, turning this distinction around so that the independent variable defines the sub-discipline is simply unmanageable. That is, it would be impossible to discuss everything that is "caused by" the family. What social fact is uninfluenced by this central social institution? Similarly, the sociology of gender, if defined by gender as an independent variable, would require its scholars to explain how gender affects almost every micro-social process and most, if not all, macro-social process. Investigating the causes and patterns of social stratification is an arduous task in-and-of-itself. Explaining how social stratification influences everything that it does would be impossible.

Thus, the sociology of drugs should focus on explaining variations in drugs. In my opinion, it should not consider the behaviors associated with or caused by using drugs. To my knowledge, drug use causes no behavior that is not also caused by a host of other factors. Drug use may "cause" crime, but poverty, low self-esteem, strain, the lack of self or social control, associating with delinquent peers, disorganized neighborhoods, or being involved in a gang may also cause crime. Figuring out the causes of crime and the relative influence of the litany of variables that do should be left to criminologists. Those people studying the sociology of drugs have enough work to do.

If we accept the definition of the sociology of drugs as the study of variations in drugs, we must now determine how drugs vary. Drugs can vary on at least six dimensions. Drugs vary on their psychoactive/physiological effects, the method of use, the reasons for use, the extent to which they are used, the social definitions of the drug, and the social definitions of their users. Although sociologists can address all of these dimensions, some of them are better suited for sociological analysis than others. Let us briefly consider each.

First, drugs vary with respect to their effects. That is, they vary on the *psychoactive/physiological dimension*. Drugs are either psychoactive or not. They also vary in the extent to which they are psychoactive. LSD, for example, is more psychoactive than is coca leaf. Within this dimension, the drug effect also varies. Drugs can stimulate or depress. They can be analgesics, relaxants, or hallucinogens. They can be synthetic, semi-synthetic, or natural. This dimension of drugs is primarily, although not totally, determined by the chemical properties of the drug, the method of ingestion, and the dose ingested. It is therefore the least social dimension of drug use. However, the social setting in which use occurs influences even this dimension. Any cigarette smoker can verify that nicotine, while chemically a stimulant, can produce relaxing effects. While marijuana has been associated with an "amotivational syndrome" in most western societies, it is used to stimulate workers in Jamaica (see Rubin 1975; Dreher 1982; Dreher 1983). Similarly, peyote consumed in a religious ritual will produce radically different effects than if a partying adolescent consumes the same drug in the same amount and manner. Thus, the social dimension of drug use even influences the physiological dimension of drugs.

Drugs also vary in their *method of use*. People ingest drugs in several different ways. They can smoke, drink or eat marijuana, for example. Heroin can be injected, smoked or sniffed. Tobacco users smoke, drink or eat their drug. In an extreme method of use, the Mayans used tobacco enemas (see Furst 1976). This dimension is determined not only by the physical properties of the drug (one cannot smoke a liquid drug), but also by social factors. First, the desired effect can produce variations in the method of use. For example, street-lore held that placing LSD in the eye instead of ingesting it orally produced a "better trip" and more vivid hallucinations (personal communication). Smoking cocaine, as either freebase or crack, produces a more immediate and intense "high" than sniffing it does. Second, social control myths can also influence the method of use. For example, some erroneously hold that smoking or sniffing heroin does not produce a physical addiction like injecting it does. Regardless of the validity or invalidity of such claims, they nevertheless pattern the method of use. Third, differing cultural traditions can lead to differences in the methods of use. Moroccan Jews, for example, who used marijuana were unlikely to smoke it in pipes since this practice was associated with Muslim Arabs (see Palgi 1975). This practice allowed the Jews to maintain a social distance between themselves and the Arabs.

A third manner in which drugs can vary is the *reason for use*. Why people consume a given intoxicant is a complicated question. The motives for use are probably as varied as those who use drugs. Generally speaking, however, motives for use can be classified as being either physiological, psychological, or social. Physiological reasons for drug use would include relieving symptoms associated with a disease, guarding against the on-set of a disease, or satisfying the pangs of physical addiction. Psychological reasons for drug use would include the variety of motives often listed to explain adolescent drug use: to relieve boredom, as a source of excitement, to escape reality, etc. Social reasons for drug use include such uses of drugs as religious or ceremonial use, to mark a rite of passage, to mark social distinctions, and recreation.

Drugs also vary in the *extent to which they are used*. The absolute number of persons using a particular substance, as well as the percentage of the population who uses a drug, varies over time and space. Indeed, rates of use vary considerably within cultures or nations over time. Several accounts have traced these variations throughout U.S. history (see, for example, Hawdon 1996b; Morgan 1981; Musto 2003). Rates of use also vary cross-culturally (see, for example, Emdad-ul Haq 2000; OGD1996; Ruggiero and South 1995; Rubin 1975; UNODC 2004). The causes of the extent of use are primarily, although not totally, social. Undoubtedly, the cultural traditions of a people, either created by them or imposed on them, largely determine if the use of drugs is common or relatively rare. The degree to which formal and informal sources of social control restrain drug use will also pattern the extent to which they are used. Yet we must recognize that some drugs are more immediately enjoyable than others are. Cocaine, for example, allegedly has the highest immediate sensory appeal of any drug (Erickson et al 1987). All things being equal then, if a culture "discovers" cocaine, it is likely that more people will use cocaine than a drug that has a lower sensory appeal. Nevertheless, *since the chemical properties of a drug are relatively constant, social and cultural factors must be the primary causes of variations in rates of use over time and space.*

Another way in which drugs vary is in the *social definitions of the drug*. The social definitions of drugs range from innocuous to "the most dangerous substance on earth." These definitions are in part due to the toxicity of the drug. While at some point in time social commentators have labeled heroin, cocaine, methamphetamine and a host of other drugs as "the most dangerous substance," it is extremely unlikely that we as a society will ever define caffeine as "the most dangerous substance on earth." Yet social definitions of drugs vary due to social factors also. It is widely recognized that tobacco and alcohol are physically dangerous drugs, yet their social standing in western cultures insulates them from the "most dangerous" label. Indeed, the correlation between a drug's social disapproval, as represented by its legal status, and the objective harm caused by the drug is weak, at best, and possibly even inverse. Moreover, the definition of a single drug varies over time. For example, the percentage of 12th-graders in the *Monitoring the Future* survey (Johnston, O'Malley and Bachman 2003) who say using marijuana once or twice poses a "great risk" ranges from a low of 8.1 percent in 1978 to a high of 27.1 percent in 1991. The percentage began to decline again after 1991 and reached another low of 13.7 percent in 2000. Assuming that the toxicity of marijuana has remained relatively stable over this time, toxicity alone cannot explain the variations in perceived harmfulness and the social standing of the drug.

Not only do social definitions of drugs vary over time and across drugs, *definitions of users* also vary. In the United States, opiate users of the late 19th century were typically considered useful members of society. By the 1920s, however, they were "diseased." By the 1940s, opiate users were fully marginalized and largely defined as outcasts (see Morgan 1981). Tobacco users were not ostracized in the 1930s or even in the late 1970s, long after we were aware of the dangers of smoking. Today, however, many view tobacco smokers as "weak-willed" or "stupid." As will be discussed later, the social status of the users largely determines our definitions of them.

I should note that I do not consider the cultivation, production, and distribution systems of drugs to be central to the sociology of drugs, per se. To be sure, there has been outstanding work conducted by historians (Courtwright

2001), anthropologists (Bourgois 1995) and sociologists (Adler 1985) that discuss and describe the distribution systems of various illegal drugs. There are also several accounts of the distribution systems of legal drugs (see, for example, Greener 2001). Although valuable to the study of drugs, this work addresses issues and questions that are best suited for the areas of organizations and criminology. This body of work discusses variations in the organization of these production and distribution systems, not drugs. That is, these systems vary in terms of their market concentration, geographic centers of production, organizational structure, career opportunities and paths, etc. Variations in these variables follow similar patterns of other systems that produce and distribute illegal commodities. That is, these systems are organized in the manner they are *because* they produce and distribute drugs. In this case, drugs are one of the *independent variables* that may or may not cause variations in the organizational structure of the system. It is true that the production and distribution systems influence the extent to which drugs are used, the method of use, the social definitions of the drug and the social definitions of the users. Yet, here again, these systems are *independent variables*. Therefore, I will not consider the production and distribution systems as *dependent variables* in this analysis.

Thus, the primary means by which drugs vary include psychoactive/physiological effects, the method of use, the reasons for use, the extent to which they are used, the social definitions of the drug, and the social definitions of their users. The social context in which drugs are used influences all of these dimensions. Therefore, sociology can contribute to our collective understanding of what influences the variation in all of these dimensions. Indeed, sociology has contributed to our understanding. However, across the six dimensions, the social context becomes increasingly more important while the chemical property of the drug and psychology of the user become less important. Although all of these areas are open to study by sociologists, I will concentrate primarily on the extent of use, the reasons for use, the social definitions of drugs, and the social definitions of users.

A Note on Methodology

The methodology used throughout this book is primarily secondary analysis. Historical and cross-cultural accounts of drug use support the various theoretical propositions presented here. When possible, I offer quantitative support for the propositions. Before discussing the world's drug consumption patterns and attempting to explain them, it is necessary to provide a cautionary note about the data. Despite the widespread use of intoxicants, the data available regarding how many people use specific types of drugs is dreadfully lacking. Reasonably accurate data exist in most developed nations, however, as a general rule, as a nation's GDP decreases, so does the reliability and validity of the nation's drug-use data.

There are several reasons for the lack of data on drug consumption. Primarily, much drug use is illegal and therefore hidden from authorities and researchers. Research indicates that self-report data concerning illegal activities are reasonably accurate if collected under certain conditions, but the validity of the data tends to vary by social class (see Katz et al 1997; Del-Boca and Noll 2000). In addition, there are other problems associated with self-report data such as telescoping and providing socially desirable or trivial answers. Even if we avoid such problems, we still need relatively large samples to provide accurate estimates of population parameters. Collecting data from numerous individuals is often cost-prohibitive, especially in developing countries.

A second source of insufficient drug data is the non-representative nature of many of the samples used for our estimations. Even well designed data collection efforts frequently miss the more marginalized sections of society such as high school dropouts, incarcerated individuals and homeless persons. Under representing these populations generally does not adversely affect the estimates of overall levels of drug use or the use of common drugs such as marijuana; however, the estimates of use of the more "problem drugs" such as cocaine and heroin are often dramatically under-reported. For example, the US National

Household Survey estimated that 3.8 million Americans used cocaine at least once in 1998 while the US Office of National Drug Control Policy (ONDCP) estimated the total number of cocaine users to be 6.5 million Americans (see UNODC 2000: 69, note n). Thus, the household survey only accounted for approximately 60% of the cocaine users in the nation. Similarly, the household survey accounted for approximately 20% of the U.S. heroin-using population. The differences in the estimates were due to the household survey's exclusion and ONDCP's inclusion of marginalized populations. Similar discrepancies were found in many European studies (UNODC 2000: 69, note n). Given the more widespread use of telephones and other communication devices in developed nations relative to developing nations, we can assume the general population in developed nations is more accessible for data collection efforts than the population of lesser-developed countries (LDCs). Thus, we can also assume that the under-representation of the marginalized populations is more pronounced in LDCs.

A third reason we lack good data on drug use is that, despite the longstanding tradition of criminalizing drug consumption, governments and researchers made little systematic effort to determine the extent of use until relatively recently. Even in the United States where federal drug legislation is nearly a century old (see Morgan 1981; Musto 2002) and resources for data collection efforts are available, widespread systematic efforts to collect data on drug consumption were not prevalent until the early 1970s (see Hawdon 1996). At the world level, while information about drug cultivation, production and trafficking have been collected for nearly a century, drug consumption information has traditionally been considered a domestic issue and not open for international assessment (UNODC 1999). A systematic attempt to collect consumption data was not undertaken until a Special Session of the United Nations General Assembly passed resolution S-20/2 and S-20/3 in June 1998. During that session, called to address the world's "drug problem," the General Assembly adopted the *Declaration on the Guiding Principles of Drug Demand*

Reduction, which established standards for international data collection efforts as part of UNODC's "balanced approach" to drug reduction. It was decided that any drug reduction activities had to be developed based on objective and scientifically valid assessments of the drug problem (UNODC 2000: 87). Although a number of countries have developed comprehensive drug monitoring systems, most countries have not. Without comprehensive surveys to estimate use patterns, the United Nations' Office of Drug Control and Prevention (UNODC) relies heavily on "observations by professionals in treatment institutions or health authorities" to assess drug-use trends (UNODC 2000: 61).

A fourth reason for the dearth of good drug data is that drugs are a political reality. As such, it is not always in the best interest of authorities to know how many people are using drugs. For example, nations may "miscalculate" rates of use because developed nations such as the United States often tie international aid to drug reduction efforts.[2] Given such pressures to decrease the cultivation and production of drugs within their boarders, LDCs have an incentive to under-report the nation's drug patterns. Conversely, other nations may over-estimate the extent of the "drug problem" in hope of attracting additional international aid to fight the "problem" (Fazey 2002). In nations with hugely profitable drug trades, reporting may result in unwanted international interference (Fazey 2002). Under reporting, or failure to report at all, can also be due to differing cultural definitions of what constitutes a "drug problem." For example, despite the widespread use of marijuana in Jamaica (see Dreher 1982; Jutkowitz and Eu 1994; Rubin 1975), Jamaica did not report a "marijuana problem" to the UN in 2002 (UNODC 2002). Finally, during times of "drug scares" or moral panics, estimates of use are often exaggerated (see Goode and Ben-Yehuda 1994; Hawdon 1996). For example, Harry Anslinger's attempt to combat marijuana use in the 1930s was complete with sensational stories of

---

[2] For example, in 1994, Bolivia agreed to eradicate 3,200 hectares of coca fields in exchange for $20 million in credit to help steady Bolivia's balance of payments (see OGD 1996: 160 - 161).

insanity and violence caused by marijuana use (see Anslinger and Cooper 1937). His efforts failed to cause prolonged concern however because rates of use were so low (see Goode 1999; Hawdon 1996; Morgan 1981). In short, the politics of drugs obscures the information we have concerning use patterns around the world. Whether misreporting or non-reporting is due to deliberate decisions, prioritizing other problems, international considerations, or simply the lack of resources, the result is "we have very little true knowledge about what is happening regarding drug consumption in the world" (Fazey 2002: 95).

With these problems in mind, we must still attempt to understand the world's drug consumption patterns. This project primarily relies on data from the United Nations' Office of Drug Control and Prevention (UNODC). UNODC's *Global Illicit Drug Trends* and *World Drug Report* provide the bulk of the data. UNODC's data are obtained from the Annual Reports Questionnaire that is distributed to all UN member countries each year.[3] These data are supplemented with other sources such as other governmental reports and data from Interpol and the World Customs Organization (for a complete discussion of sources and limitations of the data see UNODC 1999: 13; UNODC 2002: 281).[4] These data, flawed and limited as they may be, are, nevertheless, the best available. The UN uses several techniques to assure the quality of their data.

When possible, I supplement the UNODC data with independent, non-governmental researchers' survey data or ethnographic accounts. Despite the UN's and my attempts to verify use estimates through the triangulation of various data sources, the estimates provided below are approximations at best. Because of this lack of exactness, any direct comparisons between countries should be considered as suggestive and tentative, not "gospel truth." What is assumed here

---

[3] Unfortunately, the Annual Reports Questionnaires did not include "demand side" information prior to 1999.

[4] For a list of sources used by UNODC, see *Global Illicit Drug Trends, 2002*, pages 281 - 285.

is that the relative ordering of consumption patterns is reasonably valid. While precise estimates of the number or percentage of the population who uses a given drug may not be entirely accurate, the relative ordering of nations with respect to use should be more conforming to facts. When discussing the limitations of the consumption data they present UNODC (2002: 283, emphasis mine) states,

> These measures -- the pooling of national results, standardization and extrapolation from subregional results in the case of data gaps -- does not guarantee an accurate picture. But it should be sufficient to arrive at *reasonable orders of magnitude about the likely extent of drug abuse in the general population.*

Thus, while we cannot directly compare estimates of cocaine use, for example, across nations and say with confidence that the United States has *twice as many youth* who have tried cocaine (6%) as does Germany (3%) (see UNODC 1999: 132), we can probably safely conclude that *more US youth have used cocaine at least once in their life than have German youth.* The confidence of the relative rankings of use across areas also increases as the geographic scope under consideration increases. That is, although UNODC's *Global Illicit Drug Trends* (1999: 133) reports a lifetime prevalence rate of heroin abuse among youth of 1.9 percent for Slovakia and only 0.7 percent for Swaziland, I am hesitant to conclude that Slovakia truly has higher rates of youthful heroin use than Swaziland. I can, however, based on these and other data, confidently state that Europe has a higher rate of heroin use than does sub-Saharan Africa.

With these cautions in mind, we can now turn to the empirical patterns of drug consumption.

CHAPTER TWO

THE EMPIRICAL PATTERNS OF "LEGAL" DRUG USE

Before any attempt to offer a sociological theory of drugs is made, we must first review the empirical patterns that are to be explained. Specifically, what are the patterns of contemporary drug use in the world at the beginning of the twenty-first century? What regions and which countries have relatively high rates of drugs use? What drugs are popular where?

I do not intend to provide an overly detailed discussion of these patterns. Interested readers should consult the UN's various documents on illegal drug trends for country-by-country data. Specifically, *Global Illicit Drug Trends*, which is published annually, and the *World Drug Report*, most recently published in 2002, report the data referred to throughout much of this work. I do intend to paint a general picture of global drug patterns of the late 20th and early 21st centuries. On occasion, I will focus on specific countries to provide a greater sense of the variations that occur. However, most of the discussion will be at the regional or sub-regional level.

While most discussions of drugs focus solely on illicit drugs with occasional mention of tobacco and alcohol, I will consider illegal and legal drugs. I will even consider the consumption of pharmaceutical drugs. While the legal status of a drug is an important independent variable when trying to explain why some drugs are used more than others, the legal/illegal distinction should not exclude us from considering some drugs that are widely used in some parts of the world. It is my position that the legality dimension is to be explained, not simply taken-for-granted. In one sense, morphine is morphine regardless of the conditions under which it is used. Moreover, the heavy reliance on pharmaceutical drugs to combat disease reflects structural and cultural characteristics of a society that can provide clues about the use of illicit drugs in that society. Thus, ignoring the use of legal drugs can blindfold us in our attempts to explain the use of illegal ones.

I will discuss the trends in "legal" drugs first. It should be recognized that alcohol, tobacco and pharmaceutical drugs are also consumed illegally. Alcohol is outlawed in some nations and many nations prohibit its use by youth. Despite these prohibitions, it is consumed where it is outlawed and by those who are forbidden to use it. Similarly, the U.S. Drug Enforcement Agency estimated that $25 billion worth of prescription drugs were sold on the black market in 1993. This amount is close to the $31 billion Americans spent on cocaine that year (Bellenir 2000). Although these drugs are consumed illegally at times, they are legal for most people to use in most countries at most times. The consumption of illegal drugs will be discussed in Chapter Three.

The Nearly Ubiquitous Drugs

Before comparing regions, we should mention the world's ubiquitous or nearly ubiquitous drugs. There are three such drugs: caffeine, alcohol, and tobacco. All of these drugs are widely used in most parts of the world. Their use is so common that many people still do not consider them drugs despite their psychoactive properties. The nearly universal use of these drugs is undoubtedly due to their legality in most societies. In addition -- and in part why these drugs are legal -- these drugs were among the favorites of Europeans during the 15th through 19th centuries while they colonized the world (see Szasz 1974; Courtwright 2001).

Caffeine

Caffeine is the world's most widely used psychoactive substance. Approximately 70 milligrams per person is consumed around the globe each day

(Courtwright 2001). Yet, depending on the culture and nation, average caffeine consumption ranges from 80 to 400 mg per person, per day (Daly and Freedholm 1998: 199). Westerners are undoubtedly familiar with caffeine from their favorite stimulating drinks coffee and tea. Indeed, by the late 1990s, the world's hot-drinks sales reached $53 billion (*The Economist* 1998). Yet, caffeine is also present in cocoa, kola nuts, guarana and over sixty other plants. Several pharmaceutical preparations, including some analgesics, also contain caffeine (htdocs 1991; Hanson, Venturelli and Fleckenstein 2002). Caffeine containing foods and drinks are used "at least weekly by nearly all adults" (Pollak and Bright 2003). Known mostly for its mild stimulating effects, caffeine has been used for medicinal purposes and has several health benefits. Of course, caffeine consumption can also pose health risks (see Parliament, Ho, and Schieberle 2000; Noonan 1995; Hanson et al 2002; McKenna 1992; Daly and Freedholm 1998).

Worldwide, and in most nations, coffee is the preferred means of ingesting caffeine (*The Economist* 1998; htdos 1991). Coffee apparently originated in what is now the highlands of Ethiopia. From there it speared eastward and northward, and was popularized in the Arab world through Sufi religious practices (Jamieson 2001). It arrived in southern Europe in the early 17th century and quickly became popular throughout Europe. In England alone, for example, there were thousands of coffeehouses by the end of the 17th century (Jamieson 2001). Throughout Europe, coffeehouses were the focal institution of the emerging bourgeois class of urban professionals (Jamieson 2001; Hattox 1985). Once the Europeans grew fond of coffee, they took control of and monopolized the existing Asian and Near East trade of the commodity. Although the European coffee trade was initially limited to the Near East, they rapidly expanded their markets and spread coffee's use throughout the world (Jamieson 2001). As Courtwright (2001: 19) notes, "it was Europeans who made coffee into a worldwide drink and global crop." Today, coffee is truly a global commodity (see Courtwright 2001; Jamieson 2001; Ponte 2002) and a multi-billion dollar industry. The retail value of U.S coffee sales alone was $18.5 billion in 1999 (Cosgrove 2000).

Coffee, a relatively inexpensive drug, is popular everywhere. According to marketing-research statistics, the Swiss consume 8.7kg of coffee per capita each year. This would make them the leading coffee consumers (the *Economist* 1998). In comparisons, Italians, known for their love of coffee and espresso, consume only 2.8 kg of coffee each year. Americans also consume a great deal of coffee, and consumption rates have increased by over 1/3 during the 1990s. Over 161 million Americans, or approximately 79 percent of adults over 18, drink coffee at least occasionally. Nearly 54 percent of those Americans who drink coffee do so daily (Cosgrove 2000).

Coffee is consumed in both private and public settings. As in earlier times, coffeehouses are primary consumption centers (Cosgrove 2000; Hattox 1985; Jamieson 2001). Although considerable amounts of coffee are consumed in the home, coffeehouses can be found in most large cities throughout the world. They have become increasingly popular in small towns and rural areas, at least in developed countries. These shops purposefully offer a comfortable setting to entice users to stay for long periods of time (Cosgrove 2000). Along with bars, coffeehouses are among the few places that openly and legally distribute drugs. Of course, one can also purchase prepared coffee in most convenience stores, fast food outlets, and restaurants in the developed world and many larger cities of the developing world. Coffee is truly a ubiquitous drug.

Tea has a similar history in that Europeans spread its use across the globe. Originally cultivated in China and India, and noted for its medicinal properties, tea spread throughout southeastern Asia and into Japan by the 9th century. The Dutch East India Company introduced tea to Europe in 1610 (Jamieson 2001). The Dutch also introduced tea in the Americas (Hanson et al 2002). Although access to tea was limited throughout much of the 17th century, by the end of that century it had become widely popular throughout Europe (Jamieson 2001). Tea especially gained popularity in England, Wales and the Netherlands (Courtwright 2001). Europeans then introduced the plant to their colonies where it also became popular, especially in English America (Jamieson 2001).

Most nations drink more coffee than tea (the *Economist* 1998). However, worldwide tea consumption is increasing. One reason that tea consumption has risen lately is the growing body of evidence that suggests tea is a healthy drink (Cosgrove 2000). Although long known for its health-promoting properties, tea is now widely promoted throughout the west as a healthy drink (Noonan 1995). Recent evidence indicates that drinking five cups of tea per day is equivalent to eating two vegetables. Tea contains antioxidants and is associated with a decreased risk of heart disease, stroke, and cancer. Both green and black teas produce these benefits (Noonan 1995).

Tea remains popular throughout Asia, especially in India and China. It is also widely consumed throughout Africa and Europe. Ireland is the leading tea-consuming nation in the world. The Irish brew "3.6kg each year, far more than the second-place British" (*The Economist* 1998: 84). Americans have recently increased their tea consumption, consuming just over 7 gallons per year, per capita by the mid-1990s (see Frank 1995). Specialty teas represent a "small but vibrant" segment of the American tea market (Cosgrove 2001: 16), and ready-to-drink tea helped per capita tea consumption surpass juice consumption in the United States in 1994 (Frank 1995). Although obviously present and consumed there, tea has never become a leading cash crop in South America (Courtwright 2001).

While most caffeine consumption comes through drinking coffee and tea, caffeine also naturally occurs in cacao, the source of our beloved chocolate. Cacao was originally cultivated in the rainforests of the Americas where the Aztecs had maintained large-scale production centers since the fifth century. Aztec and Mayan elites frequently used cacao (Courtwright 2001). Throughout the Aztec empire, cacao served as a form of currency and was an important commodity in the empire's trade networks (Jamieson 2001). Although originally confined to the upper class, chocolate was especially popular in Spain, Italy and France during the 17th and 18th centuries. It became widely used throughout Europe in the 19th century when industrialized production and widespread cultivation made it increasingly affordable. By the early 20th century, "cocoa became a breakfast drink for children, chocolate candies tokens of middle-class affection" (Courtwright 2001: 25).

Today, cacao is primarily grown in West Africa (Coe and Coe 1996), while the United States and the Netherlands are the primary grinding countries (ICCO, 2000). Consumed in nearly every part of the world, it is estimated that the worldwide confectionery industry, including sugar and chocolate confectionery, is worth around £51billion a year (ICCO, 2000). The United Kingdom spends more per-head per-year than any other nation on earth. The per-head per-year expenditure on cocoa in the UK was 90 U.S. dollars in 1999. The Irish and Swedes also have above average per-head per-year expenditures on cocoa products (ICCO 2000).

In 1999, world cocoa consumption was 0.525 kg per-person per-year (ICCO 2000), but rates vary considerably. Generally speaking, developed nations consume more cocoa than developing countries. In many northern European countries, residents consume between 7 and 10 kg per year (Seligson, Krummel, and Apgar 1994). The Swiss have the highest consumption rate (10.18 kg per person per year). Belgium, Germany and Austria also consume higher than average amounts of chocolate (ICCO 2000). While Europe consumes the most chocolate per person, it is also extremely popular in the Americas. Per capita consumption of chocolate in the United States is approximately 4.6-4.8 kg per year (Seligson et al 1994). Asia and Oceania have the lowest per person consumption rates at .093 kg per-person per-year, although consumption in Japan and Australia is relatively high. In 1999, the average rate of cocoa consumption in Africa was .146 kg per-person per-year (ICCO 2000).

Caffeine is also naturally found in kola nuts. Indigenous to West Africa, kola nuts contain more caffeine than coffee and traces of theobromine, another mild stimulant. Despite its popularity in West Africa, kola did not become popular worldwide until it was added to "soft drinks" (Courtwright 2001). In the late 19th century, Europeans began mixing kola with wine. John Pemberton's

Coca-Cola originally included cola, coca, and alcohol. When he removed the alcohol to appease prohibitionists, he was able to market his product as a temperance, or "soft," drink. Coca-Cola spawned numerous imitators and competitors, including Pepsi-Cola. Cola-containing drinks became a popular child's drink by the early 20th century. After World War II, Coca-Cola became global. By 1991, Coca-Cola was sold in 155 nations across the globe (Courtwright 2001: 26), and Coca-Cola products are now sold in over 200 nations. According to the Coca-Cola Company, there are nearly 10,450 Coca-Cola soft drinks consumed every second of every day (Coca-Cola web page, April 14th 2003). While not all Coca-Cola products contain caffeine, most do. Caffeine containing sodas are the primary source of caffeine for most children in most countries (see, for example, Pollak and Bright 2003; Hanson et al 2002). Although natural kola has been replaced by synthetic cola-extract in most soft drinks (Daly and Freedholm 1998), the caffeine content of these drinks still ranges from 30 to 60 milligrams per 12-ounce serving (Hanson et al 2002: 287).

Kola nuts are still widely used in West and Central Africa. They are widely consumed in Nigeria, Ghana, and the Congo region (see Gureje and Olley 1992). The nut is broken into small pieces and chewed. It is used for its stimulating qualities and believed to be an aphrodisiac. Although children will eat kola nuts on occasion, the practice is primarily limited to adults. Kola nuts are frequently consumed after meals to stimulate conversation and are often served with a sweet palm wine to counteract the bitterness (personal observation). Because kola nuts dry easily, the practice of chewing raw nuts is largely limited to Western Africa where the crop grows.

Although caffeine is found in over 60 plant species, coffee, tea, cacao, and kola are the four primary sources of the drug. Still, there are other noteable products that offer a dose of caffeine. In Japan, for example, there is a thriving market for caffeine chewing gum and lozenges marketed under the names "Strong Man" and "Sting" (Vaughn, Huang, and Ramirez 1995). Regardless of its form, caffeine-containing substances are legal and consumed by adults and children.

While once products of the elites, coffee, tea, chocolate and kola-containing beverages are now as "class-less" as any product is. The young and old, males and females, and rich as well as the poor consume caffeine. The legality, stimulating effects, and relative safeness of these drugs undoubtedly contribute to caffeine's popularity and its dominance of the world's drug consumption habits.

Tobacco

Tobacco is indigenous to the Americas and had been used there for centuries before the Europeans arrived. At the time of Columbus' arrival in Haiti, there was "virtually no Indian population, from Canada to southern South America, to whom one or another of the major species of tobacco was not sacred and that did not either cultivate it or obtain it by trade from their neighbors" (Furst 1976: 27). Tobacco was used by numerous tribes in a number of religious ceremonies (Furst 1976; French 2000; Hirschfelder and Molin 1992; Lyon 1998; McKenna 1992; Pego et al 1995; Perrin 1992; Thompson 1970). The most common means of use was smoking; however, tobaaco was also drunk, snuffed, licked, sucked, eaten, and used in enemas (Furst 1976).

Tobacco was a sacred drug and rarely, if ever, used for simple pleasure by indigenous Americans (Furst 1976; French 2000; Mooney 1891). The importance of tobacco to Native American culture is apparent in numerous myths and creation stories. Carib Indians, for example, believe a good shaman defeated a bad shaman by producing more varieties of tobacco. Quiche Maya of highland Guatemala hold that the Hero Twins were able to fool the rulers of the underworld by placing fireflies at the tips of their cigars and only pretending to smoke them. They were then able to light the unused cigars in the morning and meet the challenge of keeping their cigars lit all night (Furst 1976). Numerous tribes have similar legends that involve tobacco. In fact, tobacco probably served a greater variety of sacred purposes for a greater variety of groups than any other plant in the New World (see, Brown 1953; Furst 1976; Hirschfelder and Molin 1992; Lyon 1998; Mooney 1891).

Tobacco use changed radically once Europeans became aware of it. Although its use spread slowly at first, it would, in time, cover the globe. The Spanish introduced tobacco to the Philippines in the late 16th century where it quickly became a cash crop. Portuguese sailors introduced the habit, and tobacco seeds, to much of the world (Hanson et al 2002; Courtwright 2001; Gately 2002; Goode 1999). As the European colonial powers began to cultivate it in numerous colonies, tobacco became a major cash crop that helped fuel European colonial expansion and, eventually, the industrial revolution (see Goodman 1993). Even though its expense limited consumption, tobacco found admirers wherever it was introduced. As Courtwright (2001: 15) notes,

> Like ripples from a handful of gravel tossed into a pond, tobacco use and cultivation spread by secondary and tertiary diffusion: from India to Ceylon, from Iran to Central Asia, from Japan to Korea, from China to Tibet and Siberia, from Java to Malaysia to New Guinea. By 1620 tobacco was, by any definition, a global crop.

When colonial tobacco production expanded and the price of tobacco fell, use became widespread, despite prohibitions against its use in several countries.[5] The English and Dutch led the way in Europe's early use of tobacco. In Asia, the Chinese were known to be fond of the drug (see Courtwright 2001; Gately 2002). Tobacco use continued to increase throughout the world in the 18th and 19th centuries. In the mid-18th century, pipe and cigar smoking surpassed snuffing and chewing in Europe. Cigarette smoking was rare until the late 19th century when, in 1881, the cigarette-rolling machine was patented. After that technological development, cigarettes rapidly gained popularity. For example, "between 1918 and 1963, the number of cigarettes sold in the United States increased from 45.6

[5.] Anti-tobacco sentiments are not new. Tobacco users have faced punishments, from mild to severe, throughout history. Russian smokers, for example, were beaten or exiled. Some of the more severe punishments found throughout history included the punishment of Turkish smokers who "under the reign of Ahmed I endured pipe stems thrust through their noses; Murad IV ordered them tortured to death" (Courtwright 2001: 16). Although never overly severe, the United States has attempted to regulate tobacco smoking for over a century. For a discussion of tobacco legislation in the U.S., see Jacobson, Wasserman and Anderson (1997).

to 523.9 billion; the per capita consumption jumped from 697 to 4,345" (Goode 1999: 198). By the late 1950s, Americans were "purchasing upwards of 15,000 cigarettes a *second*" (Courtwright 2001: 18, emphasis in the original). By the mid-20th century, cigarettes had become the primary means of consuming tobacco in most countries of the world (Courtwright 2001; Gately 2002).

Rates of use continued to increase until the 1970s after which they stabilized or declined in most developed nations. For example, according to OECD data, over half of the United Kingdom's population smoked daily in 1960. By 2000, only 27 percent of the population smoked daily. Likewise, in the Netherlands, 59 percent smoked daily in 1960 while only 33 percent did so in 2000. Similar trends are evident throughout much of the developed world. Twelve of fourteen of the most developed nations have seen the percentage of the population who are daily smokers decrease since the 1970s.[6] Since 1990, per adult consumption in the developed world has declined by approximately 1.4 percent (WHO 2000).

While tobacco use has stabilized or decreased in many developed nations, rates of use have continued to increase in many developing nations (WHO 1997). For example, per capita consumption rose by 3.9 percent per year between 1983 and 1992 in China (WHO 2000). By the early 1990s, it was estimated that sixty-eight percent of adult males and seven percent of adult females in China smoked (WHO 1997). Similarly, per capita consumption rose by 2.0 percent per year between 1983 and 1992 in India. Over half of all adult Indian males now smoke (WHO 2000; 1997). In Kenya the smoking rate among primary school children was estimated at forty percent in 1989. This rate was only at ten percent in 1980 (WHO 2003). In Bolivia, Argentina and Chile, smoking among 13 to 15 year olds ranges from 31 to 39 percent (WHO 2002). In general, while consumption in

---

[6.] The 12 developed nations that have witnessed declines in smoking are Australia, Belgium, Canada, Denmark, France, Italy, Japan, the Netherlands, Sweden, Switzerland, the United Kingdom, and the United States. Austria and Germany's consumption rates have increased slightly (OECD 2002).

the developed world declined by approximately 1.4 percent during the 1990s, it increased in the developing world by 1.7 percent during that decade (WHO 2000).

Today, tobacco is the third most widely used drug in the world in terms of number of people who use the drug. It ranks behind both caffeine and alcohol. However, it is probably the most *frequently used* drug in the world since heavy users consume 20 to 80 cigarettes per day, every day. Even the most hard-core caffeine addict would be hard pressed to drink 60 cups of coffee per day. Having 60 alcoholic drinks per day would likely kill the user in a relatively short period; probably on the first day such quantities were consumed. Thus, although they are outnumbered, tobacco users use their drug a lot.

The World Health Organization (2000) estimates that there are about 1.1 billion regular smokers worldwide. These smokers smoke 5.5 trillion cigarettes annually, or "a pack a week for every man, woman, and child, smoker or non-smoker, on the planet" (Courtwright 2001: 19). The vast majority of smokers, approximately 800 million, are in developing nations. However, per capita daily consumption is still higher in the developed world. In the early 1990s, the highest average per capita consumption was in Europe. Europeans smoked 2,290 cigarettes per-adult per-year (WHO 2000). Smoking is especially prevalent in Eastern Europe (WHO 2002b; WHO 1997). For example, among 13-15 year old students surveyed for WHO's *Global Youth Tobacco Survey*, over thirty-three percent of youth in the Ukraine and the Russian Federation smoked. Moreover, nearly thirty percent of urban Polish youth indicated they smoked (see Warren et al, 2000). The next highest per capita consumption rates were in the Western Pacific at 2,000 cigarettes per adult per year. Africa, at 540 cigarettes per adult per year, had the lowest consumption rate.

While residents of developed nations still smoke more than those in developing countries, the gap in tobacco consumption is narrowing. In 1970, per adult consumption was 3.25 times higher in developed countries than it was in developing countries. By 1980, this ratio had narrowed to 2.38. The 1990 ratio was only 1.75 and it has likely narrowed more since then (WHO 2000). Today, it is estimated that 50% of the adult male population in developing countries smokes tobacco, and consumption rates are increasing (WHO 1999; Shenon 1998). Per adult cigarette consumption in the developing world has been increasing since 1990. For example, consumption has increased by 7.0% each year since 1990 in China where over 300 million people now smoke (Shenon 1998). Per adult cigarette consumption in the developing world is expected to surpass that of the developed world sometime between the years 2005 and 2010 (WHO 2000). Thus, global trends are beginning to mirror those within many developed nations. While the "class gap" in smoking was almost non-existent in the United States in the 1950s, today, smoking is inversely correlated with social class. The emergence of this class gap was due to wealthier persons quitting their smoking habit while more among the poor began to smoke. Globally, the wealthier nations are reducing their use of tobacco while the poorer nations are increasing their tobaccoconsumption. Eventually, if these trends continue, smoking will be a habit of the poor, both within nations and among them.

Tobacco use is highly gendered, at least in the developing world.[7] According to WHO (2000), 700 million males smoke in developing nations while only 100 million females do. In India, for example, while nearly 65% of adult males smoke, only 3% of adult females do so (WHO 1999). Similarly, in South Korea, 68% of males and 7% of females smoke (Levinthal 2002: 251; WHO 1999). In Sri Lanka, the gender gap is even more pronounced. While 54% of Sri Lankan males smoke, less than 1% of females smoke (WHO 1999). These gender gaps in smoking are likely to continue in many developing countries.[8] For

---

[7] Nepal may be an exception to the general pattern of large gender "smoking-gaps" in developing countries. In rural Nepal, it was estimated in the early 1980s that nearly 85% of males and 72% of females smoked tobacco. However, in urban Nepal, nearly 65% of males smoke while only 14% of females do so (WHO 1999).

[8] Bangladesh may be an exception to this trend. While over 60% of men and only 15% of women in Bangladesh smoke, smoking among women increased from 1% to 15% in the early 1990s (WHO 1999). Since rates have remained relatively stable among males, the gender smoking-gap in Bangladesh appears to be decreasing.

example, while smoking rates among male Korean teenagers was 30% in 1993, only 9% of Korean female teenagers smoked. In Tashkent, Uzbekistan, 22.5% of teenage males smoke, but less than 1% of teenage females do so (WHO/EURO 1997). Similarly, among Nigerian secondary school children in Lagos, 40% of the boys and 8.4% of the girls smoked tobacco (Gureje and Olley 1992). Given that the majority of smokers begin while in their teenage years or earlier (DiFranza and Tye 1995; Warren et al 2000; WHO/EURO 1997), these numbers indicate that smoking in developing nations will likely continue to be a male-dominated past time for the foreseeable future.

The story is much different in the developed world where the gender gap tends to be much smaller. In total, 200 million males and 100 million females are regular smokers in developed countries (WHO 2000). The gender gap in smoking is quite narrow in some nations. In Australia, for example, 29% of men and 21% of women smoke. In the U.S., 28% of men and 24% of women are regular smokers. A similar ratio is seen in the United Kingdom. These relatively small gender gaps will likely reduce more in the near future. In some nations, they may even reverse. In Sweden, for example, females are actually more likely to smoke: 24% of females and 22% of male Swedes smoke regularly (Levinthal 2002: 251). According to WHO's European regional office's (WHO/EURO) Health Behavior in School-Aged Children survey, young women are smoking more than young men in eight of the twenty countries that were surveyed in 1993/1994. Fifteen-year-old women were more likely to smoke than their male counterparts in Austria, Denmark, France, Germany, Norway, Spain, Sweden, and the United Kingdom. The ratio was almost 1:1 in Greenland, Belgium, Finland, and Israel. Only in the lesser-developed nations of the former eastern bloc (Estonia, the Czech Republic, Hungary, Latvia, Lithuania, Poland, the Russian Federation, and Slovakia) did the gender ratios in smoking among fifteen-year-olds heavily favor males (WHO/EURO 1997).

In a matter of 500 years, tobacco has been transformed from a rarely used, tightly controlled ceremonial drug of the Americas to a frequently used, largely uncontrolled, recreational drug of the world. Its primary users were once religious elites. Now, the primary users of tobacco are the lower classes of the developed world and those who can afford it in the developing world. Its preferred method of use went from chewing and sniffing to smoking. Despite harsh restrictions on its use, tobacco continued to find consumers. Even the threat of death, torture or exile did not halt the spread of tobacco. Despite frequent warnings about tobacco's ill effects, millions -- indeed over a billion -- use the drug. It is one of the world's favorite drugs.

Alcohol

Humans have consumed alcoholic beverages for over 8000 years. Archeologists have found beer jugs dating back to the Stone Age. Fermentation is a natural process. Thus making alcohol is easy. In fact, natural sources of alcohol that require no human action exist. Palm wine, for example, is extracted directly from the tree. Animals have been known to "get drunk" from eating rotting fruits. It is therefore very likely that even our earliest ancestors experienced the intoxicating effects of alcohol, if only on occasion and by accident.

As a species, humans got serious about alcohol with the advent of viticulture, the making of wine from grapes. Viticulture can be traced to the Armenian region as early as 6000 to 4000 BC. Over the next 4000 years, wine drinking spread throughout the Mediterranean world (Courtwright 2001; Unwin 1991). Wine was extremely popular in ancient Greece, so much so that regulations were passed to dilute wine with water and supervise drinking at meals (Fort 1973). From Greece viticulture spread to Russia, India and China (Courtwright 2001), although the Chinese had been drinking fermented rice and millet long before the arrival of wine in the second century B.C. The process of distilling spirits was advanced by the Arabs. Despite Islamic law prohibiting its use, Arabs introduced distilled spirits to Europeans in the 11th century (Courtwright 2001; Fort 1973). The consumption of alcohol flourished in medieval Europe. Beer and ale were the drinks of the commoner while the aristocracy preferred wine.

Alcohol reached the western hemisphere during the 16th century with the arrival of Europeans. By mid-century Jesuit priests had spread viticulture as far south as Chile and across the continent of South America. English colonists brought distilled spirits to North America. Indeed, colonial ships often carried more alcohol than water. By the mid-17th century, the Dutch had introduced viticulture to the southern tip of Africa, and the distilling industry was booming throughout Europe. Viticulture reached Australia when the British established a penal colony there in 1788 (Courtwright 2001; Unwin 1991). Thus, Europeans clearly stimulated the growth in alcohol production and consumption. Although the use of alcohol was known to most cultures (with the possible exceptions of arctic dwellers and North American Indians), Europeans raised it to an art form.

Global alcohol consumption has increased in recent decades, with most of this increase occurring in developing countries (WHO 2002b). Still, drinking patterns vary dramatically between sub-regions. Table 2.1 reports the liters of pure alcohol, beer, wine, and spirits consumed per adult by region. The data are from WHO statistics for 172 countries. They report the adult per capita alcohol consumption for the year 2000 (WHO 2003b).

Table 2.1: Alcohol Consumption by Region

REGION	Total alcohol liters per capita age 15+	Beer liters per capita age 15+	Wine liters per capita age 15+	Spirits liters per capita age 15+
Africa	3.18	1.45	0.42	0.62
North Africa/Middle East	1.23	0.39	0.15	0.52
North America	6.89	3.94	1.17	1.77
South America	5.74	2.00	0.83	2.88
Eastern Europe	10.25	2.79	2.50	4.86
Western Europe	11.47	3.98	5.26	2.10
Oceania	4.46	2.62	1.26	0.56
Asia	2.69	.47	0.33	1.70
World	5.24	1.83	1.33	1.80

Source: *World Health Organization*

As can be seen from Table 2.1, the world consumed 5.24 liters of pure alcohol per adult in 2000. This consumption is nearly equally divided between beer, wine and distilled spirits. As WHO warns, there are the typical problems with these data; however, these data should provide a reasonable estimate of the world's alcohol consumption.[9]

Not surprising, Europeans are the leading alcohol consumers in the world today. As seen in the table, the people of Western Europe consume 11.47 liters of pure alcohol per adult each year while Eastern Europeans consume 10.25 liters per person. The top seven alcohol-consuming nations of the world, and twenty-three of the top thirty (76.7%), are European. France leads all nations with an annual consumption rate of 20.28 liters of pure alcohol per person. Other heavy-drinking European nations include, in descending order, Spain, Moldavia, Portugal, Luxembourg, Ireland, and the Czech Republic. While most European nations consume relatively large amounts of alcohol, there is variation across Europe in the preferred type of alcohol. While almost half of Western Europe's consumption is in the form of wine, almost half of the alcohol consumed in Eastern Europe is from distilled spirits. Western Europe's wine consumption is astounding. The Western European nations consume, on average, 5.27 liters of pure alcohol in the form of wine per person each year. This is more than double the amount consumed by Eastern Europeans, the second-leading wine consuming region of the world. Western Europe also edges out North America for the leading beer-consuming region of the world (3.98 versus 3.94, respectively). Similarly, no one truly comes close to drinking as much distilled alcohol as the Eastern Europeans. Eastern Europeans consume a full 2 liters of pure alcohol in the form of spirits more than the second leading spirit-drinking region of South America. To put this in perspective, those two *additional liters* is more than the

---

[9.] The category beer includes data on barley, maize, millet and sorghum beer combined. The amounts from beer, wine and spirits do not necessarily sum to the total. The total alcohol data also include other beverage categories such as palm wine, vermouths, cider, fruit wines, etc. (WHO 2003b).

total distilled spirit consumption rates of Africa, the Middle East, North America, Oceania and Asia. It nearly matches Western Europe's rate of 2.10 liters per person.

At 6.89 liters per person, Eastern Europe's total alcohol consumption is over thirty percent more than North America's, the third leading alcohol-consuming region. At 9.08 liters per adult, the United States ranks fortieth in pure alcohol consumption. Canada is 47th (7.58 liters per adult) and Mexico is 83rd (4.01 liters per adult). South and Central America, including the Caribbean, is the fourth leading alcohol-consuming region. Consumption of pure alcohol in this region is 5.75 liters per person. However, some notably heavy-drinking nations dramatically increase the region's mean. The Bahamas, for example, is the eighth leading alcohol-drinking nation at 15.26 liters per person, and St. Lucia is ninth at 14.56 liters per person. In fact, if the Caribbean is separated from Central and South America, the Caribbean has a higher rate of alcohol consumption than North America. Central and South America have similar consumption rates (3.32 and 3.35, respectively). The leading South American nation in terms of alcohol consumption is Argentina, which ranks 38th out of the 171 nations. While wine is relatively unpopular in the region, beer and distilled spirits account for nearly equal shares of the alcohol intake.

The residents of Oceania consume, on average, 4.46 liters of pure alcohol per-person per-year. Australia, New Caledonia, and New Zealand have similar, and relatively high, consumption rates (10.29, 10.08, and 9.80, respectively). However, there is a high degree of variation within Oceania. While Australia ranks 29th at 10.29 liters of pure alcohol per person, The Federated States of Micronesia has a consumption rate of only 0.64 liters per person and the Solomon Islands ranks 161st at 0.16 liters per person. The preferred type of alcohol in Oceania is clearly beer.

Africa, which as a region ranks sixth in average alcohol consumption, has a per adult consumption rate of 3.19 liters per person. Uganda has the highest consumption rate among all independent African nations at 13.10 liters per person. South Africa, Burundi, Gabon, and Nigeria also have relatively high alcohol consumption rates (also see Gureje and Olley 1992). Conversely, the nations of Western Africa, which are predominately Muslim or heavily influenced by Islam, consume relatively little alcohol. For example, the consumption rate in Senegal is only 0.43 and Mali has a rate of only 0.28 liters per person per year. As in Oceania, beer is the most common source of alcohol in Africa.

Alcohol is not as popular in Asia as it is in most other regions. In Asia, the annual consumption rate is 2.69 liters of pure alcohol per adult. Distilled spirits are the preferred type of alcohol in Asia. Slightly more than sixty-three percent of the total alcohol consumed in the region is in the form of distilled spirits. Thailand is an exception to Asia's relatively low rates of consumption. The average per adult consumption of pure alcohol in Thailand is 13.59 liters. This ranking rates Thailand the 13th spot in the world's alcoholconsumption list. Generally speaking, more alcohol is consumed in northern and eastern Asia than in central or southern Asia. Among the Asian nations, the top five alcohol consumers are all in the northern and eastern parts of the continent. In descending order, these nations include Thailand, Korea, Japan, Laos, and China. Similarly, the Philippines, Mongolia and Singapore consume more alcohol than the average for the region as a whole. In general, many of the former Soviet republics of Central Asia (Georgia, Kyrgyzstan, Azerbaijan, and Kazakhstan) also consume above average amounts of alcohol for the region (ranging from 2.76 to 4.32). Conversely, less alcohol is consumed, on average, in the South Central Asian nations of India, Sri Lanka and Nepal. The alcohol consumption rate in India is 1.01 liters per person, and only 0.07 liters per person in Nepal. Those Asian nations with a significant Islamic population also have low rates of alcohol consumption. In Indonesia, whose population is nearly 90 percent Muslim (CIA 2001), the rate of consumption is only 0.08 liters per person per year. Similarly, the heavily Islamic nations of Pakistan and Afghanistan have the lowest consumption rates (0.03 and 0.02 liters per person, respectively) of any nation for which data are available.

Not surprisingly, the nations of Northern Africa and the Middle East consume very little alcohol. This region's consumption rate is 1.24 liters per person per year. In most of these nations, the majority of the population is Muslim, and Islam condemns the use of alcohol. One can predict alcohol consumption rather well simply by knowing if the nation is heavily influenced by the Islamic faith. The bi-variate correlation between alcohol consumption and percent of the population who is at least nominally Muslim is -.552 (p < .01).[10] Thus, the Qu'ran's four prohibitions against alcohol seem to be effective at controlling its use.

Among the nations of the North African / Middle Eastern region, the United Arab Emirates (UAE) has the highest alcohol consumption rate (2.75 liters per person). However, less than 20 percent of the UAE's population are citizens since scores of foreign laborers come to work in its oil industry (CIA 2001). A similar situation exists in Bahrain, which has the second highest rate of alcohol consumption in the region. The only non-Muslim state in the region, Israel, has the third highest alcohol consumption rate, but this too is relatively low at 1.84 liters per person. Saudi Arabia and Yemen have among the lowest alcohol consumption rates in the world at 0.07 liters per-person per-year.

Although the use of alcohol produced social concern everywhere it has flourished, it is still a popular pass-time. Despite religious constraints against its use in Islamic, Buddhist and Hindu societies, and Christian and Jewish norms of moderation, alcohol is the world's second most popular drug. There are three primary reasons why alcohol became popular. First, it is easy and relatively inexpensive to produce. Second, it was often safer to consume alcohol than water. Contaminated water is "possibly the single greatest menace to human health since the advent of civilization" (Courtwright 2001: 10), and alcohol

---

[10] Data for percent Muslim were taken from the CIA's *World Factbook*. A Spearman's rank-ordered coefficient, or rho, is used since it is a non-parametric test and does not make assumptions about the distribution of the data. As discussed in Chapter One, given the questionable quality of drug data, even that on alcohol consumption, relative rankings are believed to be better suited for analysis than absolute values.

provided a germ-free alternative. Third, alcohol was one of the favorite drugs among Europeans. Their colonial expansion spread the use of alcohol and their imperial rule made sure it was legal, and profitable, in all of their colonies (see Szasz 1985).

Pharmaceutical Drugs

Many sociological discussions of drugs do not consider pharmaceutical drugs. Those that do tend to do so only briefly. These substances are often overlooked because they are legal and many pharmaceutical drugs are not psychoactive. Indeed, it should be noted that the most frequently consumed pharmaceutical drugs are not psychoactive. For example, the top selling pharmaceuticals are cholesterol and triglyceride reducers, and anti-rheumatic non-steroidals are the second most frequently used class of pharmaceuticals (IMS 2001). These drugs are not psychoactive. Nevertheless, to ignore these substances produces a biased view of drug consumption patterns.

First, many pharmaceutical drugs are psychoactive. For example, the third leading class of pharmaceuticals is anti-psychotics, which have powerful psychoactive effects. Anti-psychotics accounted for 4.2% of global pharmaceutical sales in 2000 when it reached the $13.4 billion mark (IMS 2001). Similarly, antidepressants are among the fastest growing classes of pharmaceutical drugs, increasing nearly 20 percent per year since the late 1990s. Second, many pharmaceutical drugs are obtained and used illegally. For example, $25 billion worth of prescription drugs were sold on the U.S. black market in 1993 (Bellenir 2000). Given the amount of pharmaceutical drugs used in the U.S., legally and illegally, it is not surprising that prescription sedatives, stimulants, painkillers, and tranquilizers are frequently mentioned in drug-related emergency room episodes (Bellenir 2000). Diverted pharmaceutical use is also prevalent in the United Kingdom, India, the Philippines, Australia and numerous other nations (see, for example, Emdad-ul 2000). Third, pharmaceutical drugs produced legally in one country are often smuggled and used illegally in another.

For example, phensedyl (a codeine-based cough linctus) is smuggled from India into Bangladesh and has become popular among youth there. According to Emdad-ul (2000: 238), this practice is widespread enough that "the problem has now become acute." Phensedyl is also popular in India, Nepal, Myanmar and the Philippines (Emdad-ul 2000). As Courtwright (2001: 66) notes, psychoactive drugs developed by multinational pharmaceutical companies "find their way into the drug underworld." Finally, the consumption of pharmaceutical drugs is not random. As will be seen shortly, there is considerable regional variation in the use of pharmaceuticals. Although these patterns reflect variations in wealth among nations, they also speak to cultural differences and other structural characteristics that shape the modern world system. Thus, in my opinion, ignoring the consumption of pharmaceutical drugs paints an incomplete picture of the world's drug use. We will therefore consider these patterns.

While caffeine, tobacco and alcohol are widely produced and used throughout the world, pharmaceutical drugs have a much more limited market. Humans have undoubtedly used plants, minerals and animal products to heal the sick since the earliest periods of human existence (see Greener 2001; Jackson 1988). However, the modern use of drugs for medicinal purposes has changed considerably since shamans, medical men, and physicians gathered and mixed their own ingredients. Today, the pharmaceutical industry has evolved into a concentrated, global, multi-billion dollar industry (see Greener 2001; Borgner and Thomas 1996; Tucker 1984). The top ten pharmaceutical firms account for approximately 48% of the world's market (IMS 2003), and the top twenty firms account "for just over sixty-four percent of the world market" (Greener 2001: 21). U.S. and European firms account for approximately 76% of the industry's output and Japanese-based firms an additional 20%. The world's remaining nations produce approximately 4.0% of the pharmaceutical drugs consumed in the world each year (see Tucker 1984; Greener 2001).

While humans have always consumed drugs to maintain or restore their health, the world's consumption of pharmaceuticals continues to grow. Total global pharmaceutical sales were $317.2 billion in 2000 (IMS 2001), but grew to $430.3 billion in 2002 (IMS 2003).[11] In two years, the world's medicine tab grew by over 35%. Based on industry marketing forecasts, this trend should continue, at least in the near future. The Asian market, mostly fueled by China, is expected to reach the $30 billion mark by 2005 (IMS 2001c). Similarly, sales are projected to increase through at least 2004 in South and Central America (IMS 2000). The strongest growth, however, "is expected to be in the Middle East and the Pacific rim" (Greener 2001: 123).

Although the reach of the pharmaceutical industry is global and the markets are growing throughout the world, North America, Europe and Japan are, have been, and will likely continue to be the primary pharmaceutical consumers. For example, in the early 1970s, Valium ranked as the most frequently prescribed drug of any type. Over 100 million Valium prescriptions were written worldwide in 1975 alone. The majority of those, 85 million, were for Americans. It was estimated that during the early 1970s, 10 to 20 percent of adults were taking benzodiazepines on a regular basis in Western Europe and North America (Lickey and Gordon 1991). While the use of benzodiazepines has decreased in Western Europe and North America since the 1970s, the overall use of pharmaceuticals has not declined in these areas. These markets accounted for more than 88% of worldwide pharmaceutical consumption in 2002 (IMS 2003). North America, the world's largest market for pharmaceuticals, accounted for approximately 51% of global sales. Europeans spent $101.9 billion on pharmaceuticals in 2002, or 25.4% of the world's consumption of pharmaceuticals. Nearly 89% of Europe's pharmaceuticals were bought by European Union nations. The rest of Europe spent slightly more than $11 billion. Japan accounted for 11.7% of pharmaceutical consumption. To put these numbers in perspective, Asia, Africa and Australia combined to buy approximately 8% of the world's pharmaceutical

---

[11.] IMS Health is an industry-leading market research firm. IMS Health's *World Review* tracks actual sales of approximately 90% of all prescription drugs and certain over-the-counter (OTC) products in more than 70 countries.

drugs. All of Latin America accounted for only 4.1% of the world's pharmaceutical sales in 2003 (IMS 2003). Table 2.2 lists the global pharmaceutical sales by region for 2002.

Table 2.2: Global Pharmaceutical Sales by Region, 2002

Region	2002 Sales (billion U.S. dollars)	Percent Global Sales
North America	203.6	50.8
Europe	101.9	25.4
European Union	90.6	22.6
Rest of Europe	11.3	2.8
Japan	46.9	11.7
Asia, Africa and Australia	31.6	7.9
Latin America	16.5	4.1
Total	400.6	100.0

Source: IMS *World Review 2003*. Sales cover direct and indirect pharmaceutical channel purchases in US dollars from pharmaceutical wholesalers and manufacturers. The figures above include prescription and certain OTC data and represent manufacturer prices.

Based on per capita sales of pharmaceuticals, the United States is the world's leading pharmaceutical consuming nation by far. In 1999, the United States spent $406 per person on pharmaceutical drugs. This amount is double what eighth-ranked Austria spent per person on pharmaceutical drugs, and nearly three times what twelfth-ranked Netherlands spends. The Japanese, who are in second place at $340 per perspon, spend 83% of what Americans do on pharmaceutical drugs. The next 10 leading pharmaceutical consuming nations were European. France, Belgium, Switzerland and Germany were the leading consuming nations in Europe. Table 2.3 reports the top 12 nations in per capita sales of pharmaceuticals for 1999. There is little reason to believe these relative rankings have changed since these data were collected.

Table 2.3: Per Capita Sales of Pharmaceuticals, 1999[12]

	Per capita pharmaceutical sales (US dollars)
United States	406
Japan	340
France	296
Belgium	252
Switzerland	231
Germany	226
Sweden	216
Austria	202
Italy	187
United Kingdom	173
Spain	154
Netherlands	145

Source: Greener 2001: 24.

In Central and Eastern Europe, Turkey and Poland are the leading pharmaceutical consuming nations as measured by sales. Turkey spent $3.2 billion on pharmaceutical drugs in 2000 and Poland spent $2.5 billion (IMS 2001b). Hungary, the Czech Republic, Slovenia, Slovakia and Bulgaria spent less than $1 billion each for pharmaceutical drugs (IMS 2001b). Assuming market projections are accurate, the Eastern European market for pharmaceuticals should continue to grow in the near future.

Although they cannot compete with North America, Europe or Japan, South America is the fourth leading pharmaceutical consuming region in the world. The people of this region spend slightly more on pharmaceutical drugs

---

[12.] Greener (2001: 24) reports the data in British pounds. The data were converted to U.S. dollars using the average exchange rate for 1999 which was 1.6178 dollars per pound.

than the people of Asia, Australia, and Africa combined. And, the South American pharmaceutical market is expected to continue its recent expansion (IMS 2000). Mexico, Brazil and Argentina are the prime consuming nations in South and Central America.[13] Mexico and Brazil each account for approximately 30 percent of the pharmaceutical sales in the region. Argentina holds 22.2% of the region's market. There is a considerable gap between these three nations and their neighbors with respect to pharmaceutical sales. The next largest market share is held by Venezuela at 7.0% and Colombia at 5.2% (IMS 2000).

In comparison to North America, Europe, Japan and South America, Asia -- or that part of it excluding Japan -- consumes relatively few pharmaceutical drugs. However, pharmaceutical sales are expected to reach $30 billion in Asia by 2004 (IMS 2001c). Again excluding Japan, China is the leading pharmaceutical market, accounting for 34 percent of the sales in Asia. India, South Korea and Taiwan each account for approximately 15 percent of the Asian pharmaceutical market. Indonesia (6 percent), the Philippines (5 percent) and Thailand (4 percent) all account for similar market shares.

South Africa is the leading pharmaceutical consuming state in Africa. Although the nations of Africa represent a small fraction of the total pharmaceutical market, forecasts call for continued sales growth. Throughout Africa, sales increased by 10% in 2000 and similar growth rates are anticipated for the next several years (IMS 2003). Despite this growth in sales, Africa, as a region, is the smallest market for pharmaceutical drugs in the world.

As is evident by the above discussion, the use of pharmaceutical drugs, like the use of other drugs, varies across regions of the world. Although lesser-developed nations are projected to increase their consumption of pharmaceutical drugs, it is doubtful they will come close to matching the rates of use found in the United States, Western Europe or Japan.

---

[13] I include Mexico in Central America in this section only. I do so because the rates of pharmaceutical use in Mexico pale in comparison to those found in the U.S. and Canada.

46

Summary

Caffeine, alcohol, tobacco and pharmaceutical drugs are nearly ubiquitous. These drugs are found and are widely used on every continent and in most, if not all, nations. While alcohol and tobacco are primarily used by adult males, the use of these drugs has increased among women and, in many nations, children. Women and the elderly disproportionately consume pharmaceutical drugs; however, males and youth also consume considerable doses of pharmaceuticals. Caffeine is the egalitarian drug. Men and women, rich and poor, old and young consume it.

With respect to the most widely used drugs in the world, Europe consistently leads the way. Europeans consume the most caffeine and alcohol. In terms of per capita consumption, Europeans also consume the most tobacco. They are among the leaders in the consumption of pharmaceutical drugs. Americans also consume large amounts of caffeine and alcohol, and they lead all nations in the consumption of pharmaceutical drugs. Australia and Canada are also leading drug-consuming nations. These patterns should not be overly surprising since Europeans are primarily responsible for spreading the use of caffeine, alcohol and tobacco across the globe. Moreover, their colonial governments profited from the production and sale of these products and, to the best of their abilities, made sure these products remained legal to use. America, Canada and Australia -- the most "European" of any of Europe's former colonies since the indigenous populations were largely eliminated there -- have carried on their mostly-European ancestor's drug-using habits.

Outside of Europe and North America, Japan is also a heavy drug-using nation. This is especially true for pharmaceutical drugs and tobacco. The use of alcohol is in the "moderate" range in Japan. The rest of Asia consumes moderate amounts of alcohol and relatively few pharmaceutical drugs. Asians do smoke heavily. South Americans are relatively heavy drinkers and lag behind only Europe, Japan and North America in terms of pharmaceutical consumption. As with the consumption of other goods, Africa typically trails the rest of the world in consuming legal drugs. North Africa, and especially the Middle East, consume little alcohol and, despite recent increases, pharmaceutical drugs.

As striking as these patterns are, they become more pronounced when "illicit" drugs are considered. We will consider regional variations in the use of opiates, cocaine, marijuana, amphetamines and other illegal drugs in the next chapter.

CHAPTER THREE

THE EMPIRICAL PATTERNS OF ILLEGAL DRUG USE

As was seen in Chapter Two, legal drugs are widely used across the globe. Although there is variation in rates of use by country, caffeine, alcohol, tobacco and pharmaceutical drugs are used almost everywhere. This is not the case for illicit drugs. While the majority of the population in most countries consumes some caffeine-containing substances and nearly a quarter of all people use tobacco, there are no nations whose majority of the population uses illegal drugs. In fact, according to 2003 UNODC estimates, only 3.4% of the world's population, or 4.7% of the world's population age 15 or older, use any illicit drug (UNODC 2003: 101). It is therefore necessary to note that we are discussing a behavior in which relatively few people participate, despite the concern these users cause. Nevertheless, the variation in the use of illicit drugs is, as we shall see, more dramatic than the variation in the legal drugs discussed in Chapter Two.

Turning our attention to the use of illicit drugs, the discussion is organized differently than most presentations of drug patterns. The discussion groups by region rather than type of drug. Organizing the discussion by region -- a major independent variable that will be used to explain these observations -- permits general trends and patterns to be gleaned. These trends and patterns then become what need explaining. That is, these patterns become the dependent variable for the remainder of our discussion.

After a brief discussion of use patterns at the global level, we will consider patterns of use in Africa, Asia, Europe, North America, Oceania, and South America. Patterns of use in specific countries will be discussed when doing so serves an illustrative purpose. The reader is again reminded that these estimates are "educated guesses." While these data are the best available, they should not be considered perfectly accurate. Despite their limitations, however, these data should suffice for observing general trends and patterns.

Global Patterns of Illicit Drug Use

As mentioned above, approximately 3.4% of the world's population or 4.7% of the population over the age of 15 are believed to use some type of illicit drug.[14] By far, the most commonly used illegal drug is cannabis.[15] During the period between 1998 and 2001, approximately 163 million persons used some type of cannabis substance annually (UNODC 2003). This figure represents 3.9% of the world's adult population. Marijuana's use is as old as it is varied. For centuries, humans have used marijuana[16] as a fiber, in religious practices, as medicine, as food, and for recreational purposes (see Able 1982; Benet 1975: Goode 1999). Today, marijuana is the most widely used illegal drug in every region of the world (see UNODC 2003; UNODC 2004).

The second most frequently used types of illicit drugs are amphetamine-type stimulants (ATS). Worldwide, over 34 million people use amphetamines. An additional 7.7 million use ecstasy. Together, nearly one percent of the world's adult population uses an amphetamine-type stimulant illegally each year

---

[14.] In the future, I will refer to the population over the age of 15 as "the adult population."

[15.] Throughout the remainder of the chapter, "drugs" will refer to "illegal drugs." Usually, "drugs" will mean cannabis, amphetamine-type stimulants, cocaine, and opiates. These are the drug categories that UNODC cover and for which reasonably reliable data exist.

[16.] I am using the term marijuana to generically refer to all cannabis substances including ganja (the flowering buds), bhang (the leaves), and charas or hashish (the resin).

(UNODC 2004). ATS rank second in number of users in Africa, Asia, Europe, South America and Oceania. They are the third most frequently used illegal drugs in North America.

Opiate drugs are rank third in terms of number of users worldwide. Worldwide, 14.9 million people use opiate drugs illegally each year. Of these, 9.5 million use heroin. Relatively speaking, narcotic use in general and heroin use specifically is rare. Less than half of one percent of the world's adult population use narcotic drugs illegally. Only 0.22% of the world's adult population used heroin during the 1998 - 2001 period (UNODC 2002; UNODC 2003). Only in Eastern Europe does more than one percent of the adult population use opiates.

Finally, cocaine, despite its fame, is limited by geography. Globally, 14.1 million people, or 0.3% of the adult population, used cocaine annually during the 1998 - 2001 period. There are more cocaine users in North America and South America than there are ATS users. Conversely, cocaine use is relatively rare in Africa, Eastern Europe and, especially, Asia (UNODC 2004).

Table 3.1 reports the global use of cannabis, ATS, cocaine and opiates. It should be noted that the total number of drug users is not the sum of the separate drug categories because many drug users take more than one substance. Based on American data, for example, slightly over 60% of illicit drug users use marijuana alone. An additional 21% of illicit drug users use marijuana and some other drug. Thus, over 80% of the America's illicit drug users use marijuana. Only 19% of illicit drug users use drugs other than marijuana without also using marijuana (SAMHSA 1999). It appears that a similar pattern of use exists worldwide. As can be seen in Table 3.1, which reports the estimated number of users and percentage of the adult population who use the major types of illegal drugs worldwide, marijuana users comprise nearly 82% of the world's illicit drug users. Approximately 37 million people across the globe use any illicit drug without also using marijuana. This figure represents approximately 0.4% of the world's population.

Table 3.1: Types of Illegal Drugs Used (1998 - 2001 period)

	Illicit Drugs	Cannabis	ATS	Ecstasy	Cocaine	Opiates	Heroin
Global use (million people)	200.0	162.8	34.3	7.7	14.1	14.9	9.5
Percent adult population	4.3	3.9	0.8	0.2	0.3	0.4	0.2

Source: UNODC, *Global Illicit Drug Trends*, 2003: 101

REGIONAL PATTERNS

While it can generally be said that, regardless of the region of the world, marijuana is the most commonly used illicit drug, there is, nevertheless, considerable variation in rates of use by region. For example, the rate of marijuana use in Western Europe is double what it is in Eastern Europe. The patterns of use become even more variable when other drugs are considered. For example, generally speaking, opiates are the least commonly used drugs. However, this is not the case in Asia or Eastern Europe. Moreover, who uses drugs varies by region. For example, while marijuana is a popular recreational drug for western adolescents, its use in some nations is limited, more or less, to adults. We must therefore consider rates of use in more detail. What follows is a discussion of rates of use by geographic region.[17] The data, unless otherwise noted, are based on the United Nation's publications *Global Illicit Drug Trends* (UNODC 1999; 2002b; 2003) and *World Drug Report* (UNODC 2000; 2002; 2004). Also, all estimates, unless otherwise noted, are for the period 1998 - 2001, as reported by the U.N. publications.

[17.] Seven regions will be discussed. These regions are Africa, Asia and the Middle East, Eastern and western Europe, North and South America, and Oceania. On occasion, sub-regions will be discussed when rates of use vary between these sub-regions.

Africa

Not surprisingly, the estimates of drug use in African nations are among the least reliable that we have. For example, fewer than half of all African nations report to UNODC. In addition, nationally representative data such as the U.S.'s *National Survey on Drugs and Health* simply do not exist in African nations. Although the *World Health Organization* (WHO) is attempting to collect better data, we simply must acknowledge that our understanding of drug use in Africa is based on limited data.

With this caution in mind, we can safely say that marijuana is the most widely used drug in Africa. In fact, Africa has the second highest rate of marijuana use in the world and "it is widely acknowledged that cannabis is by far the most widely abused illegal substance" (UNODC 1999: 119). Marijuana's history in Africa is indeed a long one and its use is embedded in many African cultures (see, for example, Du Toit 1976; Du Toit 1975; Palgi 1975). Today, Africa has the second highest rate of marijuana consumption among the major regions of the world. For the period 1998 - 2001, it is estimated that 33.2 million Africans, or 8.6% of the adult population, uses some sort of cannabis substance on an annual basis. Moreover, marijuana use has been reportedly increasing in "several countries of Southern Africa, Central Africa, Northern and Western Africa" (UNODC 2000: 65).[18]

While marijuana use is prevalent across Africa, regional variations, as well as variations across countries within regions, are marked. Based on available data, rates of marijuana use are higher in West and Southern Africa than they are in Northern Africa. Eastern Africa has the lowest rates of marijuana use. The average rate of marijuana use in Western Africa is 8.6% of the adult population,

---

[18] Nigeria is an exception. Although Nigeria reported a "strong increase" in cannabis consumption in 1994, they reported "some decline" in 1998 (see UNODC 2000: 67).

while in Southern Africa it is 9.3% of the adult population. The mean usage rate in North Africa is estimated to be 4.5% of the adult population. In East Africa, the mean rate of use, at least among those nations for which estimates were made, is only 2.0% of the adult population.[19]

According to the World Health Organization's data, the top six marijuana-using nations in Africa are in either West or Southern Africa. The West African nation of Ghana has the highest reported rate of marijuana use at 21.5% of the adult population. Sierra Leone (16.1%) and Nigeria (14.4%) also have relatively high rates of marijuana use (UNODC 2003). In Nigeria, however, marijuana use is more common among the youth population than in many other African nations (see Gureje and Olley 1992). Marijuana use is also widespread in The Gambia, Niger, Cameroon, Mali, Cape Verde, and Burkina Faso (see Affinnih 1999; OGD 1996; Stares 1996). Conversely, the West African nations of Chad (0.9%), Sao Tome Principe (.01%) and Cotes d'Ivoire (0.01%) have very low reported rates of marijuana use (UNODC 2003). Unlike in other parts of Africa where marijuana use is an ancient practice, in West Africa the widespread use of marijuana is a somewhat recent phenomenon. According to Du Toit (1975: 101), marijuana was introduced to the region during and shortly after the Second World War. Despite its late start, however, West Africa now has some of the highest rates of marijuana consumption in the world.

While marijuana was introduced to Western Africa relatively recently, it has been used for centuries throughout Southern Africa. In fact, its use pre-dates tobacco. South Africa has the highest rate of marijuana use among Southern African states and the second highest rate of use among all African nations. Over 18% of South African adults use marijuana (UNODC 2003). This is not a new

---

[19.] It is emphasized that these means are tentative. They are based on available data only. Plus, what data are available are, at best, "educated guesses." However, I do feel confident that these data are reliable enough to establish general patterns of use in a relative sense. Thus, while I provide these means for illustrative purposes, their absolute magnitude should be considered with extreme caution. The relative rankings by region, however, should be reasonably accurate (see UNODC 2004).

habit for South Africans. Du Toit (1975: 96) noted that, "in a traditional Zulu community it was common for men to smoke cannabis, even daily." More recently, cannabis mixed with methaqualone has become popular in South Africa and Botswana, especially among the youth population (UNODC 2000; also see Macdonald 1996; Affinnih 1999).

Zambia, where use pre-dates the arrival of Livingston, also has a relatively high marijuana usage rate. Approximately 15.0% of Zambian adults use marijuana (UNODC 2004). The small island nation of Mauritius (7.0%) has moderately high levels of marijuana use. Zimbabwe, where an estimated 6.9% of the adult population uses marijuana, has similar rates. However, marijuana use among Zimbabwean secondary school children is much higher (Eide and Acuda 1997). Local studies suggest that marijuana use is also relatively widespread in Malawi (Stares 1996). Angola, on the other hand, has comparably low rates of marijuana use (2.1%) (UNODC 2002; UNODC 2003).

North Africans, who tend to use hashish instead of smoking cannabis leaves, also have a long tradition of using cannabis. Despite this tradition, however, rates of marijuana consumption appear to be lower in North Africa than in either West or Southern Africa. Among the reporting North African states, Morocco has the highest rate of cannabis use. Slightly over 7% of Moroccan adults use some cannabis product. Morocco has long been known as a hashish-using culture (see Palgi 1975), especially in the Rif Mountain area where cannabis is a major cash crop. As Joseph (1975: 190) said when discussing the widespread practice of using hashish among the Rif Mountain inhabitants, "the use of *kif* (cannabis) remains an integral part of the culture and is regarded as no great threat to the community or group relationships." Egypt also has "a tradition of cannabis consumption" (UNODC 2000: 61). Hashish use in Egypt dates back to around the mid-twelfth century during the reign of the Ayyubid dynasty (Khalifa 1975). Use, although relatively widespread, is more common in the lower class except in the urban areas where it is prevalent among middle-class youth. Egypt's reported marijuana-use rate for the 1998 - 2001 period was 5.2% of the adult population (UNODC 2004).

Among those East African states that reported to the UN, Kenya had the highest rate of marijuana use at 4.0% (UNODC 2003). Somalia and Uganda report relatively low rates of use at 2.5% and 1.4% of the adult population, respectively (UNODC 2004; also see Affinnih 1999). Despite its relatively low rate of use, Uganda allegedly produces a considerable amount of marijuana (Stares 1996). Although marijuana use is not overly prevalent in the general population, the Twa, who comprise a small minority of the Rwandan population, have traditionally been known for their use of marijuana (see Codere 1975). In Burundi, a relatively small segment of the urban population use cannabis heavily. Increasingly, however, Burundi youth between the ages of eight and sixteen are becoming the largest group of marijuana users in the country (OGD 1996). Marijuana is also widely used in Tanzania, although recent estimates report only 0.2% of the population using cannabis (UNODC 2004). Yet, it is well documented that cannabis smoking was prevalent there in the 1880s among the Nyamwezi, and by the mid-1900s, cannabis use was very widespread (Du Toit 1975).

Although marijuana use is widespread in Africa, the use of other drugs is relatively limited. Africa, as a region, has among the lowest rates of amphetamine, opiate and cocaine use. However, this pattern appears to be changing as Africa becomes increasingly central in the global illicit drug distribution networks (see OGD 1996). Cape Verde, Ghana, Senegal, Mali, Niger, Nigeria, Kenya, and South Africa are all major transit points in the global drug trade. As drugs pass through these nations, spillover occurs thereby increasing the rate of use among the local population. As a result, most African nations reporting to the U.N. cite increasing use of heroin, cocaine, and, to a lesser extent, amphetamine-type stimulants (see UNODC 2000).

While many African nations report increased use of heroin, cocaine, LSD and other drugs, amphetamine-type stimulants still rank second as the illicit drug of choice in Africa. Most amphetamine and amphetamine-type drugs used illegally in Africa were legal products that were diverted to the illegal market (UNODC 2000). Approximately 2.3 million Africans, or 0.5% of the adult population, use ATS (UNODC 2003). As with cannabis, ATS use appears to be highest in West Africa. Approximately one percent of the adult populations of Nigeria, Ghana, and Cameroon use ATS (UNODC 2004). Slightly lower rates of use are seen in Mali, Senegal and Chad (UNODC 2002). In Southern Africa, South Africa (0.6%), Zambia (0.4%) and Zimbabwe (0.1%) have relatively high rates of ATS use by African standards (UNODC 2004). South Africa is believed to be a main manufacturing area for ATS (UNODC 2000). In North Africa, ATS use in Egypt (0.5%), Ethiopia (0.3%) and Morocco (0.3%) is similar to those of some Southern African nations (UNODC 2003). It is also estimated that Kenya has moderately high rates of ATS use at 0.6% of the adult population (UNODC 2004). Recent observations note the increased use of ATS in the slums of Nairobi.

Recently, anecdotal evidence suggests that ecstasy use has penetrated the youth cultures of some Southern African nations, especially South Africa, Namibia and Swaziland (UNODC 1999; UNODC 2004; Gureje and Olley 1992). In South Africa, approximately 0.3% of the adult population report annual use of ecstasy (UNODC 2004). In Swaziland, 0.3% of those ages 15 - 24 are believed to have used ecstasy at least once in their life (UNODC 1999: 130). Ghana and Namibia report low levels of ecstasy use (less than 0.1% of the adult population), but the drug appears to be gaining popularity in these nations (UNODC 2003). Despite the recent inroads ecstasy has made in the southern region of the continent, ecstasy remains a rare drug in Africa. Recent estimates report that approximately 100,000 Africans use ecstasy on an annual basis. This corresponds to a mere 0.01% of the adult population.

Although twelve of the twenty-one African nations that provided data to the U.N. reported increases in cocaine use during the late 1990s and early 2000s (UNODC 1999; UNODC 2003; UNODC 2004), cocaine is still a rarely used drug in Africa. Overall, slightly less than 1 million Africans, or 0.2% of the adult population, are estimated to use cocaine on an annual basis (UNODC 2004). Nevertheless, this rate "is estimated at close to the global average" (UNODC 2000: 78). Like other drugs, cocaine consumption is particularly concentrated in West Africa and, to a lesser extent, Southern Africa. Ghana and Nigeria are the leading cocaine-consuming nations in Africa (see Courtwright 2001; UNODC 2002; UNODC 2003; Obot 1990; Affinnih 1999). Ghana reported the highest rate of use at 1.1% of the adult population, while approximately one-half of one-percent of the adult Nigerian population uses cocaine annually (UNODC 2004). Other West African nations with reported cocaine "problems" include Togo, Benin and Cameroon (UNODC 2002; UNODC 2002b; UNODC 2003). Despite reporting a "problem" with cocaine, however, no estimates of the extent to which cocaine is used were available for many of these nations. Sao Tome Principe (0.02%), Sierra Leone (0.02%) and Chad (0.01%) report low rates of cocaine use.

Outside West Africa, Southern Africa also has, by African standards, a high rate of cocaine consumption. Among the adult population, 0.5% of South Africans use cocaine (UNODC 2004). Cocaine is less prevalent in Namibia (0.2% of the adult population), Angola (0.01%) and Zimbabwe (0.05%); however, these nations reported increases in cocaine use during the late 1990s (UNODC 1999; UNODC 2002b; UNODC 2003). In East Africa, Kenya, Burundi and Tanzania reported increases in cocaine consumption. In Kenya, 4.5% of youth are reported to have used cocaine (UNODC 1999b: 132).

Like cocaine, there is a limited use of opiates in Africa. In total, 0.2% of the adult African population, or slightly less than one million people, use opiate drugs (UNODC 2000; UNODC 2003). However, rates of opiate use in Africa nearly doubled during the late 1990s (see UNODC 2000 and UNODC 2003).

Still, opiate use in Africa is approximately two-thirds the global average. Like in other regions, the vast majority of African opiate users use heroin.

Once again, Ghana has the highest rates of opiate consumption in Africa at 0.7% of the adult population (UNODC 2003). Nigeria has been known as a transshipment point for heroin since the mid-1980s (Courtwright 2001; Stares 1996), and the use of heroin has increased there since that time (Obot 1990; Affinnih 1999). Heroin use is especially pronounced in Lagos. Recent estimates are that 0.6% of the adult Nigerian population use opiates. Yet, according to World Health Organization data, opiate use has declined in Nigeria since the late 1990s (UNODC 2002). Similarly, Chad, with an estimated rate of use of between 0.3 and 0.5% of the adult population, reports a "strong decline" in opiate use. There has been a reported decline in opiate use in Cotes d'Ivoire. Conversely, Togo, Senegal, Benin, Sierra Leone and Cameroon report relatively low levels of opiate use, use appears to be increasing in all of these nations (UNODC 2002).

Opiate use is relatively high in the Southern African nations of South Africa and Zambia (0.4%) and less common in Namibia, Swaziland and Zimbabwe (UNODC 2003). All of these nations reported increases in opiate consumption during the late 1990s (UNODC 2000). In Swaziland, for example, 0.7% of youth are estimated to use heroin annually (UNODC 1999b: 133), and this rate is believed to be increasing (see Macdonald 1996). Use, although still relatively low, is also increasing in Zambia. Lower rates, less than 0.1% of the adult population, are found along the Indian Ocean in Mozambique and Tanzania. These nations reported modest increases in opiate use during the late 1990s. Neighboring Kenya has slightly higher rates of use (UNODC 2000; UNODC 2002) than other East African nations and reports continued increases in consumption. In fact, it is believed that over 10,000 intravenous drug users live in Nairobi, and the vast majority of these are heroin users (Siringi 2001). Uganda and Rwanda also saw increases in opiate use during the late 1990s. In North

Africa, Egypt has seen an increase in heroin use since 1990; however, rates of use remain comparably low (UNODC 1999b).

Other drugs that are widely used in Africa include khat (or qat) and betel. Both of these drugs are natural stimulants and are chewed. Khat is used widely and frequently throughout North Africa, and to a lesser extent in the coastal nations along the Indian Ocean (see Weir 1985; UNODC 2000). Its use is also increasing in Southern Africa (Macdonald 1996). Khat chewing has traditionally been limited to adult males, and the practice is still heavily male-dominated. It is customary to chew khat at social events to stimulate discussion (Weir 1985). Although most Islamic clergy condemn the use of khat, it remains a popular drug throughout the Muslim world, including North Africa. The use of betel is also prevalent in East Africa. Similar in its effect to tobacco, betel is a popular natural stimulant. Although widely practiced around the world -- it is believed that nearly 10% of the world's population uses betel -- its use in Africa is largely confined to the East Africa nations from Somalia to Mozambique (see Courtwright 2001). Here, however, the practice is widespread.

Traditional hallucinogenic drugs are also used in remote areas of Africa. Iboga, an indigenous plant with similar effects as LSD, is used in the religious rite of the Mbwiti cult of the Fang of Cameroon and Gabon (UNODC 1999b). Likewise, male elders in rural Zimbabwe use the hallucinogenic drink *mudzepete* (UNODC 1999b). Anecdotal evidence suggests that these traditional hallucinogens are becoming increasingly popular among small segments of the youth population in Ghana, Cameroon and Gabon.

In addition to these traditional drugs, some African nations report high rates of inhalant use. Kenya, for example, has the fourth highest reported rate of inhalant use in the world. Nineteen percent of Kenyan youth between the ages of 12 and 18 are reported to have used inhalants at least once in their life (UNODC 1999: 134). Zimbabwe (12.0% of youth) and Swaziland (11.8%) also report high rates of inhalant consumption. These rates are among the top ten reported in the world. The use of solvents is also fairly common among young people in South Africa, especially those from historically disadvantaged groups (Rocha-Silva 1998). The South African press has also reported the growing use of inhalants, as well as ecstasy, among school-aged children (*Africa News Service* 2003a; *Africa News Service* 2003b). As previously stated, methaqualone, often mixed with cannabis, is also consumed for its unique effects. South Africa reports the highest rate of methaqualone use; however, use appears to be increasing throughout Southern Africa, especially in Zimbabwe and Mozambique (UNODC 1999b; UNODC 2000; Macdonald 1996). Mandrax is "also found in some of the other African countries along the Indian Ocean" (UNODC 2000: 61; Affinnih 1999), including Kenya (UNODC 1999b). Finally, other psychoactive pharmaceuticals, including benzodiazepines, have recently been diverted to the illicit drug markets of the high-use African nations such as Ghana, Nigeria, South Africa, Zimbabwe and Kenya. LSD and other synthetic hallucinogens also appear to be limited to these relatively affluent African nations (UNODC 1999b; OGD 1996).

To summarize, illicit drug use is common in Africa, at least relative to the rest of the world. In fact, Africa has among the highest rates of illicit drug use in the world. In general, rates of use are highest in West and Southern Africa and lowest in North Africa. Most of this use, however, is limited to marijuana or other traditional plant-based psychoactive drugs such as khat and betel. The use of marijuana in Africa, especially West and Southern Africa, is widespread. As a region, Africa ranks second only to Oceania in rates of cannabis use. In more remote regions of the continent, natural hallucinogens are also used and have been for centuries. Other drugs, however, are rarely used in these areas. Synthetic drugs, although increasingly popular among youth in the more developed nations of Africa, still have a limited market; however, synthetic drug use appears to be on the rise. As Africa plays an increasingly important role in the global drug distribution system, synthetic drugs are becoming more available to, and more widely used by, the local populations.

Asia and the Middle East

As with Africa, the data on the consumption of illegal drugs for most Asian nations is sparse. Nevertheless, we can say that, despite the long history of drug use in the region and the region's notoriety for the production of illegal drugs, Asia has the lowest rate of illegal drug use among the major regions of the world. This is true despite the fact that nearly half of all opiate users, two-thirds of all amphetamine users, and nearly thirty percent of all marijuana users live in Asia. Thus, while millions of Asians use illegal drugs, rates of use, as a percentage of the total population, are relatively low throughout the region.

Not surprisingly, the most commonly used illicit drug in Asia is marijuana. An estimated 54.9 million Asians use marijuana (UNODC 2003). Although this figure represents approximately 34% of the world's marijuana users, it is only 2.2% of the region's adult population. This is the lowest rate of use among the major regions of the world. Moreover, during the late 1990s, several Asian nations reporting to the United Nations indicated that the use of marijuana was decreasing. Pakistan reported a "strong decline" in use, while Iran, Syria, India and Thailand claimed "some decline" in marijuana use (UNODC 2000: 67). China reported that marijuana use was stable. Among those Asian nations that reported to the U.N., only Myanmar, Jordan and Indonesia reported a "large increase" in marijuana use (UNODC 2000; UNODC 2002).

Central Asia, comprising a number of the former Soviet republics, has the highest average marijuana use-rate in Asia. Among those Central Asian nations reporting to the U.N., the average use rate is 3.3% of the adult population. Afghanistan has the highest consumption rate in the sub-region (7.5% of the adult population) and is closely followed by Kyrgyzstan with a rate of 6.4% of the adult population (UNODC 2004). Iran (4.2%), Uzbekistan (3.9%) and Tajikistan (3.4%) report use rates above the sub-region's average, while Kazakstan (2.4%) and Azerbaijan (1.1%) report rates slightly below the average (UNODC 2004). Marijuana is currently the major drug of choice in these areas, however, recent studies show that injectable drug use is growing (Drug Policy Alliance 2002). Although official statistics indicate marijuana use is rare in Armenia -- only 0.8% of the adult population uses marijuana -- there is anecdotal evidence that marijuana use may be more common than the Armenian government admits. In the village of Marduni, for example, "Hemp Monday" is celebrated every Easter by stuffing traditional dolmas (rice balls wrapped in vine leaves) with the flowering tops of marijuana plants. Marijuana use is allegedly prevalent in the Lake Evan region and in the rural villages near Yerevan (OGD 1997). However, the official estimate of marijuana use in Armenia is relatively low (UNODC 2000); UNODC 2002). Turkmenistan at a rate of 0.3% of the adult population, has the lowest reported rate in the region. Although the reported rates of marijuana use in Central Asia are below the global average, use is believed to be stable or showing slight increases in every reporting nation of the region (UNODC 2000). As in other Islamic cultures, marijuana use often exceeds the use of alcohol. This pattern is especially true in Tajikistan and Uzbekistan where alcohol consumption is among the lowest in the world. It is also true in Turkmenistan despite the relatively low rate of marijuana consumption found there.

South Asia has the second highest rate of marijuana use in Asia, at least based on data reported to the United Nations. The average use rate in South Asia is approximately 2.6% of the adult population (UNODC 2004). Pakistan, with a rate of 3.9% of the adult population using cannabis, leads the sub-region. Bangladesh, Nepal and India all have use rates of approximately 3.2% of the adult population. The use of marijuana on the Indian sub-continent is ancient and deeply rooted in the culture. Marijuana was probably first used as an intoxicant in India around 1000 B.C., and soon became an integral part of Hindu culture (Snyder, 1971: 125; NCMDA 1972). Even today, marijuana is used as a food, as medicine, for recreation, as an aphrodisiac and for religious purposes

(Courtwright 2001; Hasan 1975).[20] High-caste Hindus are not permitted to use alcohol; however, bhang is used at religious ceremonies and as an intoxicant at marriage ceremonies and family festivals. Laborers in India use bhang in much the same way as Americans use beer. Older, male, rural farmers use ganja during "smoke breaks." In the urban areas, marijuana use is more common among the wealthier youth and the "lower and criminal classes" (Courtwright 2001: 40). The pattern is similar in Nepal. Nepal, where cannabis grows wild and is cultivated in the mid-lands, is an international source for marijuana (Fisher 1975; OGD 1996). Although found and used in most parts of the country, the percentage of people using it for recreation is relatively small. Users tend to be male and older, although urban youth have increasingly used marijuana since the 1970s (see Fisher 1975). Sri Lanka has rates of marijuana use that are slightly below the sub-region's average. In Sri Lanka, where rates appear to be stable, 1.6% of the adult population uses marijuana (UNODC 2000). The island nation of Maldives, meanwhile, has well below average use rates at 0.5% of the adult population (UNODC 2004).

The Middle East also has a long tradition of using cannabis. Evidence of this comes from cuneiform tablet interpretations that ascribe use in Persia circa 700-600 B.C. (NCMDA 1972). The drug's popularity as an intoxicant spread throughout the Middle East and thoroughly permeated Islamic culture within a few centuries. The popularity of marijuana and other drugs such as opium in the Middle East has been attributed to the Islamic prohibition against alcohol. The use of intoxicants is proscribed by the Islamic religion in general terms; however, *The Qu'ran* makes no special reference to cannabis, or to any other drugs known at the time. The Middle East has the third highest rate of cannabis consumption in

---

[20.] Cannabis is used in three different preparations in India. *Bhang* is made from the leaves and stems of uncultivated plants and blended into a liquid drink. *Ganja* is the flowering tops of cultivated plants and is smoked. *Charas* is similar to hashish and is obtained by scraping and pressing the resin from the leaves of the cultivated plants. It is also smoked.

Asia at approximately 2.5% of the adult population using annually (UNODC 2004).

In Lebanon, approximately 6.4% of the adult population uses marijuana. Jordan (2.1%), Syria (2.0%) and Oman (1.0%) report below-average rates of cannabis use (UNODC 2004). Reported rates range from 0.1% to 0.4% in Qatar and Bahrain. Officials believe cannabis use is increasing in Kuwait, Qatar and Jordan (UNODC 2000). Israel, the only non-Arabic state in the region, reports a relatively high and increasing consumption rate of 5.7% of the adult population (UNODC 2004). *Les Observatoire Géopolitique des Drogues* (OGD 1997) reports that "hundreds" of young tourists are cannabis consumers. In Tel Aviv, these tourists are hired illegally to serve as temporary workers during the periodic closing of the border to Palestinians from the occupied and autonomous territories. The use of marijuana by tourists in Israel has been reported since immediately following the Six-Day War (see Palgi 1975). Today, drug use in Israel follows two basic patterns. First, the traditional use of hashish is found among older Israelis and Arabs living in Israel. Second, marijuana use, although limited, is found among small segments of young people. Still, anti-drug sentiments are widespread in Israel and lifetime prevalence of use of any kind of illicit drug in Israel is "low compared to many Western countries" (Javetz and Shuval 1990: 429; also see Javetz and Shuval 1982; Javetz and Shuval1984).

Compared to other regions of the world, marijuana use is rare in South East Asia. According to U.N. estimates, only 0.9% of the adult population of South East Asia uses marijuana on an annual basis (UNODC 2004). Low rates of use are reported for Lao People's Democratic Republic, Indonesia, Malaysia and Viet Nam. Rates of use in these countries range from 0.1% to 0.7% (UNODC 2004). Singapore and Brunei Darussalam report even lower rates. These nations claim to have marijuana use rates of less than 1/10 of one percent of the adult population (UNODC 2003; UNODC 2004). Myanmar (1.4%), Cambodia (1.3%) and Thailand (1.5%), although relatively high for the sub-region, report use rates well below the global average. Only the Philippines has a use rate that approaches the global average of 3.5% of the adult population (UNODC 2004).

East Asia has the lowest reported rate of marijuana use in the region. Indeed, East Asia has the world's lowest rate of marijuana use. Based on UN data, only 0.3% of East Asian adults use marijuana. The former British colony of Hong Kong has the highest rate in the sub-region at 0.6% of the adult population (UNODC 2002), and use there is apparently increasing (Laidler, Hodson and Traver 2000). Marijuana use is not extensive in China, despite the fact that hemp was widely used as food in ancient China and considered along with millet, rice, barley and soybean as a major grain (Li 1975). While its use as a medicine was once widespread, even this practice has largely declined. Li (1975: 57) noted that,

> The medicinal use of the hemp plant was widely known to the Neolithic peoples of northeastern Asia and shamanism was especially widespread in this northern area and also in China, and cannabis played an important part in its rituals. . . (Yet) while shamanism, and the use of cannabis in particular, were on the upswing in other Asiatic locales, hallucinogenic practices slowly declined in China beginning with the Age of Confucius. Only in scattered small areas did shamanistic traditions continue in China during later ages.

Apparently, marijuana is inconsistent with the Chinese Doctrine of the Mean and their strong family-based social system (see Li 1975; also see NCMDA 1972). However, there is some evidence that Chinese marijuana use increased during the 1990s (see Wang 1999).

Other parts of East Asia also have low rates of marijuana use. Like China, Taiwan also has a rate of only 0.5% of the adult population, while Korea and Japan have even lower reported rates at 0.1% (UNODC 2002). Although the data are from the early 1990s, North Korea reports a marijuana use rate of only 0.05% of the adult population. Like the rest of East Asia, Japan was a relative latecomer in the recreational marijuana market. Despite an active illicit drug scene (see, for example, Suwanwela and Poshyachinda 1986), marijuana did not become popular in Japan until the late 1960s. Even after marijuana use began to increase in the youth population, Japan has not experienced major problems with "marijuana abuse" (Vaughn, Huang and Ramirez. 1995). Both Korea and Japan, however, reported increases in marijuana use during the late 1990s. The remaining reporting East Asian nations claimed that use was stable (UNODC 2000).

While marijuana is the most widely used drug in Asia, the region is noted for the production and use of opiate drugs. Despite this reputation, however, the region ranks below the global average in the use of opiates at 0.29% of the adult population (UNODC 2002). Still, 7.5 million, or over half of the world's 14.9 million, opiate users live in Asia. Unlike most other regions of the world, opiate use in Asia is not limited to the use of heroin. While nearly all opiate users are heroin users in Africa, Western Europe, Oceania and North and South America, only 3.6 million of Asia's 7.5 million opiate users (48%) are heroin users. Thus, over half of Asia's opiate users smoke or eat opium.

The history of opiate drugs in Asia is well documented. Yet, this history appears to be continuing. Every reporting nation in Asia except Myanmar and Laos indicated that opiate use increased during the late 1990s. Myanmar reported "some decline" and opiate use was "stable" in Laos. While most Asian nations indicated there had been "some increase" in opiate use, several of the former Soviet republics in Central Asia reported a "large increase" in opiate use (UNODC 2000: 64). A number of other sources corroborate that opiate use is spreading in those Central Asian states (e.g. Kozlova 2002; OGD 1996; OGD 1997; Drug Policy Alliance 2002). It is therefore not surprising that according to the most recent data, Central Asia now has the highest rate of opiate use among the Asian sub-regions at 1.0% of the adult population. Central Asia has apparently surpassed South East (0.46%) and South Asia (0.42%) as Asia's leading opiate-using regions. East Asia (0.15%) and the Middle East (0.11%)

have the lowest rates among the Asian sub-regions, based on UNODC (2004) figures.

Iran is the leading opium-consuming nation in the Central Asia, where 2.8% of the adult population uses opiate drugs (UNODC 2002). The majority of these users smoke opium; however, as in other traditional opium-smoking countries, heroin use is becoming more common. Kyrgyzstan, Kazakhstan and Tajikistan, where annually an estimated 2.3%, 1.3% and 1.0% of the population use opiates respectively, also have high rates of use. Approximately 0.6% of the adult populations in Uzbekistan, Afghanistan and Georgia use opiate drugs (UNODC 2004). These rates are not significantly higher than in other Central Asian nations such as Armenia or Turkmenistan. Indeed, the lowest rate reported in the region was for Azerbaijan at 0.2% of the adult population. Thus, the tradition of smoking opium is well established throughout the Caucasus (OGD 1997). While heroin use was practically unknown a few years ago, recent evidence indicates that intravenous drug use, in particular injecting heroin, is increasing (Drug Policy Alliance 2002; Kozlova 2002). This increase in use is in part due to the greater supply of refined heroin passing through Central Asia on its way to Europe (OGD 1996).

South East Asia also has a well established opiate-using population. In particular, Laos, Myanmar, Thailand and Viet Nam have reputations for high rates of opium and heroin use. Although officials claim use is stable in Laos, some of the highest rates of opiate use in the world can be found there. According to UNODC (2004) data, 1.8% of the adult population use opiate drugs. According to *Les Observatoire Géopolitique des Drogues* (OGD 1996; 147), nearly 63 tons of Laotian poppy is converted to opium and smoked by some 42,000 Laotian consumers. Heroin use is rarer in Laos than in neighboring nations, however, "some cases have recently been reported among craftsmen in certain northern villages" (OGD 1996: 148). Myanmar is also known for its opiate production. As part of the "golden triangle," Myanmar is the world's second largest producer of opiates, topped only by Afghanistan (see Drug Policy Alliance 2002; Stares 1996; UNODC 2000). It is also a major transit area for heroin destined for Europe, Australia and the United States. It is not overly surprising then that some of the opium and heroin produced for the global market finds its way to local consumers. Although rates of use are reportedly declining (see UNODC 2002 and UNODC 2004), Myanmar still has an opiate use rate of 0.7% of the adult population (UNODC 2004). While lower rates are reported for Thailand (0.5%), Viet Nam (0.3%), Malaysia, India (0.2%), Singapore (0.01%) and Brunei Darussalam (0.01%), opiate use in these nations has been increasing. For example, there were officially 240,000 drug addicts in Viet Nam at the end of 1996 (compared to 180,000 a year earlier), and most of them were under 35 (OGD 1997). Traditional consumption in Viet Nam is limited to opium smoking, and although this practice is still prevalent, heroin use appears to be increasing. For example, in the Lam Son province, urine tests in schools revealed traces of heroin in one child out of ten. Similar rates are estimated in Hanoi. In the largest urban center in southern Viet Nam, liquid opium residue is popular. Heroin and other opiate-based drugs have become popular among well-off teenage students and members of the very poorest segments of society (see OGD 1997).

South Asia has the third highest rate of opiate use in the region, and Pakistan and India have rich opiate histories. It was once believed that nearly 3 million persons were opiate users in Pakistan (see Drug Policy Alliance 2002), however, a recent survey conducted by UNODC in 2000 and 2001estimated there were only 0.5 million "heroin abusers" in Pakistan (UNODC 2002: 223). Nevertheless, Pakistan still has a relatively high rate of opiate use at 0.9% of the adult population. Similarly, in India, with a rate of 0.4% of the adult population using opiate drugs (UNODC 2004), heroin use has effectively replaced the more traditional use of opium (Drug Policy Alliance 2002). Nepal also has an opiate-using rate of 0.4% of the adult population. Most users in Nepal smoke opium. However, as in other parts of Asia, heroin appears to be gaining popularity. Bangladesh, Indonesia and Singapore have very similar rates of opiate use. Rates in these nations range from 0.1% to 0.3% of the adult population. Brunei has the lowest reported rate in Asia at 0.01% of the adult population.

East Asia has use rates that are below the region's and the world's average at 0.15% of the adult population. However, the opiates have played a significant role in the region's history. As Courtwright (2001: 33) notes, "of all the Asian lands in which opium use and cultivation became entrenched, the most significant was China." China's opium-related history is indeed storied and includes several attempts to outlaw its use, two wars to ensure Britain's continued exportation of Indian opium to China, and three international conferences that eventually paved the way for international agreements on curtailing the drug trade.

While opium has been used medicinally in China for centuries, its recreational use is more recent. Beginning in the seventeenth century, the Chinese began smoking opium, most typically mixed with tobacco. By the early nineteenth century, use was common among laborers and the peasantry when it was estimated that over 16 million Chinese, or 6.0% of the adult population, were daily opium smokers and "perhaps half of the adult population smoked opium at least occasionally, to celebrate festivals or ward off disease" (Courtwright 2001: 34).

Today, China continues to be a major transit point for illegal narcotics and opium smoking is somewhat common (Drug Policy Alliance 2002; Stares 1996; OGD 1996; OGD 1997; Wang 1999). It is estimated that 0.1% of the adult population use opiates (UNODC 2004), but the actual rate is probably much higher (OGD 1997). This figure, according to OGD, does not include some ethnic minorities in Yunnan Province who traditionally smoked opium but who are now increasingly turning to heroin. Reports claim that the popularity of injecting drugs has also increased in other parts of China since 1990. Beijing recently estimated that there were more than 900,000 drug addicts in the People's Republic -- compared to 70,000 in 1990 -- but independent reports put the number at between 6 and 7 million. More than 50 percent of addicts are believed to be injecting drugs, and many narcotics experts estimate that within five years China will have the most heroin addicts of any country in the world (Kurlantzick 2002). The majority of Chinese opiate users are males with a low educational level or middle class youths. However, in the southwestern province of Guizhou, opium users are usually elderly people (OGD 1997; Kurlantzick 2002).

Although now part of China, the former British colony of Hong Kong owes its former existence to opium. Ceded to the British after the First Opium War of 1839 - 1842, Hong Kong served as a center for the global distribution of opium. The British also promoted opium use in Hong Kong to help finance the colonial administration there (see Yi-Mak and Harrison 2001). As a result, opiate use has been popular in Hong Kong for well over a century. In addition to opium, heroin grew in popularity during the 1940s and 1950s. In fact, it was in Hong Kong where the practice of "chasing the dragon," a means of smoking heroin, first became popular in the 1950s (see Strang, Griffiths, and Gossop 1997).

Today, according to UNODC (2002), 0.2% of Hong Kong's adult population uses opiate drugs. According to Laidler et al (2000), heroin is widely available and the most commonly used illegal drug in Hong Kong, at least among those known to the *Central Registry of Drug Abuse*. Moreover, "most users report that heroin consumption is on the rise" (Laidler et al 2000: 29; also see OGD 1997; UNODC 2000). Heroin is widely available and inexpensive. As one heroin user stated (cited in Laidler et al, 2000: 29),

> Sei Jai (number 4 heroin) is the most popular. It's also widely available. You can easily get it anytime of the day, even in the afternoon. You can get it in every district, and the price is not too expensive.

Heroin use, however, may be declining among those under the age of 21. Registry data indicate there has been a significant decline in the number of newly reported heroin users under the age of 21 since 1995 (see Laidler et al 2000).

While opium and heroin use are historically important in China and Hong Kong, use rates, at least those reported to the UN, are actually higher in Taiwan. According to UNODC (2004) data, 0.3% of the adult population use opiates in Taiwan. This is the highest rate of opiate use reported in the sub-region. Heroin is one of the major drugs of choice, almost matching the consumption of marijuana (Drug Policy Alliance 2002; see UNODC 2002). Yet, unlike most of Asia, heroin use appears to be decreasing (compare UNODC 2000 and UNODC 2004). The other East Asian nations that report to the UN, Japan and South Korea have low rates of opiate use. In both of these nations, less than 1/10 of the adult population is believed to use opiates. Unlike other eastern countries, Japan has never experienced a high rate of opium use among its citizens, in part because it was the first Asian power to enact legislation against opiate use within its boundaries (Greberman and Wada.1994). Meanwhile, in South Korea, use may be increasing. The amount of heroin trafficked into the country appears to be on the increase and most of it appears to be for local consumption (Drug Policy Alliance 2002). However, opiate use rates are still relatively low in South Korea.

Finally, opium has a long-entrenched history in the Middle East. It figured prominently in Arab medicine, and Arab traders are believed to have introduced opium to the Indian sub-continent and China sometime in the eighth century (Courtwright 2001). Its use in the Middle East, both medicinally and recreationally, occurred long before then. Despite this long history, reported rates of use are lower than one may expect. Based on UNODC (2004) data, the ten reporting Middle Eastern nations had an average adult consumption rate of 0.11%. Israel and Bahrain report that 0.3% of the adult population uses opiates. Jordan reports a rate of 0.2% of adults, and Oman reports that 0.1% of adults are opiate users. All of the remaining Middle Eastern nations report a rate of opiate use of under 1/10 of a percent (UNODC 2004). Use in the region, however, is believed to be increasing. Israel, Syria, Kuwait and Bahrain reported "large increases" in opiate use in the late 1990s (UNODC 2000); however, these increases apparently did not materialize (compare UNODC 2000; UNODC 2002; UNODC 2003; UNODC 2004).

The other illicit drug that is popular in Asia is methamphetamine (amphetamine-type stimulants or ATS). In fact, 22.5 million Asians use methamphetamine. That represents over 2/3rds of the world's ATS users and the world's second highest rate of ATS use. Throughout the region, 0.9% of the adult population use ATS-type drugs (UNODC 2002). Unlike opiate use, which is relatively constant across Asia, the widespread use of ATS is found in a limited number of countries. Generally speaking, the primary ATS-using nations are located in South East Asia and East Asia.

Thailand has not only the region's but also the world's highest rate of ATS use. According to UNODC (2004) data, 5.6% of the Thai population use ATS. Behind alcohol and tobacco, ATS are the most widely used drugs in Thailand. There is an estimated 2 to 3 million drug users in Thailand, and most of them use methamphetamine (Drug Policy Alliance 2002). Sixty percent of incarcerated drug offenders in Thailand are charged for possessing ATS. In addition, Thailand reports the highest rate of ecstasy use in Asia at 0.1% of the population (UNODC 2004). While still not widely used in Asia, ecstasy appears to be penetrating the South East Asian market, and Thailand has been the initial point of entry. Compared to Thailand, the Philippines has a relatively low rate of ATS use. However, nearly 3% of the adult population uses ATS in the Philippines, and an additional 0.01% use ecstasy. The rate of ATS use is the second highest rate in the region and fourth highest in the world (see UNODC 2004).

Methamphetamine is also popular in East Asia. After alcohol, methamphetamine is Japan's most popular drug (Drug Policy Alliance 2002), and the rate of ATS use is higher than that of marijuana. Estimates from the mid-1990s stated that Japan had 400,000 stimulant addicts as well as millions of occasional users (Vaughn, Huang and Ramirez 1995). UNODC (2002) estimates that 1.8% of the Japanese adult population currently uses ATS. The widespread use of ATS in Japan dates to the World War II when the Japanese government sanctioned the use of stimulants and used them in the war effort (Vaughn et al 1995; Greberman and Wada 1994). ATS are still popular today, and rates of use seem to be increasing (UNODC 2000). Moreover, whereas stimulant use has traditionally been limited to males in urban areas, ATS are now used in both cities and rural areas. Estimates from the early 1990s also indicated a higher proportion of women, particularly housewives, are now using methamphetamine (Greberman and Wada 1994; JICA 1992).

Taiwan and South Korea also report high rates of ATS use. Along with heroin, ATS are the drugs of choice in both nations. In Taiwan, the rate of ATS use is 1.2% of the adult population. South Korea has a rate of ATS use of 0.2% (UNODC 2004). While inhalation, snorting and ingestion are primary methods of use in Taiwan, injection remains the preferred method of taking methamphetamine in South Korea (Drug Policy Alliance 2002). Methamphetamine, known as *bingdu*, is also becoming increasingly popular in China, particularly among the unemployed youth population (Kurlantzick 2002).

Throughout East and South East Asia, every reporting nation except Japan indicated that there had been either "some increase" or a "large increase" in ATS use during the late 1990s (UNODC 2000). While Japan reported there had been "some decline" in ATS use, rates of use there remain well above the global average. Throughout the region, ATS are quickly becoming the most popular illicit, and occasionally licit, drug. As stated in UNODC's *World Drug Report* (2000: 74, 76),

> In many of these countries, including Japan, the Republic of Korea, Taiwan, the Philippines and Thailand, use of methamphetamine already exceeds that of the opiates, the traditional substance of abuse in the region. In addition to the instrumental use of methamphetamine . . . large-scale recreational use among high-school and university students has started spreading.

The Middle East reports a moderately high rate of ATS use, although consumption there appears to be much lower than in East and South East Asia. Jordan reports a rate of 0.4% of the adult population. Israel, Lebanon, Bahrain and Oman report ATS-use rates that range from 0.01% to of 0.3% of the adult population. Qatar and Syria report lower rates, yet ATS use appears to be increasingly popular (UNODC 2002).

Despite its popularity in other regions, the use of cocaine is rare in Asia. According to UNODC (2004), Asia has the world's lowest rate of cocaine use at a mere 0.01% of the adult population. Although the use of cocaine is reported to be increasing in South Korea, Israel (UNODC 2002) and China (Wang 1999), it remains a drug used by few Asians. For example, the highest reported rate of annual cocaine use in Asia is for Indonesia at 0.1% of the adult population. This rate ties them for 53rd place in the world. Of course, other illicit drugs are used throughout Asia. For example, in India, it is reported that a growing number of people are using licitly manufactured drugs, in particular codeine-based cough syrups and benzodiazepines, in combination with illicit drugs like heroin (Drug Policy Alliance 2002). Similarly, drugs such as diazepam, mogadon, and codeine-based cough syrups are widely available throughout the Middle East (OGD 1997) and in Hong Kong (see Laidler et al 2000). The sniffing of solvents, including glue, paint thinner, and gasoline, can be found among small segments of the Japanese youth population (Vaughn et al 1995). The use of Club Drugs, such as MDMA and ketamine, has also been reported recently in Hong Kong (Laidler et al 2000) and can be found in larger urban areas in East Asia, South East Asia, Central Asia and the Middle East. However, generally speaking, the major illicit drugs found throughout Asia have been, and remain, marijuana, opiates and methamphetamine.

To summarize, Asia, despite its history with intoxicants, has relatively low rates of drug use. In fact, in terms of percentage of the population who uses drugs, Asia has the lowest rates of use on the planet by far. Among the seven major regions of the world, Asia ranks 3rd in ATS use, 5th in opiate use, 6th in heroin and ecstasy use and last in marijuana and cocaine use. While more drug users live in Asia than in any other region of the world, rates of use there are consistently among the lowest in the world.

Europe

Unlike Africa, which is known for the widespread use of marijuana, and Asia, which is known for opiate production and consumption, Europe is not often associated with drugs other than alcohol. This perception is probably due to the imposition of European standards on the rest of the world (see, in particular, Szasz 1974). It certainly has little to do with empirical reality. Unlike Africa and Asia, Europe provides a substantial market for all of the major illicit drugs. Eastern Europeans are now the world's leading opiate consumers (see UNODC 2004). Western Europe, while not leading in consumption rates for any specific drug category, consistently ranks in the "middle of the pack." That is, Western Europeans have relatively high use rates of several types of drugs, including above average rates of marijuana, cocaine and ecstasy use. Overall, Western Europe ranks third with respect to drug consumption rates, at least based on UNODC data. Therefore, the perception of who uses drugs does not necessarily match the reality of drug use.

With respect to illicit drugs, cannabis again leads the way in Europe. An estimated 34.1 million Europeans, or 5.2 percent of the adult population, use cannabis annually (UNODC 2004; EMCDDA 2002; EMCDDA 2003). Yet, Western Europeans are twice as likely to use marijuana as Eastern Europeans. While approximately 7.2% of the adult population of Western Europe uses cannabis annually, only 3.3% of adults in Eastern Europe do (UNODC 2004). Although Western Europe has a higher consumption rate than Eastern Europe, the Czech Republic reports the highest annual rate of cannabis use in all of Europe at 10.9% of the adult population. The United Kingdom has the second highest rate among European nations at approximately 10.6 percent of the adult population (UNODC 2004; also see ESPAD 1997; Parker, Williams and Aldridge 2002). France, Spain, and Ireland are the next three leading cannabis-consuming states with adult use rates ranging between 9 and 10 percent (UNODC 2004; also see Mayock 2002). Switzerland, Denmark, Italy, Belgium, the Netherlands, Germany, Austria and Iceland also have relatively high rates of marijuana use, ranging from approximately 5.0% of the Icelandic adult population to 7.0% of the Swiss adult population (UNODC 2004). The next leading cannabis-consuming states in Europe, at least according to UN data, are Norway, Greece and Slovenia with approximately 4.5% of the adult population using cannabis annually. Thus, after the Czech Republic, the next fourteen leading consuming nations in Europe are in Western Europe. It is also worthy to note that the leading cannabis-using nations in Eastern Europe are the two nations with the highest GDPs in the region. The lowest reported rates of annual cannabis consumption in Europe are primarily found in Eastern Europe. Belarus, for example, reports a rate of 0.10% of the adult population. Bulgaria, Moldova, Estonia, Hungary, Poland, Albania and Lithuania all report rates of cannabis use under 2.5% of the adult population (UNODC 2004). The two notable Western European nations with low rates of marijuana use are Finland (2.9%) and Sweden (1.0%).

The fact that cannabis products are the most widely used illicit drugs in Europe is not surprising. As in Africa and Asia, marijuana is the illicit drug of choice. However, Europe differs from these other regions in one important respect. While cannabis use is largely confined to adult males in many African and Asian areas, it is a habit of the young in Europe. This pattern is especially pronounced when the more remote regions of Africa and Asia are compared to the more urbanized areas of Europe. For example, while hashish smoking is common among older males in Afghanistan, Nepal, India, Kenya, Zimbabwe and Rwanda, very few older Europeans use cannabis. Conversely, an estimated 37% of youth use cannabis annually in the United Kingdom and Ireland. Between 20 and 25 percent of young adults in the United Kingdom are regular recreational cannabis users (Parker et al 2002; also see ESPAD 1997; ESPAD 2001; Flood-Page et al 2000). Similarly, 35% of young Czechs used marijuana, according to a recent survey (Ladislav, Kubikaa, and Nociarb 2002). Denmark (28%), Spain (26%) and Switzerland (25%) all report that over one-fourth of youth use cannabis annually. Throughout the European Union (EU) countries, "more than one in four young Europeans (28.9%) have tried cannabis and more than one in ten (11.3%) have used cannabis over the last month (EORG 2002: 4; also see EMCDDA 2002). In comparison, only 12% of Kenyan youth use marijuana annually, despite a slightly lower overall use rate than Denmark (see UNODC 2002). Thus, in the developed world of Europe, the young use marijuana. In fact, the youthful recreational use of cannabis has become "normalized" in some European nations (see Parker et al 1998; Parker et al 2002). In contrast, in the lesser-developed regions of Africa and Asia, the old, particularly older males, use marijuana. In these regions, recreational marijuana smoking among the young, while apparently increasing, is far from "normalized."

The youthful use of marijuana is common in other European nations. For example, Germany, the Netherlands, Belgium and Italy report that approximately one-fifth of their youth use cannabis annually. Several Eastern European states also report relatively high rates of youthful cannabis use. Slovenia, for example, reports that 15.7% of youth use cannabis annually. Bulgaria (15.0%), the Ukraine (14.5%), Slovenia (13.0%) and Poland (12.2%) have similarly high rates. While these Eastern European states are closing the gap with the west, youthful cannabis use is still higher in Western European than in Eastern Europe. According to UNODC data, approximately 16.6% of youth in Western Europe and 11.5% of youth in Eastern Europe use cannabis annually (UNODC 2002). An estimated 27.2% of 15 and 16 year-old Western Europeans have used marijuana at least once in their life. In Eastern Europe, 15.7% of 15 and 16 year olds have used marijuana at least once. In total, nearly one-in-four European teens have used marijuana (UNODC 2002b: 216). Among young adults between the ages of 15 and 24, the numbers are even higher. For example, nearly half (47.0% and 44%

respectively) of young adults in Denmark and France, and over one-third in the United Kingdom and the Netherlands, have tried cannabis at least once (EORG 2002).

In both Western and Eastern Europe, cannabis use is more prevalent among males than among females. For example, among young adults in EU countries, males are twice as likely to use cannabis regularly as are European females (EORG 2002). This correlation, while pronounced among the very young, appears to weaken with age (see, for example, Parker et al 1998; Mayock 2002). Regular cannabis use is also correlated with occupational status. Based on survey data from 2002, compared to manual workers, managers were nearly twice as likely to use cannabis on a monthly basis (EORG 2002). Urban young adults are slightly more likely to use cannabis on a monthly basis than are Europeans from rural communities, small towns, or medium sized cities (EORG 2002; EMCDDA 2002). Thus, young professional males are the primary users of cannabis; the most widely used illicit substance in Europe. This is especially true in Western Europe. However, as in the United States and other developed nations, cannabis use is not limited to any particular group. It is relatively common in all social strata, at least among the young.

The second most widely used illegal drugs in Europe are amphetamine-type stimulants and ecstasy.[21] While Europe is among the leading consumers of illegal drugs worldwide, it is interesting to note that ATS use in both Western and Eastern Europe is below the world average. According to UNODC (2004), approximately 0.81% of the world's adult population use ATS annually. In Europe as a whole, however, the estimated number of ATS users is approximately 3.3 million and the annual rate of use is only 0.51%. In fact, as far as ATS use is concerned, Europe ranks ahead of only Africa. Western Europe, with an estimated 2.3 million ATS users, and Eastern Europe, with 2.4 million ATS users,

---

[21.] ATS and ecstasy are being considered together. If these drug classes are considered separately, opiates actually rank second in Europe (see UNODC 2004).

have nearly identical rates of use at approximately 0.5% of the adult population. This rate is only 62% of the global average for ATS use.

Yet, there are European nations with relatively high rates of ATS use. For example, Ireland and the United Kingdom, the leading European ATS using nations, tie for 7th place in the world with 1.6% of the adult population using ATS annually. Denmark (1.3%) ranks 11th and Spain (1.2%) is tied at 12th. The Czech Republic, Norway, Estonia and Hungary are also among the top 25 ATS-using nations worldwide with above average rates of use. In fact, methamphetamine is the most widely used illegal substance among "problem drug users" in the Czech Republic (Ladislav et al. 2002). Other European nations with relatively high rates of ATS use include Belgium, Germany, Iceland, the Netherlands, Latvia and Poland with between 0.6% and 0.7% of the adult population using ATS annually (UNODC 2004). As with cannabis use, ATS are primarily used by European youth. For example, ATS use is fairly common among English and Dutch youth (Ter Bogt, et al 2002; also see Parker et al 1998). In the 26 European nations that reported to the *European School Survey Project on Alcohol and Drugs*, researchers estimated that approximately 10 percent of the youth population used stimulant drugs (ESPAD 1997). Similarly, 3.3% of all European 15 and 16 year olds are believed to have used amphetamines or methamphetamine at least once in their life. Indeed, after cannabis, "ATS are the drug of choice among 15 and 16 year olds" (UNODC 2002b: 219).

With the exception of the Czech Republic, Eastern European nations have relatively low rates of ATS use. The Czech Republic, with an estimated rate of 1.1 percent of the adult population, leads all Eastern European states in ATS use. Estonia, Hungary, Latvia, Poland and Slovakia are the only other Eastern European states that have rates above the regional average (UNODC 2004). Lithuania, Croatia, Bulgaria, and the Russian Federation have average levels of use for the region. Belarus has the lowest reported level of ATS in Europe (UNODC 2004). However, ATS use appears to be increasing in many Eastern

European nations. In Poland, Estonia and the Czech Republic, for example, 15 and 16 year olds have among the highest rate of ATS use in Europe (ESPAD 2001; UNODC 2002b).

In addition to amphetamine and methamphetamine, ecstasy use is relatively common in Europe (see USDEA 2001). Introduced in Western Europe in the mid-1980s, ecstasy became popular among youth in the "house" dance culture. In 1987 and 1988, ecstasy use became increasingly popular, especially in the United Kingdom and the Netherlands (see Ter Bogt et al 2002). Its use has spread to other countries since then, and is now popular among youth involved in the dance and rave cultures throughout Europe (see USDEA 2001). In Europe as a whole, the 3.3 million ecstasy users nearly match the number of ATS users (UNODC 2004). The 2.9 million ecstasy users in Western Europe surpass the 2.3 million ATS users there. Western Europe is the third leading ecstasy-consuming region of the world with a rate of 0.78% of the adult population using it annually.

Ireland now leads all European nations in ecstasy use with an estimated 3.4 percent of the adult population using annually (UNODC 2004). The United Kingdom, Spain and the Netherlands are close behind. All of these nations have annual use rates of over 1.0 percent of the adult population (UNODC 2002b; UNODC 2004). Belgium, Iceland, Germany, Austria and Norway also have relatively high rates of ecstasy use, ranging from 0.6 to 0.9 percent of the adult population (UNODC 2004; also see Measham, Aldridge and Parker 2001). In Austria, ecstasy has replaced heroin as the drug of choice among the younger population (USDEA 2001). Most of the remaining Western European states all report rates of ecstasy use in the range of 0.4 to 0.5% of the adult population. Cyprus, Greece and Sweden have the lowest rates in Western Europe, but still report at least 0.1% of the adult population using ecstasy annually (UNODC 2004).

Ecstasy began penetrating the Eastern Europe market after the dissolution of the Soviet Union and its former empire. However, Eastern Europe still has relatively low rates of ecstasy use at 0.13% of the adult population. Yet, this rate is increasing and is now only slightly below the world average (see UNODC 2002b; UNODC 2004). As with other drugs, the Czech Republic has the highest reported rate of ecstasy use at 2.5 percent of the adult population (UNODC 2004). In fact, among the European states, the Czech Republic is second only to Ireland in reported ecstasy use. Ecstasy use also increased considerably during the late 1980s in Hungary (WHO Regional Office for Europe 2001a). Generally speaking, patterns of ecstasy use in Eastern Europe closely match the patterns of ATS use there.

While marijuana and ATS-type drugs are the most commonly used drugs in Europe, Europeans also have relatively high rates of cocaine use. Cocaine is not a new drug to Europe by any stretch of the imagination. It has been used in Europe since coca leaves were used in popular elixirs during the mid-19th century (see, for example, Morgan 1981; Courtwright 2001). When Europe "discovered" cocaine, its popularity spread quickly due to its immediate sensual appeal and the promotion of its use by European governments. For example, Prussian soldiers used cocaine during the wars of German unification and English soldiers used it during World War I. Once introduced, cocaine use increased throughout much of Western Europe until the early 1920s. Use was especially prevalent in Germany, France, Belgium, Great Britain and Austria (see Phillips and Wynne 1980; Parssinen 1983). Beginning in the 1920s, however, cocaine began to lose its appeal in Europe and virtually disappeared from the European drug scene until the 1970s (see Phillips and Wynne 1980; Ruggiero and South 1995; Parssinen 1983). Cocaine's popularity began to increase noticeably in Western Europe during the 1980s (van de Wijngaart 1990), and several countries, such as Spain, the Netherlands and the United Kingdom, reported considerable increases in use during the 1990s. In Spain and the Netherlands, for example, the annual prevalence of cocaine use increased between 1995 and 2001 by over forty percent (see UNODC 2003).

Today, Europe is the third leading cocaine-using region in the world with approximately 3.71 million Europeans using the drug annually (UNODC 2004). Like cannabis and ecstasy, cocaine use is far more prevalent in Western Europe than in Eastern Europe. In fact, Western Europe, with 3.4 million annual users and an adult use rate of 1.1%, is the second leading cocaine-using region in the world (see UNODC 2004). The United Kingdom (1.7% of the adult population), Spain (1.5%) and Ireland (1.3%) have the third, fourth and sixth leading cocaine use rates in the world, respectively. Germany, Belgium, Italy and Cyprus also have high rates of cocaine use at just less than one percent of the adult population. The Netherlands, Iceland, Austria, Portugal, Norway, Denmark, Greece, Switzerland and Luxemburg have usage rates that range from 0.4% to 0.7%. Since the global adult use rate for cocaine is only 0.33%, all of these nations have above average rates of cocaine use. Finland and France, with 0.2% of adults using cocaine annually, are the only Western European nations that have below average rates of cocaine use (see UNODC 2004). However, France has recently reported that cocaine use has been increasing (UNODC 2003).

While cocaine has re-entered the Western European market, its use remains relatively rare in Eastern Europe. According to UN data, approximately 300,000 Eastern Europeans use cocaine annually. Thus, only 0.09% of Eastern European adults use cocaine (UNODC 2004). Croatia, Slovenia and Estonia are the only Eastern European nations that report that more than 0.1% of the adult population use cocaine annually. Yet, this pattern appears to be changing. Most Eastern European nations reported an increase in cocaine use in the late 1990s (UNODC 2002b). During the late 1990s, a booming cocaine market led to decreasing prices and enticed new users in Croatia, Bosnia and Hungary (OGD 1997; also see EIS 2002). More recently, increases in cocaine use have been reported in Bulgaria, Serbia and Montenegro, Belarus, Slovakia and Poland (UNODC 2003). Hungary and Croatia continue to report a growing population of cocaine users (WHO Regional Office for Europe 2001a; UNDOC 2003). The

Russian Federation, the largest nation in Eastern Europe, still reports very low levels of cocaine use (UNODC 2004).

As with most illicit drug use in Europe, cocaine is primarily used by the young. Two percent of 15 and 16 year old Europeans have used cocaine at least once in their lifetime. Again, this rate is higher in Western Europe than in Eastern Europe. In Western Europe, 2.3% of 15 and 16 year olds have used cocaine, while in Eastern Europe, only 1.2% have (UNODC 2002b). In Spain, the United Kingdom and Germany, over 3.0% of youth are annual users of cocaine. In the Netherlands, 2.0% of youth are annual users (UNODC 2002). In all of these countries, the rate of use among youth is at least double the rate of use among all adults (also see, EMCDDA 2002; EMCDDA 2003). Unlike many other drugs that are either used by the lower social classes or by all social classes, cocaine is still a "rich man's drug" in Europe. At least through the mid-1990s, cocaine use was largely confined to the "fashionable" and more affluent members of European society (see Ruggiero and South 1995).

While cocaine has only recently re-entered the Eastern European market, opiate drugs have been there for centuries. Europeans have used opium for medicinal reasons since the 12th century. Its recreational use in Europe dates to at least the 18th century (van de Wijngaart 1990). Opium was easily available and widely used throughout Europe until the early 20th century. Like cocaine, however, rates of use decreased appreciably after World War I and remained low until the 1960s. Since then, opiate drugs, including heroin, have re-emerged in most European nations (see, for example, Ruggiero and South 1995). While rates of use remain low compared to other illicit substances, Europe has the highest annual opiate use rate of any region of the world (UNODC 2004).

Unlike most other illegal drugs, however, Eastern Europe has a much higher rate of opiate use than Western Europe. According to the most recent data, the annual rate of use among adults is 2.57 times higher in Eastern Europe than in Western Europe. Nearly 3 million Eastern Europeans use opiate drugs annually, compared to only 1.6 million Western Europeans. Those numbers translate to approximately 1.1% of the adult population in Eastern Europe and 0.4% of Western Europe's adults (UNODC 2004). The Russian Federation reports the highest opiate use rate in Europe and the third highest in the world. Russian authorities estimate that there are approximately 4 million drug users in Russia, and half of these use opiate drugs. Thus, slightly over 2% of Russian adults use opiate drugs annually. Latvia and Estonia also report use rates above 1.0% of the adult population. These three Eastern European nations are among the top seven opiate-using nations of the world. The Ukraine, Croatia, Slovenia, Lithuania, Czech Republic, Bulgaria and Albania also report annual use rates that are nearly double the global average. All of these countries report that at least 0.5% of their adult populations use opiate drugs annually (UNODC 2004). The preferred method of opiate use appears to be smoking rather than injecting, at least among the young. Nearly four percent of Eastern European 15 and 16 year olds smoked heroin or some opiate-containing substance at least once in their life. Less than one percent of these youth have injected opiate drugs (UNODC 2002b).

Opiate use, including the use of heroin, has increased rather dramatically over the last decade. For example, in the Russian Federation, the number of registered addicts doubled between 1991 and 1995 and quadrupled between 1995 and 2000 (UNODC 2004). Although less well documented, similar increases occurred in much of Eastern Europe during the 1990s. More recently, Belarus, Latvia, Romania and Albania reported increases in heroin use. However, in other parts of Eastern Europe, use has recently stabilized. After years of increases, the Ukraine, Bulgaria, Hungary and the Czech Republic reported stable use levels in 2002. Poland, Slovakia and Croatia all reported declines of opiate use in 2002 (UNODC 2004). However, even in areas where use is decreasing, rates remain relatively high by global standards. Moreover, large shipments of Afghan heroin flooded the Russian market in 2003. This increased supply drove down the price and increased the availability of heroin (UNODC 2004). The effect of the price drop has yet to be determined.

Opiates are not as popular in Western Europe. While opiates trail only cannabis in terms of rates of use in Eastern Europe, they are the least widely used major illicit drug in Western Europe. The highest levels of opiate use in Western Europe are found in Luxembourg, Portugal, the United Kingdom, Italy and Switzerland, with rates ranging from 0.6% to 1.0% of the adult population. Again, use is highest among European youth. For example, in Italy, over 2.0% of youth are opiate users (UNODC 1999). Greece, Belgium, Denmark, France, Ireland and Norway comprise Western Europe's second tier of opiate-using nations. These nations report rates of use that are near Western Europe's average and slightly above the world's average. Finland, Sweden, Monaco, Liechtenstein and Turkey report very low rates of opiate use (UNODC 2004). Rates of use in Western Europe have been generally stable over the past decade. Rates increased slightly between 1993 and 2000 and have decreased slightly since then (UNODC 2004).

Among youthful Western Europeans at least, smoking heroin is preferred over injecting it. While 1.7% of Western European 15 and 16 year olds have smoked heroin at least once in their life, only 0.7% have ever injected it (UNODC 2002b). Western European opiate users are typically polydrug users. Many opiate users also use cocaine, barbiturates, amphetamine, cannabis, ecstasy and, of course, alcohol (see UNODC 2002b; also see Parker et al 1998; UNODC 1987; EMCDDA 2002; EMCDDA 2003). Not surprisingly, opiate users in Western Europe tend to be marginalized and young. For example, heroin users in Sweden, although few in number, are found "primarily among marginalized groups, such as ex-convicts, institutionalized youth and prostitutes" (Hartnoll et al 1989: 11). Heroin users in Dublin are typically young, single, unemployed and have low levels of education (Ruggiero and South 1995). Similarly, van de Wijngaart (1990: 533) observes that Dutch users "tend to have lower levels of education and come from the lower classes." This positive correlation between opiate use and marginalizaion is found in most Western European nations (see EIS 2002; UNODC 1987).

While it is true that Eastern Europe has a higher rate of opiate use than does Western Europe, there is a dramatic difference in what opiate is used in each region. In Western Europe, approximately 83% of opiate users use heroin. By contrast, only 55% of Eastern European opiate users are heroin users (see UNODC 2004). Although this appears to be changing, heroin use was relatively uncommon in Eastern Europe until the mid-1990s (see, Ruggiero and South 1995). Throughout Eastern Europe, "*kompot*" and "*makiwara*" are commonly used. These "home brews" are derived by extracting active opiate substances from poppy straw. *Kompot* is injected and *makiwara* is orally ingested. This practice, allegedly developed in Poland in the mid-1970s, accounts for significant numbers of opiate users in Eastern Europe (see UNODC 1987; Ruggiero and South 1995; UNODC 2004; WHO Regional Office for Europe 2001b). For example, while there are an estimated 2 million opiate users in the Russian Federation, only half use heroin (see UNODC 2004). If only heroin use is considered, Eastern and Western Europe have similar rates of use. However, if the use of opiate "home brews," other opiate-containing drugs such as codeine and morphine, and synthetic opiates are considered, the East clearly has a larger opiate-using population. In fact, opiates are the only major illicit drugs that are consumed by a greater percentage of Eastern Europeans than Western Europeans.

In addition to the major illicit drugs, Europeans use a wide variety of other intoxicants. The illicit use of tranquilizers, sedatives, solvents, inhalants and hallucinogens is also relatively common in both Eastern and Western Europe. For example, in Hungary, it is estimated that approximately 100,000 people, mainly middle-aged and elderly women, are "dependent on tranquilizers and sedatives, making this one of the most common drug dependency problems" (WHO Regional Office for Europe 2001a: 25). Tranquilizers and sedatives are also popular among European youth. These drugs have been used by approximately eight percent of European teens. Among younger adolescents, solvents may be the most commonly used drug in Europe after cannabis (European Commission 2002; also see Parker et al 1998). Among slightly older adolescents, inhalants are popular. Over nine percent of Western European 15 and 16 year olds and seven percent of Eastern European 15 and 16 year olds have used an inhalant at least once in their life. In the United Kingdom, inhalant use is even more common. In a survey conducted in the early 1990s, an estimated 35.3% of British youth had used amyl nitrite at least once (Parker et al. 1998). LSD and other hallucinogens have been used by about 2.6% of European 15 and 16 year olds (UNODC 2002b). Data from the United Kingdom can give a sense of the variety of drugs used in Europe. For example, in a sample of 14 to 18 year olds, 28.0% had used LSD, 8.5% had used natural hallucinogens, 4.5% had used tranquillizers and nearly 12% have used solvents (Parker et al 1998). While these numbers may be higher than what would be found in other European countries, there is a consensus that many types of illegal and pseudo-legal drugs are used in both Western and Eastern Europe.

All things considered, Europe has very high rates of drug use. While Europe typically ranks in the "middle of the pack" on most of the major illegal drugs, all of these drugs are widely available and used there. Moreover, Europeans have above average rates of use of many of the "minor" illegal or pseudo-legal drugs. As a region, Europe ranks third in the consumption of illegal drugs. When one adds the use of legal drugs such as alcohol, tobacco, caffeine and pharmaceuticals, Europe is the world's leading drug-consuming region. While Western Europeans are more likely than Eastern Europeans to use illegal substances, rates of use in these regions appear to be converging.

North America

North America, comprising only Canada, the United States and Mexico for our purposes, is highly stratified. Of course, this stratification is across numerous dimensions, but it is also very apparent with respect to drug use. While the United States and Canada are among the wealthiest nations on earth, Mexico is clearly a "middle income," "developing," or "semi-peripheral" country.

Similarly, while the United States and Canada are in the top ten percent of drug-using nations on the planet, Mexico is among the bottom half of drug-using nations. The United States and Canada have very similar drug-using patterns. Mexico's pattern of use is considerably different. While Mexico is a major producer of marijuana and opiates and an important transit nation for cocaine, the United States and Canada are major producers of amphetamines, ecstasy, LSD and other synthetic drugs. Of course, the United States is also a major cannabis producing nation; however, very little of America's marijuana is exported. Combined, the United States and Canada are by far the world's largest and most profitable drug markets. Mexico is one of the world's biggest suppliers (see UNODC 2004; UNODC 2003). Apparently, the North American Free Trade Association (NAFTA) nations have a symbiotic relationship with respect to drugs.

Another dimension of stratification that separates the northern members of North America from Mexico is the quality of data available from each nation. The United States is the most widely studied country and, therefore, has the best data concerning illicit drug use. Large scale, scientifically sophisticated, nationally representative samples have been collected in the United States since the early 1970s. The data on American drug use are, relatively speaking, exceptionally rich and accurate. Given the similarities between the United States and Canada, we can use U.S. data to paint a very good picture of drug use in these nations. Moreover, Canadian data, also of relatively high quality, confirms that the U.S. and Canada have similar drug-using patterns. Unfortunately, the data from Mexico are not as rich nor do they inspire as much confidence as the American or Canadian data. Plus, U.S. data cannot be used to accurately reflect drug use in Mexico. Still, we are reasonably confident that Mexico is not a major drug-consuming nation in the world, and it certainly lags far behind its fellow NAFTA members.

As a region, North America ranks second in terms of overall rates of illegal drug consumption. With the exception of ecstasy, the United States edges

Canada as the leading consuming nation in the region. However, the differences between these nations, regardless of the drug in question, are usually small. In essence, the United States and Canada have remarkably similar drug-using patterns. Mexico is always a distant third. As in all of other regions, marijuana is by far the drug of choice in North America. An estimated 23.5 million North Americans, or 7.5% of the adult population, use marijuana annually. Unlike other regions, the vast majority of the cannabis consumed in North America is herbal, not cannabis resin. An estimated 11.0% of the American adults use marijuana annually and 10.8% of Canadians smoke marijuana annually. These figures place the U.S. and Canada among the top 12 marijuana-consuming nations in the world with respect to consumption rates. However, the United States is the world's largest cannabis market, and Americans smoke the most tonnage of marijuana each year in the world. Mexico, despite growing a significant cannabis crop, has extremely low rates of use. Only 0.6% of adult Mexicans use marijuana annually (UNODC 2004). Among the 137 nations that provided data on cannabis use to the U.N., Mexico ties for 110th place.

As in Europe, adolescents and young adults are much more likely to use marijuana than are older adults. In the United States, for example, 34.9% of high school seniors reported using marijuana at least once in 2003 (Johnston et al. 2004). Recall that only 11.0% of Americans over the age of 15 use marijuana annually. Thus, late adolescents and young adults are much more likely to be current marijuana users than are young adolescents or older adults. Moreover, over 21% of seniors used marijuana in the 30 days prior to being interviewed (Johnston et al. 2004). The rate of marijuana use among Canadian seniors is also substantially higher than among the rest of the Canadian population. Over 44% of Ontario's high school seniors use marijuana annually (Adlaf and Paglia 2003; also see Dell Garabedian 2003). A similar correlation between age and the use of marijuana is found in Mexico, although among much fewer marijuana users. Based on a survey conducted of urban schools in Mexico in the early 1990s, 4.0% of 18-year-old males and 0.4% of 18-year-old females used marijuana in the 12 months preceding the survey. While these relatively low rates confirm that the United States and Canada have much higher rates of marijuana use than does Mexico, they also highlight that marijuana use in Mexico, like in the U.S. and Canada, is primarily a habit of late adolescents and young adults (Villatoro et al 1998).

Cocaine is the second most widely used illegal drug in North America, and North America has the highest rate of cocaine consumption in the world by a substantial margin. An estimated 6.3 million North Americans, or slightly more than two percent of the adult population, use cocaine annually. This is nearly double the number of users and double the rate of use found in the next leading consuming region. While part of the substantial gap between North America and the rest of the world in terms of cocaine consumption can be explained by the region's proximity to the producing nations of South America, there is more to the explanation than simple proximity. First, North America has higher rates of cocaine use than South America. Second, the rate of use in the United States is five times the rate of use reported in Mexico. Thus, despite being more proximate to producing countries (and, undoubtedly where it is easier to smuggle cocaine across the border), Mexico does not come close to matching America's cocaine habit, with an estimate of only 0.5% of the adult population using cocaine annually.

America is the world's leading cocaine consuming nation by far. Even Canada, the 10th leading cocaine-consuming nation, has a much lower rate of use than does the United States. While an estimated 2.6% of adult Americans use cocaine annually, only 1.0% of Canada's adult population are annual users (UNODC 2004). Again, Mexico is a distant third. However, cocaine use may be increasing in Mexico, especially among the young (USDEA 2003). In addition to powdered cocaine, both the United States and Canada have high levels of crack use. In the United States, 2.2% of high school seniors use crack annually

(Johnston et al 2004). Similarly, 2.4% of seniors in Ontario smoke crack at least once a year (Adlaf and Paglia 2003). Crack use is not widespread in Mexico.

Cocaine, regardless of the chosen ingestion method, is a drug of the young. In the United States, 4.8% of high school seniors used cocaine at least once in 2003 and nearly 2% used it within 30 days of being interviewed (Johnston et al 2004). In Canada, nearly 7.0% of high school seniors in Ontario used cocaine in 2003 (Adlaff and Paglia 2003). Even in Mexico, where rates of cocaine use are low by regional and Western Hemisphere standards, 2.8% of 18-year-old males were annual cocaine users (see Villatoro et al 1998). In fact, officials believe cocaine use is as prevalent as marijuana use among young persons in parts of the country (USDEA 2003). Therefore, we can confidently state that young adults in North America are two to five times more likely to use cocaine than are older adults.

Next, cocaine remains a drug of the affluent. While this correlation has weakened over time, it remains nevertheless. Based on the United State's 2000 *National Household Survey on Drug Abuse*, the correlation between income and the average number of days per year one uses cocaine is moderate and positive (r = .14; p < .001). Similarly, contrary to popular perception, crack use is also positively correlated with income. In fact, the correlation between the use of crack and income is even stronger than the correlation between the use of cocaine and income. In 2000, the correlation between income and the average number of days per month one smoked crack was significant and positive (r = .18; p < .001). Thus, cocaine tends to be a rich-person's drug, regardless of the form in which it is used.

Ecstasy is the third most widely used illegal drug in North America. In terms of the percentage of adults who use ecstasy annually, North America ranks second among the major regions of the world. Slightly more than 3.5 million North Americans, or 1.1% of the adult population, use ecstasy annually. Once again, the North American ecstasy market is not in Mexico. An estimated 0.01%

of the adult population there uses ecstasy. This rate is far below the global average for ecstasy use of 0.18%. Once again, the United States and Canada drive North America's drug consumption. This time, however, Canada leads the way. At least according to U.N. data, Canada is the 7th leading ecstasy-consuming nation and the United States is the 9th. While 1.8% of adult Canadians use ecstasy annually, 1.3% of adult Americans do. Canada's lead is even more striking when comparing the youthful use of ecstasy. While 4.5% of American high school seniors use ecstasy annually, 7.2% of Ontario's seniors do (see Johnston et al 2004 and Adlaf and Paglia 2003, respectively).

ATS-type stimulants, the next most widely used illegal drug in North America, are the only major drug class in which North America is not substantially above the global average with respect to rates of consumption. Approximately 2.6 million North Americans use ATS annually. This figure translates to 0.82% percent of the adult population. The global average for ATS use is 0.81% (UNODC 2004). Thus, North Americans are only "average" in their ATS consumption. Despite being "average," North America only trails Oceania, Central America and South East Asia in ATS use. The United States (1.4% of the adult population) and Canada (1.0%) have similar rates of ATS use. Mexico (0.1%) has low levels of use, at least according to U.N. data.

Finally, 1.5 million opiate users live in North America. For the region, an estimated 0.48% of the adult population use opiate drugs annually. As a region, North America ranks behind Central Asia, Oceania and Eastern Europe -- and just ahead of South East Asia -- in terms of opiate use. However, unlike in Asia and Eastern Europe, heroin is the primary opiate used in North America. While other opiates can certainly be found and are certainly used in North America, the use of opium is negligible when compared to the use of heroin. The percentage of adults who use heroin in North America is approximately twice the global average (UNODC 2004). For opiates other than heroin, however, North America's rate is only 1.3 times the global average. Interestingly, opiate use is not substantially higher in the United States and Canada than in Mexico. An estimated 0.6% of Americans, 0.4% of Canadians and 0.4% of Mexicans use opiates annually. It should be noted that North Americans, at least Americans and Canadians, are increasingly turning to synthetic opiates. While only 0.8% of American high school seniors reportedly used heroin in 2003, slightly more than 9% of American seniors reported they used some "other narcotic" at least once in 2003. Over 4% of them used oxycontin (see Johnston et al. 2004). Data at this level of detail are not available for Canada; however, anecdotal reports suggest the trend toward synthetic narcotic use is also occurring there.

As in Europe, considerable varieties of other illegal drugs are used in North America. Plus, these drugs are used by a considerable number of people, especially in the United States and Canada. Table 3.2 reports the annual use of various substances by high school seniors in the United States and the Province of Ontario. It also reports the annual use of substances by 18-year-old males in Mexico. As can be seen from this table, youth in the United States and Canada use a wider variety of intoxicants than those in Mexico. While the numbers are not directly comparable,[22] high school seniors in both the United States and Canada report relatively high rates of use (at least 1.0%) for marijuana, cocaine, amphetamine, methamphetamine, ecstasy, narcotics, inhalants, solvents, hallucinogens, tranquilizers and sedatives. Substantial numbers of young Americans and Canadians use specific drugs like GHB, vicodin, ice, ritlin, rohypnol, and ketamine annually (see Johnston et al 2004; Adlaf and Paglia 2003). Although marijuana is clearly the drug of choice among these students,

---

[22.] The numbers are not directly comparable because the numbers for the United States are taken from a nationally representative sample while the numbers from Canada are for the Province of Ontario only. While similar rates of use can be found in some other Canadian Provinces (see, for example, Liu et al 2003), the national rates are lower than those reported. For example, in a national sample of 17 - 19 year olds, approximately 14% had used marijuana in 1995 (see Health Canada 1995). The Ontario numbers are reported because they are the most comprehensive with respect to specific types of drugs and are relatively recent. The figures from Mexico are even less comparable. First, they are reported only for males. Second, these figures are somewhat dated (from the early 1990s). However, these are the most comparable data available.

many of them are polydrug users. While Mexican adolescents and young adults also recreate with a variety of intoxicants, fewer of them do so. Plus, those who do use drugs for recreational purposes use a less varied pharmacopoeia (see Villatoro et al 1998).

Table 3.2: Annual Use of Various Substances by High School Seniors

	United States (High School Seniors)	Ontario, Canada (High School Seniors)	Mexico (18 year old Males)
Any illicit substance	39.3	47.1	NA
Marijuana	34.9	44.8	4.0
Cocaine	4.8	6.7	2.8
Crack	2.2	2.5	NA
Stimulants	9.9	7.8	3.1
Methamphetamine	3.2	3.6	NA
MDMA	4.5	7.2	NA
Hallucinogens	5.9	15.3	0.7
LSD	1.9	2.7	NA
PCP	1.3	2.7	NA
Tranquilizers	6.7	2.7	2.4
GHB	1.7	1.7	NA
Rophypnol	1.3	1.3	NA
Ketamine	2.1	3.7	NA
Sedatives	6.0	1.8	0.6
Inhalants	3.9	3.9	1.8
Heroin	0.8	1.1	0.1
Vicodin	10.5	NA	NA
Ritlan	4.0	3.1	NA

Source: United States: Johnston et al 2004; Canada: Adlaf and Paglia 2003; Mexico Villatoro et al 1998.

All things considered, North America is the world's largest drug market. This fact is true for both legal and illegal drugs. The residents of the United States smoke the most tonnage of marijuana each year. They are also the world's main cocaine consumers. Canada fares well also (or poorly, depending on one's perspective). Both of these nations are among the top 20 consumers of marijuana, cocaine, ecstasy, and amphetamines. Moreover, given the enormous pharmaceutical market in the United States and Canada, there is a substantial market for pseudo-legal drugs and legal drugs that have been illegally diverted to the street scene (see Bellenir 2000). Despite all data limitations and questionable estimates, there is little doubt that the United States and Canada are both big-time addicts, so to speak. Mexico is the only North American nation that does not rank among the top drug-consuming nations of the world.

South America, Central America and the Caribbean

South America, Central America and the Caribbean are significant drug producing and trafficking regions.[23] For example, Bolivia, Peru, Colombia, Ecuador, Venezuela and Brazil are the world's leading cocaine producing nations. In fact, South America produces all of the world's non-pharmaceuticalcocaine. The region, not surprisingly, is also the central distribution point for cocaine. Columbia, which hosts the world's leading cocaine cartels, is the primary distributing nation. In addition to cocaine, the region produces and distributes significant amounts of marijuana. Similarly, the Caribbean, given its proximity to major drug producers and consumers, is a primary drug-exporting region (see, for example, Griffith 1998; Stares 1996). According to the United Nations (UNODC 2003b), 3.4% of the Caribbean's GDP is from the exportation of illicit drugs. Similarly, illicit drug exports accounts for 2.3% of Colombia's GDP. To place these numbers in perspective, only 0.65% of Mexico's and the United States'

[23.] I will refer to these three regions as "South America" throughout this section when referring to the region as a whole.

GDP is from exporting illicit drugs. Illicit drug exports account for only 0.4% of the European Union's GDP and 0.5% of the world's GDP (UNODC 2003b). Therefore, the region's reputation appears well deserved. Despite producing significant amounts of illegal drugs, however, the region's consumption rate is rather modest. Of the world's major regions, South America ranks fifth in the consumption of the major illicit drugs.

Once again, cannabis is the most widely used illegal drug in the region. An estimated 13.1 million people, or 4.6% of the adult population, use marijuana annually (UNODC 2004). The Caribbean has a significantly higher rate of marijuana use than does either South or Central America. The top three, and four of the top six, marijuana-consuming nations in the region are Caribbean. Overall, rates of marijuana use among the adult population are approximately 1.3 times higher in the Caribbean than in Central America and 1.5 times higher in the Caribbean than in South America (see UNODC 2004; also see Jutkowitz and Eu 1994).

According to U.N. data and ethnographic accounts, Jamaica is the leading cannabis-consuming nation in the region. An estimated 8.0% of the Jamaican adult population smoke marijuana annually (UNODC 2004). However, cannabis use is much more prevalent among the rural working classes than among the more affluent in Jamaican society. Based on ethnographic work, nearly 70% of rural working class males smoke marijuana (see Comitas 1975; Dreher 1982; Dreher 1983). Throughout the Jamaican working-class, marijuana is smoked, made into a tonic with white rum, or brewed in a tea. It is used to treat a number of ailments, including teething and colic in babies, asthma, fevers, and nausea in pregnant women. It is also given to children to encourage a good appetite and help them do well in school (see Dreher 1982). According to Rubin (1975: 261),

> Multipurpose ganja use in the working-class milieu, particularly in rural areas, is a cultural regularity, used in the dietary and extensively in folk medicine. Ganja smoking is also a manifest -- if somewhat secluded --

practice, despite the stringent legislation against possession and use. . . . It is the rare working-class Jamaican male who has never had an initial experience of smoking cannabis as an informal rite de passage to "manship."

Marijuana is also used religiously in Jamaica. Rastafarianism, a political and religious movement that combines West African oral traditions with Christianity, began in the 1930s and quickly spread throughout Jamaica and the Caribbean. Rastafarians, who are disproportionately found in the working classes, believe marijuana, or ganja, is the herb of life mentioned in the Bible. The use of ganja is justified by Psalms (104:14) that says, "He causeth the grass to grow for the cattle and herb for the service of man, that he may bring forth food out of the earth." Known as "wisdom weed," marijuana serves as a sacrament and meditation aid. Ganja, which is consumed orally and by smoking, helps Rastafarians get closer to their inner spiritual self, Jah (God) and Creation (see Barrett 1988; Hebdige 2002; Mack, Ewart and Tafari 1999). While Rastafarianism is often associated with the use of marijuana, the religion is a complex philosophy and involves much more than simply using drugs. Nevertheless, given the central importance of marijuana to working class culture and religion, the use of marijuana is deeply institutionalized in Jamaica (see Dreher 1982). It is a central dimension of male working class identity.

While Jamaica is the most well-know cannabis-consuming nation in the Caribbean, several other island nations also have high rates of marijuana use. For example, in Barbados, 7.7% of adults use marijuana annually. Barbados' consumption rate is therefore twice the global average. St. Vincent (6.8% of the adult population) and the Bahamas (5.0%) also have above average rates of marijuana use. Conversely, the Dominican Republic, Montserrat, Dominica and Grenada have very low rates of marijuana use. In the Dominican Republic, an estimated 2.1% of the adult population smoke marijuana annually. In Montserrat,

Dominica and Grenada, less than 1.0% of adults are annual marijuana users (UNODC 2004; also see Jutkowitz and Eu 1994).

In Central America, Belize has the highest reported rate of use where nearly seven percent of adults consume marijuana annually. In Guatemala, with a use rate near the global average, marijuana remains the most widely used illicit drug and is readily available throughout the country (UNODC 2004; USDEA 2003b). Nicaragua, Panama, El Salvador, Honduras and Costa Rica have below average rates of marijuana use that range from 1.3% to 2.6% of the adult population. Although rates of use are low in these countries by global and regional standards, marijuana is still the most widely used illegal drug (UNODC 2004). Moreover, it is easy to obtain marijuana in most areas of Central America. In Costa Rica, for example, marijuana is readily available in nightclubs, discos and from smalltime street dealers. As is the case throughout the region, the majority of the marijuana consumed in Costa Rica is locally produced, primarily in the southwestern and Talamanca areas (USDEA 2003c).

Chile leads all South American nations in marijuana consumption with approximately 6.8% of the adult population using it annually. Colombiaclosely follows with an estimated 4.3% of adults using marijuana annually. Argentina, Venezuela and Ecuador report adult use rates that are close to the global average. The remaining South American nations (Guyana, Bolivia, Suriname, Peru, Paraguay and Brazil) report rates of use that are thirty percent lower than the global average (see UNODC 2004). Despite being used in Brazil for over 500 years (see Hutchinson 1975), Brazil has the lowest reported rate of use in South America. According to the most recent U.N. data, only 1.0% of the adult population uses marijuana annually in Brazil.

As is in most nations, South American youth are disproportionately more likely to use marijuana. In Chile, for example, while slightly over 5.0% of adults use marijuana, over 20.0% of youth do so. In Jamaica, 17.0% of youth have used marijuana. The Bahamas report a similar percentage (UNODC 1999; BDIS

2002.). In Brazil, while only 7.6% of youth are believed to be marijuana users, adolescents are approximately seven times more likely to use marijuana than older adults are. Also, South American males are more likely to use marijuana than are South American females.

Despite its reputation for cocaine production and use, the second most popular illicit drug in the South American region is amphetamine (see UNODC 2004). However, this is not consistent across the sub-regions of South America. While ATS drugs are the second most popular for the region as a whole, this is largely due to their relatively widespread use in Central America. The nations of South America and the Caribbean report relatively low rates of ATS use. In both of these regions, and especially in South America, cocaine is the second most widely used illegal drug (see UNODC 2004).[24] It should also be noted that the difference between ATS and cocaine use for the region is relatively small. According to U.N. data, approximately one-half million more people use ATS drugs annually than use cocaine. Nevertheless, based on the available data, ATS drugs edge cocaine as the region's second-choice illegal drug.

Over three million of the region's citizens, or slightly more than one percent of the adult population, use ATS-type drugs annually. Central America has the highest rate of ATS use in region by far. Based on U.N. estimates (UNODC 2004), rates of ATS use are approximately four times higher in Central America than they are in either South America or the Caribbean. For those South American nations that provided data to the U.N., Honduras, Guatemala, Panama and Costa Rica are the leading ATS-consuming nations in the region. Honduras, with approximately 2.5% of the adult population using ATS drugs annually, has a rate of use nearly 2.5 times higher than the region's average. The other reporting

---

[24.] Only one Caribbean nation, the Dominican Republic, reported ATS as a problem drug to the U.N. Based on this lack of mentioning by the other Caribbean nations, we must assume that ATS use there is relatively infrequent.

Central American nations have rates above the regional and global average (see UNODC 2004).

As mentioned above, South America and the Caribbean have relatively low rates of ATS use. Venezuela leads all South American nations in ATS use with slightly more than 0.8% of the adult population using annually. Venezuela's rate of ATS use is nearly identical to the global average. Colombia, Argentina, Chile and Bolivia have slightly lower reported rates, ranging from 0.5% to 0.7% of the adult population. However, in Colombia at least, rates of ATS use are almost double the national average among those between the ages of 10 and 24 (see UNODC 2003). Brazil, Ecuador, Uruguay and Suriname report rates of use less than half the global average and one-quarter the region's average. The only Caribbean nation that reported an ATS problem to the U.N. was the Dominican Republic. Approximately 0.4% of the adult Dominican population use ATS drugs annually (UNODC 2004).

South America is known for its production and use of cocaine. The history of coca in the Andean region of South America is long, dating back to at least the Incan Empire (see, for example, Phillips and Wynne 1980). While the Incaic use of coca was primarily religious, cocaine is now primarily a recreational drug. Of course, South Americans are the world's growers and producers of illicit cocaine. Bolivia, Peru, Argentina and Ecuador have been leading coca-growing nations for decades. More recently, Colombia, Venezuela and Brazil have increased their coca production. For the last 30 years, Colombian cartels have controlled the world's illegal cocaine industry. Despite being the world's cocaine producers, however, South America ranks behind North America, Western Europe and Oceania in rates of use (see UNODC 2004). Still, throughout the region, some 2.7 million persons, or 0.94% of the adult population, use cocaine or some coca product annually.

South America has the highest rate of cocaine use in the region, followed closely by Central America. Generally speaking, cocaine use is most prevalent in transit countries such as Argentina, Colombia, Chile and Venezuela. For example, Argentina has the highest reported rate with nearly 2.0% of the adult population using cocaine annually. Colombia, with 1.2% of adults using it annually, has the second highest rate of cocaine use in South America (UNODC 2004), and rates of use are even higher among Colombian youth and young adults. In 2001, 4.5% of 10 to 24 year olds used cocaine (UNODC 2003). Peru (1.0%), Venezuela, Chile, Bolivia (0.9% in each), Brazil and Ecuador (0.8%) also have relatively high rates of cocaine use. All of these nations have rates of use at least 2.5 times the global average.

It should be noted that throughout South America, the use of cocaine is not limited to cocaine hydrochloride. In addition to cocaine hydrochloride, coca leaf chewing remains a widespread practice throughout the Andean region. Coca leaf chewing has been an integral part of Andean culture for at least 5,000 years (see Allen 1988; Plowman 1984; Batchelder 2001). Millions of people, especially in Peru and Bolivia, chew 4 to 5 grams of coca leaf throughout the day. Coca chewers also consume the leaves in the form of a hot infusion (*mate de coca*). In Peru alone, it is estimated that over 13.0% of the population chew coca. This percentage is even higher in the mountain towns, where children, adults and the elderly consume coca (Jeri 1984; Batchelder 2001). Coca is popular among Andean miners to stimulate them during work and is served as an after dinner treat in remote areas (Phillips and Wynne 1980). It is also used ritualistically and is highly esteemed in the cultural traditions of the Quechuan people (Allen 1988). Although technically illegal, medical research indicates that coca chewing may be beneficial to the chewer's health and an important means of adapting to high altitude living (see Batchelder 2001).

While coca chewing is an ancient practice, smoking coca paste, or *basuco*, is a more recent means of ingesting cocaine. Coca paste is an intermediary product in the chemical extraction of cocaine from coca leaves. The paste is smoked in a dried form, which contains between 40 to 90 per cent cocaine (see

Jeri 1984). Since the 1980s, coca paste smoking has become widespread in many South American countries, particularly Bolivia, Colombia and Peru. In Lima Peru, for example, a household survey conducted in 1979 found that 1.3% of the population, or approximately 390,000 people, smoked coca paste, and coca paste smoking has probably increased considerably in Lima since that survey was conducted (see Stares 1996; also see Jutkowitz and Eu 1994). In Colombia, 1.2% of 10 to 24 year olds smoked coca paste in 2001. In Chile, 0.7% of 12 to 64 year olds smoked coca paste (UNODC 2003). Unlike coca chewing, there are no health benefits to coca paste smoking. On the contrary, coca paste smoking is a very dangerous habit (see Jeri (1984) for reviews of some early medical studies of coca paste smokers). When we consider all of the various means by which coca and its derivatives are consumed in South America, it is little wonder why it is the second most popular drug in that sub-region.

Cocaine products are also popular in Central America. Honduras, a major transit state, is the leading cocaine-consuming nation in Central America. Over 1.2% of the Honduran adult population use cocaine (UNODC 2004). Guatemalaand Panama also report rates of cocaine use over three times the global average. El Salvador, Belize and Costa Rica report lower rates (ranging from 0.3% to 0.6% of the adult population), however, use there is reportedly increasing (UNODC 2003; UNODC 2004). Much of the cocaine use in Central America is not in the form of cocaine hydrochloride, coca leaf or coca paste. Instead, cocaine is consumed by smoking crack. The increasing use of crack in Central America is largely due to the region's status as a transit site for the cocaine destined for the United States. Colombian cocaine traffickers frequently pay Guatemalans for their logistical support with cocaine instead of money. Guatemalan traffickers then convert the cocaine into crack and sell it locally (USDEA 2003b). As a result, crack use rose sharply in the late 1990s in a pattern similar to that witnessed in the urban United States during the mid-1980s (USDEA 2003b). This same phenomenon occurred in Costa Rica (see USDEA 2003c). The primary users of cocaine and crack in Central America are young, urban males (USDEA 2003b).

Cocaine use has also recently increased in several Caribbean nations. Jamaica and the Dominican Republic have the highest rates of cocaine use in the Caribbean with slightly less than 1.0% of the adult populations using the drug annually (UNODC 2004). As with the Central American states, Jamaica and the Dominican Republic serve as transit points in the international cocaine distribution system (see Griffith 1998). Aruba, another, albeit lesser, transit nation, also reports a higher-than-average rate of cocaine use. Interestingly, cocaine is the only "problem drug" Aruba reports to the United Nations (see UNODC 2003; UNODC 2004). The remaining Caribbean nations that report to the United Nations indicate that cocaine use in their nations is either at or below the global average of 0.33% of the adult population.

Marijuana, ATS and cocaine are the primary illegal drugs of choice in South America. Opiate drugs, on the other hand, are relatively rare. Approximately 0.12% of the region's adult population use opiate drugs annually. This percentage translates into slightly over 300,000 adults (see UNODC 2004). Brazil has the highest estimated rate of opiate use in the region with an estimated 0.6% of the adult population using opiates annually. Venezuela, Chile and Colombia have rates near the global average. All other nations in the region that reported to the U.N. have opiate-use rates that are less than or equal to 0.1% of the adult population (see UNODC 2004). While the availability of heroin in many transit countries such as Costa Rica has increased recently, the use of heroin throughout most of South America is minimal and rates of use appear to be stable (see UNODC 2003; UNODC 2004; USDEA 2003b).

Although opiate users slightly outnumber ecstasy users in South America, the difference is small and apparently shrinking. As with opiate drugs, approximately 0.12% of the region's adult population use ecstasy annually (UNODC 2004). In Guatemala, the leading ecstasy-consuming nation in the region, 0.4% of the adult population use ecstasy. Despite being a relatively new phenomenon in Guatemala, ecstasy is readily available in nightclubs and at raves in larger cities such as Guatemala City and Antigua (USDEA 2003b). The same is reportedly true in the clubs of San Jose, Costa Rica, which has become the region's primary distribution point for ecstasy (USDEA 2003c). Although Central America has become the region's distribution center for ecstasy, there is little difference between it and the other sub-regions of South America with respect to ecstasy use. Colombia, South America's leading ecstasy-using nation, and Barbados, the leader among Caribbean nations, have rates of use that approximate Guatemala's. In both of these countries, approximately 0.3% of the adult population use ecstasy annually (UNODC 2004). The only other South American nations that report an ecstasy "problem" to the U.N. are the Bahamas, Nicaragua, Venezuela, Chile, El Salvador and Suriname. All of these nations report that between 0.1% and 0.2% of their adult populations are annual ecstasy users (UNODC 2004; also see BDIS 2002). Authorities blame the influx of young vacationing Western Europeans and Americans for the recent increase in ecstasy availability and use in South America (USDEA 2003c).

South Americans also use other illegal drugs. Natural stimulants are abundantly available and widely used (UNODC 2003). In addition, inhalants are one of the most widely used drugs by South American youth (Jutkowitz and Eu 1994). In Brazil, 13.8% of youth use inhalants, and 10% of Bolivian youth use inhalants. Colombia (5.9%), the Dominican Republic (3.6%), Chile (3.4%), Panama (3.2%), Peru (3.0%) and the Bahams (1.2%) also report significant numbers of youth who use inhalants (see UNODC 1999; BDIS 2002). Like in other regions, pharmaceutical drugs are diverted to the underground market for recreational use. While the use of diverted pharmaceuticals and other "minor" illegal drugs occurs throughout South America, the number of people who use these drugs and the variety of drugs used do not approach that seen in North America. For example, based on a recent survey of 12th graders in the Bahamas, only 2.2% of these youth had used tranquilizers in the year prior to the survey and

0.9% had used stimulants. Hallucinogens were used by only 0.2% of the youthful respondents (BDIS 2002).

Despite low rates of hallucinogen use among Bahaman youth, a discussion of drug use in South America would not be complete without at least a brief description of hallucinogenic use. South Americans have been using hallucinogenic plants in religious rituals since the pre-Columbian Aztecs (Schultes and Hofmann 1979). While the shamans used hallucinogenic plants in numerous regions of the world, the practice was the most developed in South America where hundreds of hallucinogenic plants have been identified and used. The use of hallucinogens was common among several South American groups such as the Desana, Matses, Siona and Guahibo of the Amazon region (Vitebsky 2001). The Matsigenka of western Brazil and eastern Peru use *ayahuasca*, a potent hallucinogen, religiously (Baer 1992). The Yanomamo of Brazil and Venezuela and the Catimbó of northern Brazil also use powerful hallucinogens in religious ceremonies (Vitebsky 2001; La Barre 1975). The use of natural hallucinogenic drugs is not prevalent in the more modernized areas of South America and therefore overall rates of use are negligible. However, in remote areas, the practice of using natural hallucinogens continues.

To summarize, South America, Central America and the Caribbean are "modest" consumers of drugs. With the exception of ATS-type stimulants, the region ranks in the "middle of the pack." It is last in the world in the use of opiate drugs. There is, however, considerable variation among the sub-regions of South America. While the Caribbean has relatively high rates of marijuana use, for example, South America has modest rates of use. Conversely, South America has much higher rates of cocaine use than does the Caribbean. Still, despite being the world's producer of cocaine, South America ranks fourth among the major regions in terms of the use of cocaine. With respect to cocaine use, we see additional variation across the sub-regions of South America. Even today, the chewing of coca leaf remains common in the Andean region. The smoking of coca paste is also common in South American countries. However, crack is more common in Central America. Compared to most regions, therefore, South Americans have moderate rates of use but they use a relatively wide variety of drugs.

Oceania

Oceania is the world's leading drug-consuming region. While the absolute number of users is limited, rates of use far exceed those seen in other regions. In fact, there really is no comparison. Oceania has the world's highest consumption rates for marijuana, ATS-type stimulants, ecstasy and heroin. They trail only Eastern Europe in overall opiate use and rank third in cocaine use (see UNODC 2004). Other drugs, including LSD and betel, are also widely used throughout much of Oceania. If Oceania had a population similar to Asia's and rates of use similar to their current rates, it would be difficult for drug traffickers to keep the region supplied.

Not surprisingly, cannabis is the most widely used "major" illegal drug in Oceania where nearly 4 million people use marijuana annually. This is an astonishing 16.9% of the adult population (UNODC 2004). The rate of marijuana use in Oceania is nearly twice that of Africa's, the second leading cannabis-consuming region. It is 2.24 times the rate found in North America. According to U.N. data, of the nine leading cannabis-consuming nations in the world, four are in Oceania. Papua New Guinea and Micronesia have the world's highest rates of marijuana consumption. Almost one-third (29.5% and 29.1%, respectively) of the adult population uses marijuana annually in both of these nations (UNODC 2004). This is approximately 2.7 times the rate in the United States. Australia ranks sixth in the world in marijuana use. Approximately 15.0% of the adult population uses marijuana annually, and nearly 50% of Australian university students have used marijuana at least once in their lifetime (Davey, Davey and Obst 2002). Over thirteen percent of adults in New Zealand use marijuana annually. This percentage ranks New Zealand 9th in the world. While these Oceanic nations have extremely high rates of marijuana use, the other, much smaller, nations of the region have very low rates. In New Caledonia, for example, only 1.9% of the adult population uses marijuana annually. In Fiji and Vanuatu a mere 0.2% and 0.1% of the adult population use cannabis products annually.

Oceania also has the highest rate of ATS, ecstasy and heroin use in the world. Over 600,000 adults use ATS-type stimulants. While the absolute number is relatively low, this translates to 2.8% of the adult population. Ecstasy is used by 2.2% of adults and heroin is used by approximately 0.6% of the adult population. To place these rates in perspective, Oceania's rate of ATS use is 2.67 times that of South America's, and their ecstasy use rate is 2.75 times that of Western Europe's. As a reminder, South America is the world's second leading ATS-consuming region and Western Europe is the world's seconding leading ecstasy-consuming region. Oceania's rate of heroin use is slightly higher than Eastern Europe's.[25] Oceania does not lead the world in cocaine consumption. It ranks third. It is far behind North America, but nearly matches Western Europe. Cocaine is used by 1.03% of the adult population in Oceania and by 1.06% of the adult population in Western Europe.

While Oceania's rates of drug use are extremely high, these rates are driven almost exclusively by Australia and New Zealand. In fact, the other Oceanic nations do not report any data on the use of drugs other than marijuana (see UNODC 2004). Thus, we will concentrate on Australia and New Zealand. For most drugs, Australia's rate of use is higher than New Zealand's.

Based on national surveys, 46% of Australians had used some illicit drug at least once in their life in 1998 (Australian Institute of Health and Welfare 1999). In addition to the "major" illegal drugs, Australians also have relatively high rates of illegal tranquilizer use (1.1% of the adult population use annually),

---

[25] Oceania's rate of opiate use is lower than Eastern Europe's; however, only about 60% of Eastern Europe's opiate users use heroin.

sedative and pain killer use (over 3%) and LSD and other hallucinogen use (1.1%) (Australian Institute of Health and Welfare 2002; also see Davey et al 2002). It should be noted, however, that rates of use appear to be decreasing in Australia. Based on the 2001 national household survey, only 37.7% of adults had used an illicit drug at least once in their lifetime (Australian Institute of Health and Welfare 2002). In fact, the use of most types of drug decreased between 1998 and 2001. The use of ecstasy, however, increased. Between 1993 and 2001, the percentage of Australians over the age of 14 who used ecstasy annually increased threefold, from 0.9% to 2.9%.

As in most nations, rates of illicit drug use are typically higher among youth than among older residents. Among university students between the ages of 18 and 21, for example, 35% are annual marijuana users while only 16% of those students over the age of 36 use marijuana annually. Similarly, 11% of 18 to 21 year olds use ATS-type stimulants annually but only 3% of those over the age of 36 do so (Davey et al 2002). Rates of use are also significantly higher among male university students than among female students. Australian males are more likely to use marijuana, ATS, ecstasy, LSD, cocaine and heroin than Australian females. Females have higher rates of use of benzodiazapines (Davey et al 2002). Yet, the differences between males and females are more pronounced in older age cohorts. Among the very young, differences in lifetime use and regular use are negligible (see Australian Institute of Health and Welfare 2002). Finally, rates of use are higher in the large metropolitan areas than they are in the more remote regional areas. However, the gap between metropolitan and rural areas has been decreasing (Headley 2001).

New Zealand has similarly high rates of drug use. While rates of use are generally lower in New Zealand than in Australia, they still rank among the leading drug-consuming nations of the world. New Zealand is the third leading ATS-consuming nation in the world and has the fourth highest rate of ecstasy use. Over 3% of the adult population use ATS annually and 2.2% of the adult population use ecstasy annually (UNODC 2004). As in Australia, ecstasy use has increased in recent years.

Table 3.3 reports the percentage of the adult population in Australia and New Zealand who annually use the "major" illegal drugs. In addition, the rankings of each nation ranks with respect to use rates of the various substances are included.

Table 3.3: Percentage of Adult Population Using Various Substances Annually in Australia and New Zealand and Ranking in World

	Marijuana	Cocaine	ATS-type Stimulants	Ecstasy	Opiates
Australia	15.0 (3rd)	1.4 (5th)	4.0 (2nd)	3.4 (1st)	0.6 (21st)
New Zealand	13.4 (9th)	0.4 (37th)	3.4 (3rd)	2.2 (4th)	0.7 (17th)

source: UNODC 2004

As can be seen from Table 3.3, Australia consistently ranks among the leading drug-consuming nations in the world. It is in the top five for every major drug except opiate drugs. Even for opiate drugs, however, Australia ranks among the leaders in the percentage of the population who are heroin users. New Zealand ranks high on marijuana, ATS and ecstasy use. Moreover, heroin use is also higher than average in New Zealand. While cocaine is not widely used in New Zealand, its use is reportedly increasing (UNODC 2004).

There is one drug that is more widely used in the rest of Oceania than in Australia and New Zealand. Betel, a relatively mild natural stimulant, is popular throughout most of Oceania and known as "the coffee of Oceania." Its widespread use in Oceania and Asia make it the most widely used stimulant in the world (see Drug Scope 2004). In most Oceanic cultures, both men and women consume betel. While older residents are the primary users, it is also popular among young children. It serves both ritualistic and recreational purposes. For example, both the Tikopia of the Solomon Islands and the Trobriands of northeastern Papua New Guinea use betel in initiation rituals and other ceremonial settings (see, for example, Firth 1936; Scoditti 1989).

To summarize, Oceania is the world's leading drug-consuming region. Rates of marijuana use are extremely high among the populations of Papua New Guinea, Micronesia, Australia and New Zealand. However, other than the natural stimulant betel, the use of other illegal drugs is relatively uncommon in most of Oceania. This is not the case in Australia and New Zealand, however. These nations, and especially Australia, are among the world's leaders with respect to the use of illegal drugs. While rates of use may be decreasing in Australia, they remain well above the world's average.

CHAPTER THREE SUMMARY

As this chapter's discussion illustrates, there is considerable variation in drug consumption patterns when viewed from a cross-cultural perspective. With respect to cannabis use, for example, the range of use is impressive. Whereas only 2.2% of Asian adults use cannabis annually, nearly 17% of adults in Oceania partake in its use. While an estimated 0.002% of adult Africans use ecstasy, over 1% of adults in North America and 2% of adults in Oceania experience the effects of ecstasy at least once a year.

In general, the regions of the world can be classified into three categories based on rates of drug use. Asia and Africa occupy the "low use" category. With the exception of ATS-type stimulants, Asia ranks near the bottom in use rates for all major illegal drugs. Even for ATS, Asia ranks only third. Similarly, while Africa ranks second in average rates of marijuana consumption, they are at or near the bottom on all of the other "major" illegal drugs. The "moderate use" category includes Eastern Europe and South America. Both of these regions have one major illegal drug that is widely used relative to the world's other regions and at least two others that are used moderately relative to the other regions. Eastern Europe ranks first in opiate use, second in heroin use and fourth in ecstasy use. South America ranks second in ATS use, fourth in cocaine use and fifth in both marijuana and ecstasy use. However, rates of use of the other major illegal drugs are relatively low in each of these regions. Finally, the "high use" category includes Western Europe, North America and Oceania. These regions have relatively high rates of use of all of the major illegal drugs. Not only do significant numbers of people use illegal drugs in these regions, wide varieties of drugs are used there. Western Europe has relatively high rates of both cocaine and ecstasy use and moderate rates of use of the other major illegal drugs. North America ranks in the top three on all the major drugs except ATS-type stimulants, for which it ranks fourth. Oceania has the highest rates of use for marijuana, ATS-type stimulants, ecstasy and heroin. They are second for opiate drugs in general and third for cocaine. Moreover, all three of these regions have relatively high rates of use of the "minor" illegal drugs such as inhalants, tranquilizers, sedatives, barbiturates, PCP and hallucinogens. Table 3.4 reports the percentage of the adult population who use each of the "major" illegal drugs annually for each of the major regions of the world. The variation in rates of use can easily be seen in this table.

While regional differences remain, consumption patterns are converging. All of the major regions in the "low" or "moderate" categories report that the use of a drug that had previously been unknown is increasing. Meanwhile, rates in the nations that comprise the "high" category have apparently stabilized, at least for the moment. We have known for sometime that the illicit drug distribution system is global. It should be little surprise then that, as this system broadens and deepens, consumption patterns will become increasingly similar. While the use of illegal drugs is deeply ingrained in some cultures, it is shunned, at least in a relative sense, in others. In some cultures, older generations use drugs, even those deemed illegal. In other cultures, using illegal drugs is a symbol of youth and youth culture. Yet, these long-time differences appear to be disappearing. The world's converging drug consumption patterns seems to be especially pronounced among the young (see UNODC 1999; EIS 2002). What can account for these patterns of drug consumption? Let us now try to explain it.

Table 3.4: Illegal Drug Use by Region (1998 - 2001 period)

	Percent of Adult Population Using Annually and Region's rank					
	Cannabis	Amphetamine	Ecstasy	Cocaine	Opiates	Heroin
Asia	2.17 (7)	0.89 (3)	0.01 (6)	0.01 (7)	0.29 (5)	0.14 (6)
Africa	8.60 (2)	0.50 (7)	0.002 (7)	0.20 (5)	0.20 (6)	0.20 (5)
Eastern Europe	3.29 (6)	0.50 (6)	0.13 (4)	0.09 (6)	1.08 (1)	0.60 (2)
South America	4.58 (5)	1.04 (2)	0.12 (5)	0.94 (4)	0.12 (7)	0.12 (7)
Western Europe	7.16 (4)	0.50 (5)	0.78 (3)	1.06 (2)	0.42 (4)	0.35 (4)
North America	7.53 (3)	0.82 (4)	1.11 (2)	2.03 (1)	0.48 (3)	0.48 (3)
Oceania	16.89 (1)	2.78 (1)	2.15 (1)	1.03 (3)	0.63 (2)	0.63 (1)
Global	3.88	0.81	0.18	0.33	0.35	0.22

Source: UNODC, *Global Illicit Drug Trends*, 2003

CHAPTER FOUR:

THEORETIC FRAMEWORK FOR A

SOCIOLOGY OF DRUG USE

In the previous chapters, the "what" and "where" of drugs were discussed. That is, the empirical patterns of consumption described *what drugs are used where*. Some general patterns of global drug consumption became apparent from those discussions. Generally speaking, legal drugs are used more frequently and by more people than illegal drugs. Caffeine, alcohol and tobacco are the world's most widely used drugs by far. With respect to illegal drugs, cannabis is the most popular. Yet, there is tremendous variation in use rates across regions. Marijuana, for example, is widely used in Africa, while its use is relatively limited in much of Asia. In Asia, opiate and amphetamine use is relatively high. Cocaine and cannabis use is prevalent in South America. Eastern Europeans like opiates, alcohol and tobacco. The people of North America, Western Europe and Oceania, especially Australia, appear to like drugs, *all drugs*, regardless of the substance's effect or legal status. What can account for these patterns?

At first glance, these patterns seem easy to explain. As a commodity, drugs, like all commodities, are used where they are available. That is, the *supply* of drugs determines the *demand* for them. The fact that drugs are used relatively heavily in the areas of the world where they are produced verifies this point. The use of opiates and heroin in Asia is not surprising when one remembers that the major poppy growing and opium/heroin production centers of the world are located there. Similarly, South America has a relatively high rate of cocaine consumption in part because the Andean region is where the world's coca is grown.

When we examine drug-trade routes, we also see supply influencing demand. The increasing use of heroin in the former Soviet Republics of Central Asia is due, in part, to their place along the heroin distribution system that runs from the production centers in Afghanistan to the lucrative markets of Russia and Western Europe (see, for example, OGD 1997). Similarly, many Central American nations such as Honduras have relatively high rates of cocaine use because of their proximity to the cocaine production centers. Caribbean nations like Jamaica have high rates of cocaine use because they are trafficking centers on the way from Colombia to the United States. Africa, as a region, currently has among the lowest rates of amphetamine, opiate and cocaine use; however, this pattern appears to be changing as Africa becomes increasingly central in the global illicit drug distribution networks (see OGD 1996). Thus, the supply of drugs does influence the demand for them.

Yet this is only part of the story. While geography, climate, and socio-political conditions can explain why certain areas of the world produce certain drugs, these do not account for other aspects of the drug distribution system. Why do drug traffickers take the routes they do? Why do the world's drug distribution systems, regardless of the legal status of the drug and where the drugs originate, so often end in Europe, North America and Australia? The answer is simple, at least on one level. The people of these areas *demand* the products.

Just as supply influences demand, demand determines supply. Put more sociologically, structure and process are complementary. That is, social interaction, if repeated over time, generates social structure. Yet, once established, structure determines, at least in part, future interaction. Blau (1977: 4) states,

This original etiology in which processes of social associations produce differences in social positions can only be observed in newly formed groups, because once distinct social positions have become established they channel further role relations and associations.

In other words, while process creates structure in the first place, structure determines process after that. Thus, patterns of drug consumption determined, at least in part, the structure of the distribution system. Once this structure of drug distribution was established, however, it influenced, and continues to influence, the process of drug consumption. Heroin traffickers go through Tajikistan not because the nation provides scenic vistas. They go through Tajikistan because it lies between where the heroin is produced and where it is wanted. Given this route, however, they might as well sell as much of their product along the way as they can. If the good people of Tajikistan want to buy some heroin, why would the traffickers, who happen to be carrying heroin, deprive them?

If we remember that drugs are commodities in the same sense that automobiles, shoes, refrigerators, and porcelain dolls are commodities, we avoid the all-too-common mistake of treating drugs, especially illegal drugs, as being somehow special. Economically speaking, drugs are not very special in the least. They respond to the same laws of supply and demand that govern the production and distribution of automobiles, shoes, refrigerators and porcelain dolls. An entrepreneur would not send her parkas to Senegal since the parka market there is far too limited. Similarly, not much cocaine is shipped to Asia since there is not a large demand for it there. In short, drugs, regardless of their legal status, are simply commodities sold on an open, capitalistic market. In fact, one could argue that what is most unique about the illegal drug distribution system is that it is one of *the most capitalistic systems* remaining on earth. This system is unfettered by governmental regulation (other than the occasional seizing of the commodity by state officials). There are no tariffs that benefit one nation and reduce the commodity's competitiveness in others. Neither the production nor sale of illicit drugs is taxed. The system's workers are not unionized. The systems of production are highly concentrated, and there is tremendous profit involved. Few other commodities are produced and distributed on such an open, capitalistic, laissez-faire market.

The point of this discussion is that since drugs are commodities distributed in a capitalistic-style system, understanding the modern world system, how it evolved, and the corresponding social processes that evolved with it may help us understand the contemporary patterns of drug consumption. Thus, drug trade routes often lead to consumers in Europe, North America and Australia because that is where *most trade routes lead*. Americans, Europeans and Australians consume the most drugs because these people consume most of almost *everything*. They are the leading nations in the modern system of capitalism and, as such, their citizens enjoy the benefits of that system the most. One of these "benefits," apparently, is access to a wide variety of substances with which to alter their consciousness.

Of course, this somewhat simplistic explanation begs the question to a large part. Saying that drugs are used where drugs are wanted is tautological. We must be able to determine why drugs are wanted where they are. It was the desire and willingness of the people of Europe and America to use heroin -- coupled with their general affluence that permitted them to afford it -- that caused the distribution systems to target these countries. We must therefore attempt to answer why this desire and willingness developed in these nations more than in others. This question appears to lead us to the age old "why do people use drugs." Which, it would seem, would likely lead to the psychological and social-psychological explanations we are accustomed to.

There are undoubtedly dozens, if not hundreds, of psychological reasons people use drugs. The usual suspects come to mind, including compensating for an inadequate personality, escaping reality, relieving stress, rebelling from authority, gaining acceptance from friends, enhancing self-esteem, feeling better, dulling emotional pain, or relieving boredom. Let us assume, for example, that boredom leads to drug use. It could be that youth in developed nations suffer more boredom than other youths. If this were so, it would explain why they are more likely to use drugs. But, explaining cross-cultural variations this way leads us to ask "why are American, English, Canadian and Australian youth so bored?" Compared to youth in other parts of the world, it would seem, at least at first glance, that the youth in developed nations would have *more ways* of keeping occupied. When they tire of their computer games, they have TV. When they grow bored with riding their ATV's, they can have sister take them for a ride in the family SUV. They can go to soccer practice, baseball games or dance class. They have the Cub Scouts, Girl Scouts, Boys and Girls Club, YMCA and YWCA. They have a wealth of objects, organizations and toys at their disposal. They have more than children in most countries throughout most of history could even begin to fathom. Why, then, are they so bored?

Alternatively, we could assume some personality trait makes youth vulnerable to becoming drug users. Let us assume a combination of low self-control, low self-esteem, deviant prone personalities and associations with deviant peers lead one to use drugs. Although investigating these factors can help us explain why specific individuals use drugs, they do not account for cross-cultural variations in use. That is, these explanations may suffice to explain why one American youth uses drugs while another does not. They, however, beg the question as to why a greater percentage of American youths than Japanese youths use drugs. Again, we ask, "Why are American youth more plagued with these problems than youth from Japan?" Why do American youths have such lower levels of self-esteem and self-control? Why do they have more deviant-prone personalities?

It may be that genetics has something to do with it. It is possible that American youth have "bad genes." After all, many of the heavy drug-using nations share a common genetic stock. Europe, Canada, America and Australia were all populated by Greek-Roman-French-Anglo-Saxon ancestors. Yet, assuming that it was their common genetic background that led to the "bad genes" that drive these nations' youth to use drugs, it must have been the same genetic background that led to the high levels of self-control needed to achieve the military and technological success that propelled these cultures to the front-and-center of the modern world system. Could the same genetic stock lead to a disproportionate number of drug addicts and Nobel Prize winners? It is, of course, possible and someday we may find that these groups have a greater range of genetic variation in their stock than the peoples of other nations.

Yet even if we assume the "common genetic" background argument is valid, we still face a problem. While this can plausibly explain variations *between* nations and groups of people, rates of drug use also vary *within* nations historically. The history of drug use in the United States can provide an example. We know that rates of drug use were much higher in the 1970s than they had been in the 1950s or came to be in the 1980s or 1990s (see, for example, Musto 1987; Morgan 1981; Hawdon 1996b). Did the youth who came of age in the 1960s have more "bad genes" than those who came of age in the 1950s? Did this genetic pool become even worse in the 1970s? Did those who matured in the 1980s regain "good genes?" Population genetics does not work that way. Something, therefore, still seems to be missing.

To explain cross-cultural variations in drug consumption, looking at individual characteristics may be helpful, but, ultimately, it is likely to generate more questions than it answers. To say that American youth are bored more than Israeli youth begs the question of "why is this so." The answer is unlikely to lie in the individual psychology or biology of the users. Instead, it likely lies in variations in the collective psychology, if you will, of the culture in which the individuals live. The low self-control of American youth relative to Japanese youth probably has more to do with differences in the two countries' normative expectations, institutionalized parenting practices and structured systems of social control than with individual-level variations in biology or psychology. Rather than their common ancestors, the cultural similarities of the United States, Canada, Western Europe and Australia are the likely the reason for these nations' similar drug-use patterns. It is something about the collectives' cultures and structures that shape the individual psychologies of their respective members that cause variations across collectives. We therefore will not address the psychological reasons for using drugs. Instead, we should ask, "What are the *social reasons* that people use drugs?" We thus return to cultural processes and social structures.

However, which cultural processes and social structures matter? It is true in a relative sense that all of these nations have some cultural and structural similarities. They are all liberal democracies that base their political legitimacy on a rational-legal system. They all have norms of egalitarianism and individualism. They are all advanced capitalistic societies. Although all of these factors likely play a part, it can be argued that the "causes" of all of these factors are the processes of modernization and rationalization (see, for example, Weber 1958; Weber 1978). The role that drugs played in helping fund the European mercantile empires and how these empires helped spread and popularize various plant-based drugs has been discussed elsewhere (see, for example, Yi-Mak and Harrison 2001; Courtwright 2001). How various multinational corporations market and distribute such drugs as tobacco to penetrate developing markets has been discussed elsewhere (e,g, Shenon 1998). These are all valuable accounts and play a role in the globalization of drugs. Yet, how modernization and rationalization shape consumption patterns has not been discussed. I will demonstrate that this line of reasoning can be fruitful.

Before addressing this task, however, it is necessary to understand that all drug use occurs within a broader, more general social context. Regardless of the specific reason a drug is used, the social context in which use occurs will help pattern that use. Thus, depending on the context of use, the nature of drug use is likely to change. The critical context in which drug use occurs is that of *sacred versus profane.* With the exception of religious use, which is always sacred, drugs can be approached as either sacred or profane objects, and therefore use occurs in either sacred or profane settings. That is, recreational drug use, for example, can occur in a sacred context, such as the Cherokee's black drink used during the Bouncing Bush festival (see Mooney 1891; French 2000), or in a profane context, such as drinking beer with friends in a local bar. The extent to which a drug is considered either sacred or profane will help shape how use occurs, the setting in which it occurs and how frequently it occurs. Let us consider this dimension more closely.

The Sacred and the Profane

Durkheim begins his definition of religion by noting that all religious phenomena are arranged in the two fundamental categories of beliefs and rites. Rites differ from other behaviors, such as moral practices, solely by the nature of the object at which they are directed (Durkheim [1915] 1968). The nature of the object, he contends, is determined by the belief about it. It is therefore the belief concerning the targeted object that distinguishes rites from other human practices. As Durkheim ([1915] 1968: 51) notes,

> A moral rule prescribes certain manners of acting to us, just as a rite does, but which are addressed to a different class of objects. So it is the object of the rite which must be characterized, if we are to characterize the rite itself. Now it is in the beliefs that the special nature of this object is expressed. It is possible to define the rite only after we have defined the belief.

He then contends that religious thought has universally divided the world into the two opposing domains of sacred and profane. All things, "real and ideal" are classified into these opposing groups. The division between these domains is so complete that they are conceived "as two distinct classes, as two worlds between which there is nothing in common" (Durkheim [1915] 1968: 54). He then concludes by stating that,

> But the real characteristic of religious phenomena is that they always suppose a bipartite division of the whole universe, known and knowable, into two classes which embrace all that exists, but which radically exclude each other. Sacred things are those which the interdictions protect and isolate; profane things, those to which these interdictions are applied and which must remain at a distance from the first. Religious beliefs are the representations which express the nature of sacred things and the relations which they sustain, either with each other or with profane things (Durkheim [1915] 1968: 56).

Based on this logic, drugs and drug use, like any object or behavior, can belong either to the world of the sacred or the world of the profane. Given the powerful psychoactive effects of many drugs, especially some of those commonly used by our earliest ancestors such as cannabis, opium, and a host of hallucinogenic plants including the fly agaric and peyote, it is understandable why so many human societies placed drugs in the realm of the sacred. They are powerful objects whose effects can be awe inspiring. Yet, as Durkheim ([1915] 1968: 261) points out, "the sacred character assumed by an object is not implied in the intrinsic properties of this latter: it is added to them. The world of religious things is not one particular aspect of empirical nature; *it is superimposed upon it*" (emphasis in the original). Moreover, it is superimposed upon it by a group. Religious beliefs are determined by the collective and are eminently collective in their origin and continued existence. Durkheim ([1915] 1968: 59) continues,

> The really religious beliefs are always common to a determined group, which makes profession of adhering to them and of practising the rites connected with them. They are not merely received individually by all the members of this group; they are something belonging to the group, and they make its unity.

Thus, it is the collective definitions and beliefs about the drug and drug use that make it sacred or profane, not its chemical properties or the psychological effects it engenders.

Therefore, the belief system surrounding a drug makes it either sacred or profane. And, the beliefs concerning any specific drug can vary across time and space. This variation is seen in the Native American Church's definition of peyote as sacred and the Anglican Church's belief that peyote is profane (see French 2000; Vitebsky 2001). It is not a matter of objective reality; it is a matter of group membership and adherence to the group's collective beliefs. As Szasz (1974: 40) argues while making a similar point about sacred and profane drug use, "acceptance or non-acceptance of an identity between ceremonial symbol and ceremonial referent is a matter of membership in a community, and not a matter of fact or logic." Once defined as sacred or profane by the group, the treatment of the object by group members is determined. Sacred objects are treated with reverence and respect; profane things are treated with irreverence or indifference. The sacred object is special, the profane mundane. Moreover, defining the object as either sacred or profane determines the behaviors involving the object. If the object is sacred, the behaviors involving it become rites, which "are the rules of conduct which prescribe how a man should comport himself in the presence of these sacred objects" (Durkheim [1915] 1968: 56).

The astute reader will note that there are also interdictions placed on the use of profane drugs. These interdictions range from the extremely informal, such as peer reactions, to the extremely formal that are promulgated by political authority. The informal interdictions are enforced by methods of informal social control including jeers, gossip and whispers. The state sanctions violations of formal prohibitions. However, these restrictions, which certainly have and do exist and undoubtedly influence human behavior, are of a different nature than religious dictates. Rites involving sacred drugs are both positive and negative.

That is, they call for both the use of the drug while simultaneously limiting the drug's use. Catholics *should* take wine when celebrating Holy Communion. The Huichol *should* use peyote during their annual ceremonies. The Rastafarian *should* smoke marijuana before praying. Although the sacred rites of drugs restrict drug-using behavior, they also demand drug-using behavior. While limiting who can use what when, religious rites involving drugs call for something to be used by someone, somewhere, at some time. Conversely, secular laws concerning drugs are always and only negative. That is, they limit drug use; they never mandate the use of drugs. Drug laws say the law-abiding *should not* use heroin, they do not require one to use another substance instead. While drinking alcohol is legal in Russia, the Russian state does not force anyone to drink. While smoking tobacco is common in Asia, it is not necessary to smoke to be a citizen of any Asiatic nation. The negativity of legal prohibitions is even present in the enabling aspect of prescribing drugs by physicians. *A medical prescription removes the negative sanction from the use of the drug. It does not require the patient to use it.* That is, a physician may write a prescription for oxicontin that permits the patient to use a formerly forbidden drug. However, this prescription does not mean the patient must use the drug, only that she can.

This distinction is critically important to remember when we are discussing *drug use*. When a drug is used sacredly, it signifies that the user has *accepted the group's definition* of the drug as sacred. Once this is done, the user has bound him or herself to the interdictions concerning use that are established in the rites involving the sacred object.[26] When a profane drug is used, however, it signifies that the user is either taking advantage of the drug's legal status or has ignored the legal restrictions on its use. In the first instance, the restrictions on

---

[26.] It is, of course, possible that the individual does not necessarily believe in the sanctity of the drug and is using it for some other reason, such as simply to achieve intoxication. However, if the drug is being used in the religious setting, the non-believer still must follow the religious dictates regarding the use of the drug or risk being exposed as a heretic. Once exposed, it is likely that the group would chastise the individual and the "heretic" would no longer be able to engage in the sacred behavior.

use are likely to be extremely general or even non-existent. For example, once an American reaches the age of 21 and he is permitted to legally consume alcohol, there are few restrictions on its use. Those restrictions that do remain, such as not being allowed to drink alcohol on the street, do not restrict who can use or how the drug can be used, simply where it can be used.[27] In an extreme example, there are no restrictions on the consumption of caffeine in the United States. Anyone can use it at any time for any reason. In the case of ignoring the legal restrictions and using a prohibited drug, the user has *rejected the group's definition* of the drug as harmful, dangerous, or morally wrong. This rejection is evident in the simple fact that use is occurring. Since the legal restriction concerned only the use of the drug (i.e. it cannot be used legally), once it is used, there remain no other state-sanctioned restrictions. That is, we pass laws that say, "the use of marijuana is forbidden." We do not pass laws that say, "marijuana use is forbidden, but if you do use it, you must use it in this way." Once the legal restrictions are rendered mute by the decision to ignore the legal dictates and use the forbidden substance anyway, the only restrictions that remain are either religious -- which would bring us back to the sacred use of drugs that would be governed by rites -- or informal ones defined by the drug-using group. While the drug subculture does have norms that regulate the user's behaviors (see Partridge 1973; Agar 1977; Cleckner 1977; Cooper 2001; Hawdon 2003), these tend to be extremely imprecise and generally permissive when compared to rites or legal codes. Even when the informal codes of behavior are precise and restrictive,

---

[27.] There may also be legal prohibitions on when a drug can be legally purchased. For example, many states in the United States prohibit purchasing alcohol after 2:00 a.m. However, these restrictions rarely apply to when the drugs can be consumed. Thus, you may not be able to purchase alcohol after 2:00 a.m. legally, but you can legally consume it after 2:00 a.m. The same holds true for states that prohibit alcohol purchases on Sunday. You may not be able to buy alcohol on Sunday, but you can legally drink it. We should also note that restrictions on other behaviors *while consuming drugs* do not negate the general principle. Laws prohibiting drunk driving, for example, regulate driving, not drinking. It is legal to be sitting in a car drunk, you just cannot drive it. Similarly, public drunkenness or drunk and disorderly laws prohibit being drunk in the wrong place or acting inappropriately while one is drunk. Again, you are allowed to be drunk, you just cannot act in a publicly injurious way while you are.

there is no authority to enforce them. While jeers, gossip and whispers can be effective means of social control (see Black 1998), these methods of informal social control lack the obligatory nature of rites (see Durkheim [1899] 1975) and the authoritative power of the state.

The use of a sacred drug, therefore, will be governed by well-defined, widely accepted rules that are embraced by the group members who have defined the drug as sacred and those members who are using the drug. Conversely, profane drug use is governed by either general rules or no rules at all. When use occurs, it is either because the secular authorities have deemed restrictions unnecessary or the user has rejected the secular authorities' definition of the situation. In either case the moral rules, laws or norms concerning profane drugs tend not to be as restrictive as rites, at least once drug use has occurred. Secular forces that are relatively easy to avoid enforce moral rules. Religious forces that, from the perspective of the devout, are impossible to avoid enforce rites.

Since behaviors involving sacred objects are defined by rites, there are clearly established, collectively agreed upon, prohibitions governing the use of sacred drugs.[28] The use must occur in the prescribed manner or the sacred drug has been desecrated. Failing to perform the ritual properly makes the drug profane and therefore does not serve the sacred purpose it was meant to. For example, simply drinking a sip of wine and eating a sliver of bread does not constitute a Christian Holy Communion. At least in the Catholic Church, before the consumption of wine is religiously meaningful, it must be transformed; it must be consecrated by a series of precise behaviors and utterances. Moreover, not just anyone can perform these behaviors and utterances. There are dictates concerning who has the power to perform the transformation. There are also dictates about when Communion should be given, who can participate in it, how it is to be

---

[28]. It is again emphasized that while the prohibitions on profane drug use may be restrictive, systematic and widely held, *these call for non-drug use, not use.* These interdictions are therefore different from rites, as previously explained.

accepted by the recipient, and what else is to be consumed with it. Of course, such interdictions and dictates do not apply solely to the Christian rite of Holy Communion. Similar restrictions that vary in content but not in form govern all sacred drug use. These restrictions exist because there is a need for consistency in sacred use to minimize variations in the setting of use. This necessity is not present in most profane drug use. Given the need for consistency, we can deduce that sacred use is more highly regulated than profane use. As Durkheim ([1899] 1975: 93, emphasis added) notes, religious matters "consist in obligatory beliefs, connected with *clearly defined practices* which are related to given objects of those beliefs." At least for the people with whom we are concerned, those who use drugs, sacred drug use will have stricter prohibitions governing it than will profane drug use. Therefore, sacred drug use will occur in the same manner over time and, hence, there will be less variation in the use of sacred drugs than in the use of profane drugs. The consistency of sacred drug use and relative inconsistency of profane drug use will be seen across a number of dimensions of use. These regulations address what can be used, when, how, and by whom. All else being equal, we will see less variation on these dimensions in sacred use than in profane use.

First, there will be less variation in *what is used in combination with the drug* when the drug is considered sacred as compared to when the drug is deemed profane. When a drug is sacred, profane objects are kept from it. Thus, if other drugs are to be used, they too must be sacred. The Huichol, for example, would use both tobacco and peyote during the peyote ritual (Furst 1976; La Barre 1975; Schaefer and Furst 2000); however, for the Huichol, both tobacco and peyote were sacred drugs. They would not use any other drug during the peyote ritual (see Furst 1976; French 2000; Schaefer 2000). Similarly, while the tobacco mix used in the Sioux Sacred Pipe Ceremony included other plants, "the plants are picked and prepared in a specific sacred manner" (Buhner 2001: 107). Conversely, the use of profane drugs can be accompanied by the use of any other profane drug. For example, it is common for contemporary British youth to use alcohol and some other drug, such as marijuana, amphetamine, ecstasy, LSD or amyl nitrite, simultaneously (Parker et al 1998: 71; Drug Scope 2000; Drug Scope 2002). Similarly, American heroin addicts often use cocaine and heroin simultaneously (see Goode 1999; Inciardi 1992; Inciardi, Horowitz and Pottieger 1993). Such polydrug use would not likely be seen with a sacred drug.

Next, there will be less variation in *when* a sacred drug is used than if this same drug was considered profane. Sacred drug use will occur at precise times. Religious celebrations that involve drugs are often tied to seasons or specific days. Religious initiation rites involving drugs occur on specific occasions. Drugs used in the context of religious healing or to communicate with spirits are used sparingly and only when the need arises. The rites dictate when the drug is used. Using the drug at other times would be blaspheming. For example, when discussing the use of tobacco among the Athapaskan, Sioux, and Cherokee, Mooney (1891: 423) noted that while

> Tobacco was used as a sacred incense . . . It was either smoked in a pipe or sprinkled upon the fire. Never rolled into cigarettes, as among the tribes of the Southwest, neither was it ever smoked for the mere pleasure of the sensation.

In fact, many aboriginal groups who used tobacco sacredly did not use it casually (see French 2000). Most societies place similar limitations on sacred drugs (see, for example, Furst 1976). Conversely, profane drug use can occur anytime the drug is available. While there may be informal "rules" concerning when one should or should not use a given drug (e.g., one should not use LSD when he or she is "feeling negative," one should not drink alcohol if pregnant), these are typically suggestions by experienced users or someone otherwise knowledgeable about the drug. In either case, there exists no "higher authority" to enforce the dictates. In addition, because of the restrictions on when a sacred drug can be used and the relative lack of restrictions concerning profane use, *sacred drug use will occur less frequently than profane use*, generally speaking.[29] While the Siberian reindeer herdsmen use the sacred fly agaric only on special religious occasions, some Americans smoke marijuana daily.

We also find less variation in *how* a sacred drug is used compared to when the same drug is defined as profane. That is, the method of how the drug is to be consumed is clearly established when a drug is considered sacred. This is not necessarily the case when a drug is defined as profane. Again, for a ritual to be religiously meaningful, it must be performed in an ordained manner. If the ritual calls for the use of a sacred drug, the rites concerning the sacred object will likely include interdictions concerning how it is to be prepared and ingested. For example, the Sioux Sacred Pipe Ceremony is highly ritualized with detailed rules concerning the preparation of the tobacco and the smoking of the pipe.[30] The ceremony involves offering the pipe to the creator, the earth, and the four directions. After each has smoked, "then, and only then, does a person smoke the pipe" (Buhner 2001: 114). With profane use, however, only physics, biochemistry, custom and occasionally the state limits the manner of use. The Incas' sacred use of coca involved only chewing it (see, for example, Phillips and Wynne 1980). The profane use of coca includes chewing it, transforming into coca-paste and smoking it, further transforming it into cocaine and sniffing it, altering it for injection, or transforming it even more and smoking it as freebase or crack. While it is true that tobacco was consumed by a variety of means -- including smoking, chewing, drinking its juice and through enemas (see Furst 1976; Baer 1992; Winter 2000) -- by those groups who used it religiously, the

---

[29] This general statement has its exceptions. For example, Warao shaman smoked "incessantly" to provide the gods with abundant tobacco smoke (see Furst 1976).

[30] See Buhner (2001: 11 1 - 114) for a detailed description of the Sacred Pipe Ceremony. Brown (1953) and French (2000) also describe the ceremony in detail.

method of use for a *particular ritual* was typically fixed. Thus, the variation in the method of sacred use is mostly *between* groups instead of *within groups.*[31] Methods of profane use, however, may very well vary within the same group and even within the same group during the same drug-using episode. A recreational user of cocaine, for example, may snort, smoke or inject. He or she may even use the drug in several or all of these manners in one prolonged coke binge.

The precise nature of sacred drug use will also limit *who* is allowed to use the drug more than the imprecise nature of profane use. When a drug is sacred, only the holy or purified can use it. Users must achieve the "holy" status by proving their worth as a shaman, passing through an initiation rite, confirming their belief in the religious dictates, or purifying themselves through some religious act. For example, among the Warao of the Orinoco Delta in Venezuela, the use of tobacco was restricted to certain rituals and used only by the shaman. Laypersons were prohibited from using the drug (French 2000). Such restrictions are common when drugs are sacred. It is true that laws regulate who can and cannot use what profane drug. However, as discussed above, laws regulate use negatively by stating a drug cannot be used. Illegal use occurs only after the user rejects these laws. So, once use occurs, the prohibitions on who can use are weakened or rendered completely impotent. Conversely, when sacred drug use occurs, the prohibitions are strengthened.

Thus, *sacred drug use is more highly regulated than profane drug use.* This proposition does not imply that profane drug use is unregulated. However, the regulations concerning sacred use will typically be more precise and more widely accepted *by those using the drugs.* Sacred users accept the religious group's prohibitions. These prohibitions therefore effectively regulate the users'

[31.] It is true that some groups who use a particular sacred drug for a number of different rituals will use different methods of ingestion depending on the ritual. However, this within-group variation is crossing different religious functions that will be outlined in the next chapter. One does not find much within-group variation in the method of use within the same group *and* within the same ritual.

behaviors. Profane users either accept the dominant group's definition of the drug as legal or reject the group's prohibitions and use it illegally. In the first case, the legal status of the drug renders its use relatively unregulated. In the second case, the users' rejection of the dominant group's definition has effectively removed all restrictions on its use. In both cases, the user's behavior is relatively unregulated by the group's dictates. In short, with sacred use, the group has tremendous influence on the manner of use. With profane use, the individual has considerably more freedom to use the drug as he or she sees fit.

Next, religion is "an eminently collective thing," and, as such, it is associated with a definite group and celebrated by the group (Durkheim [1915] 1965: 59; 63). Sacred objects, regardless of their form, are collectively conceived and come to represent the group itself. Rituals that celebrate the collectively defined sacred object also celebrate the group. Its existence, beliefs, values, actions, and moral code are declared and extolled. Religious use is proclaimed proudly and publicly as a sign of devotion and affiliation with the religious group. Thus, sacred drug use celebrates the group and integrates the user into the larger collective. For example, the male puberty rites of the Cahuilla of Southern California require the boys to use datura (Jimsonweed) to produce visions. The visions, which last for several days, are highly patterned and conform to group expectations and Cahuilla mythology. These visions "enable him to glimpse the ultimate reality of the creation stories in the Cahuilla cosmology" (Bean and Saubel 1972: 62). Dobkin (1975) makes a similar observation concerning female initiation at puberty among the Shagana-Tsonga of northern Transvaal Africa, and Vitebsky (2001) and Dobkin (1975) note this characteristic of religious rites-of-passage and visions in general. Sacred use, governed and regulated by rites, is therefore integrative. As Durkheim [1915] 1968: 432) notes, "before all, rites are means by which the social group reaffirms itself periodically."

Conversely, profane use is often disintegrative, at least to the larger collective. Although profane use may integrate the user into some sub-group, it simultaneously separates this sub-group from the larger group. This bifurcation of the group especially applies if the profane drug is deemed illegal by the larger collective. In such a case, a relatively small percentage of the population will violate the legal dictate and use the drug (as we saw in Chapter 2, only 3.1% of the world's population uses illicit drugs). People who use a drug despite its illegality will likely conceal their use from the broader, law-abiding collective. This results in much illegal use occurring in relatively small groups. Thus, for example, while smoking ganja serves an integrative function among the Jamaican working-class, it is simultaneously a polarizing issue in the larger Jamaican society and serves as a means of effectively dividing the social classes (see Dreher 1982). While illicit drug use may create a bond among youthful users, it simultaneously casts them as "deviants" and "outsiders" to the larger society.

This argument does not imply that everyone will accept all sacred use. Indeed, the sacred drug of one group can be considered profane by another. The dominant Anglo-American culture does not accept the use of peyote by the Native American Church as "legitimate." Muslims do not tolerate the Christian use of wine, despite the sacredness of the drug for Christians. However, in these instances, the sacred use is dividing two different groups, not dividing one large group. Of course, one could argue that in some respects all use divides some larger group. If we consider American Muslims and American Catholics as being divided by their acceptance of wine as a sacred drug, for example, this division divides the larger group of Americans. However, this requires the scope of the group to be changed. That is, in the above example, we are changing the dimensions of heterogeneity from religion (Catholic versus Muslim) to nationality (Americans). With sacred use, the entire group accepts its use as legitimate. Thus, all Huichol accept the use of peyote, even if they personally do not use it. Similarly, all devout Catholics accept the use of wine during Holy Communion, even if they do not partake in the ritual. This widespread acceptance is not typically the case with profane drug use. Even though many working class Jamaicans accept the use of marijuana as being legitimate and use is a source of solidarity for the working class, not all working class Jamaicans use marijuana or accept its use. Use is not considered "wise" or even "acceptable" by those who desire upward social mobility (see Dreher 1982). Similarly, not all youth approve of their fellow youths' involvement in the drug subculture. Thus, profane use divides the group (i.e. working-class Jamaicans or youth) without changing the scope of the group or unit of analysis. Thus, generally speaking and regardless of the function it is fulfilling, *sacred drug use will integrate the larger collective more than profane drug use.* Relatedly, *profane drug use will be more divisive of the larger collective than sacred drug use.*

Noting these differences allows us to understand drug use better. Even if the same drug is being used for the same reason, the pattern of use may differ dramatically from one setting to the next depending if the person and group using the drug considered it a sacred or profane object. Recreational use, for example, while serving the same function, will likely take on a very different air if the drug is sacred versus profane. Although the drug is being used for recreational pleasure in both cases, there will likely be greater controls over who uses the drugs, what other drugs are used, how much of the drug is used and the method by which the drug is ingested when the drug is sacred compared to when it is a profane substance.

These general principles concerning the differences between sacred and profane drug use will influence use regardless of the reason drugs are being used. As with most generalizations about social behavior, there are exceptions to these rules. They do apply in most cases however. Table 4.1 summarizes the influence of considering a drug as sacred or profane. As mentioned throughout the above discussion, the patterns for profane drug use are contingent on the extent to which secular interdictions in the form of laws apply to the drug. Therefore, Table 4.1 divides use into three categories: use when the drug is defined as a sacred object; the use of a profane drug that is legal to use; and, the use of a profane drug that has deemed illegal by the state.

Table 4.1: Relative Variations in Use Patterns between Sacred and Profane Drug Use

Dimension of Use	Drug as Sacred object	Drug as Profane object, but legal	Drug as Profane, but illegal
Restrictions over use	Use is highly restricted. Rites define who can use the drug, when it can be used, how it should be used, and why it is used.	Restrictions are general and often permissive.	Restrictions are on *non-use*. Once use occurs in defiance of non-use restrictions, few restrictions apply.
Who uses	The "holy" members of the larger group. Qualifications for use are specific, widely known and enforced.	All members of the larger group meeting relatively permissive qualifications.	Members of "deviant" sub-groups.
How used	Highly ritualized and used in a limited number of manners.	Ritualized but highly variable manners of use.	Highly ritualized but with highly variable manners of use.
How often used	Rare	Frequently	Frequently
Other drugs used in combination	No other drug, unless the drug is also sacred.	Other profane but probably legal drugs.	Any other drug
Integrative result of use.	Use integrates the entire group and reaffirms group values.	Use integrates sub-groups but does not divide sub-group from the larger group.	Use integrates sub-groups while dividing sub-group from the larger group.

Applying these rules allows us to see that when a drug is sacred, its use is typically highly regulated. Although use may be widespread throughout the group that defines the drug as sacred, its use will be more highly patterned and controlled than if the drug was profane. The social control exercised over sacred drugs limits their use in casual settings. We would therefore expect less "abuse" of drugs. Conversely, when a drug is defined as profane, drug use will be less highly regulated and greater variation will occur on several dimensions of use including who uses, what is used in combination with the drug, when the drug is used, the method of ingestion and the frequency of use.

If we accept the general accuracy of this argument, we can return to the original question posed earlier in this chapter. Why do Americans, Western Europeans, Canadians and Australians use drugs more frequently and use a greater variety of drugs than others around the world? The answer is because these cultures are more likely to define use as profane rather than sacred. Of course, this answer leads to another question, "Why do these cultures define drugs as profane while other cultures are more likely to define drugs as sacred?" More generally, it leads to the question, "What determines the sacredness or profaneness of a drug?" Of course, it would be nearly impossible to determine how each culture comes to define each substance as either sacred or profane. Indeed, defining the general principles concerning how any object is defined as sacred or profane is a daunting task (see, for example, Durkheim [1915] 1968). Fortunately, this task is largely unnecessary for us to proceed. Instead, we can address the process that has been "specific and peculiar" and has influenced occidental cultures more than other cultures (see Weber [1904-1905] 1958). The process leads to a general tendency for humans to shrink their sacred world and expand their profane world. As numerous theorists have observed, the extent to which the world and the objects in it are considered sacred has diminished over time due to the processes of rationalization and modernization. Let us now consider this process in general terms. We will then consider the use of drugs in light of the processes that have occurred over the centuries.

Rationalization and Modernization

Much of the discipline of sociology has been dedicated to documenting the cultural and structural changes brought about by the rationalization and modernization processes. Despite the considerable effort expended on this question, what led to this process is still debated. Whether rationalization was the cause of modernization or modernization was the cause of rationalization is unclear. It is debated whether rationalization was a result of changes in belief systems, institutions and social systems, cultural systems, or worldviews. Whether modernization was due to a historic materialism, changes in cultural values and worldviews, or simply the increase in the density of social associations is unclear. This debate is far beyond the scope of the project and my abilities to resolve. Yet, while there is debate over what ultimately caused the process of modernization, it is undeniable that this process has occurred and has resulted in well-documented changes. While it would be impossible to summarize all of the work that bears upon our question, I will nevertheless synthesize some of the major insights to weave an explanatory narrative that can be useful in accounting for contemporary patterns of drug consumption. I apologize to the authors of these great works in advance for glossing over some of the details, subtleties and nuances of their work. While these details make these works outstanding, it is the "big picture" that these scholars so wonderfully explained that make their works classics. This "big picture" interests us now.

I should warn the reader that it will be several pages before we return to the issue of drug consumption. I believe, however, it is necessary to outline the breadth and depth of social changes spurred by rationalization and modernization because these changes transformed drug use from being sacred to being profane. To use Howard Parker and his associates' word, these forces have "normalized" recreational drug use (Parker et al 1998). Moreover, as the reader will eventually see, all of these changes become pertinent for explaining the variations in contemporary patterns of drug consumption. So, please bear with me.

Early Pre-Modern Societies: Culture

In the earliest stages of development, [32] the realms of nature and culture are undifferentiated to the extent that little distinction is made between natural events

---

[32.] Parsons calls this stage "primitive" societies; Habermas calls them "neolithic."

and social phenomena. Culture and nature are one. This "leveling" of nature and culture gives rise to a mythical worldview in which powerful spirits can control the human and natural world. Habermas (1979: 47) makes this point by noting that,

> What we find most astonishing is the peculiar leveling of the different domains of reality: nature and culture are projected onto the same plane. From this reciprocal assimilation of nature to culture and conversely culture to nature, there results, on the one hand, a nature that is outfitted with anthropomorphic features, drawn into the communicative network of social objects, and in this sense humanized, and on the other hand, a culture that is to a certain extent naturalized and reified and absorbed into the objective nexus of operations of anonymous powers.

The failure to differentiate between nature and culture creates the appearance that divine and human actions are similar. The forces of nature and culture, which seem to be spontaneous, are therefore endowed with human characteristics. Nature and culture are attributed the same traits of consciousness, will and power that humans possess. However, these forces are able to assert their will and use their power in manners beyond human abilities. Given the undifferentiated nature of these societies, the extraordinary powers that direct and influence social life are seen as residing in tangible incarnations such as objects and people. Those objects that possess the extraordinary powers are endowed with sacred authority. Moreover, the means by which these powers are exploited are also endowed with sacred authority.

Thus, in the earliest stages of this mythological world view, the leveling of nature and culture results in most objects being considered sacred, so secular and sacred values are not differentiated (Parsons 1977). All worlds, the social, natural and supernatural, are understood as a totality. The whole world makes sense, everything is significant and everything can be explained within this all-encompassing, mythic order. Habermas (1984: 46) notes,

The deeper one penetrates into the network of a mythical interpretation of the world, the more strongly the totalizing power of the 'savage mind' stands out. Abundant and precise information about the natural and social environment is processed in myths. . . . this information is organized in such a way that every individual appearance in the world, in its typical aspects, resembles or contrasts with every other appearance. Through these contrasts and similarity relations the multiplicity of observations is united in a totality.

Since any rock, tree, stream, breeze, cloud, bone, animal, or person could potentially possess extraordinary power, it was best to approach everything with reverence for fear of earning its wrath. Although objects and their powers were to be treated reverently, considerable effort was made to manipulate these through magic. Yet, because there were few distinctions made between nature and culture, there was little distinction between the natural and supernatural world. Everything occurred in this world, there was no exclusive realm of the divine (see Weber [1922] 1964). Thus, magic was directed toward manipulating this world, including both its natural and social phenomena, ostensibly for obtaining worldly goods. As Weber ([1922] 1964: 1) argues, "the most elementary forms of behavior motivated by religious or magical factors are oriented to *this* world" (emphasis in the original).

Structure

In early pre-modern societies, human groups were typically small and bound together through kinship and religious ties, and the undifferentiated mythical worldview was reflected in a similarly undifferentiated social structure (see, for example, Parsons 1977). The political, economic and religious institutions are highly interconnected and mutually reinforce the social order. There was little stratification, and that which existed was primarily based on ascribed characteristics such as age and gender (see Lenski 1966). Due to their relative isolation, simple division of labor and egalitarian distribution of collective goods, the similarities among the group members were striking. They looked the same physically, they dressed the same, they shared similar beliefs and, as a result, they behaved in similar ways. The solidarity of the group was based on this homogeneity (Durkheim [1915] 1964). With this "mechanical solidarity," or solidarity based on likeness, almost every individual member shared a "collective conscience." There was a general normative consensus because everyone was well aware of the group's normative standards (see Durkheim [1915] 1964; Habermas 1984b; Parsons 1977; Luhmann 1982). The collective sentiments "(were) strong because they (were) uncontested" and "universally respected" (Durkheim [1915] 1964: 103). A unified worldview secured the cohesion of social life (see, for example, Luhman 1982). The relative homogeneity, low rates of social mobility[33] and uniformity in worldviews and normative systems limited the number of associations of early pre-modern society members. That is, there were simply few groups for people to select from in pre-modern societies. The undifferentiated social structure and group's isolation from other groups limited the member's interactions to fellow group members.[34] Since the individual was relatively isolated from other groups, non-conformity placed the individual at risk of being ousted and denied access to the group's collective power. Group membership was therefore highly valued and the exit costs associated with leaving the group or being forced from it were extremely high (see Hechter 1987;

[33.] Following Blau (1977) social mobility is defined broadly to encompass all movements of persons between structural positions. Blau (1977: 3) states,

> To speak of social structure is to speak of social differentiation among people, for social structure . . . is rooted in the social distinctions people make in their role relations and associations with one another. What is meant here by social structures is simply the population distributions among these differentiated positions.

I use Blau's definition of social structure. Therefore, social mobility is the movement of persons between differentiated positions. It would include occupational mobility and migration, of course. However, it would also include such changes in position with respect to religious denominations, political affiliations, educational levels, or family status (see Blau 1977).

[34.] This follows from Blau's (1977) assertion that when rates of social mobility are low, the saliency of structural parameters is high and inter-group contacts are therefore reduced.

Hawdon 1996b). With high exit costs, the group's ability to enforce its moral standards was enhanced. Unless one was willing to risk social isolation, the individual had to conform.[35]

The high levels of social control exercised by the group meant that violations of the moral code were met with strict punishment. This punishment was necessary to maintain the unified collective sentiments. As Durkheim ([1915] 1964: 103) states,

> But crime is possible only if this respect is not truly universal. Consequently, it implies that they are not absolutely collective. Crime thus damages this unanimity which is the source of their authority. If, then, when it is committed, the consciences which it offends do not unite themselves to give mutual evidence of their communion, and recognize that the case is anomalous, they would be permanently unsettled.

Thus, repressive law marked the mechanical solidarity of early pre-modern societies. Given the strength of the collective conscience and the high levels of social control enjoyed by the group, the individual had very little freedom and their behavior was severely restricted. As Durkheim ([1915] 1964: 130) notes,

> Solidarity which comes from likenesses is at its maximum when the collective conscience completely envelops our whole conscience and coincides in all points with it. But, at that moment, our individuality is nil. It can be born only if the community takes smaller toll of us.

Therefore, there was very little individuation in early pre-modern societies. Like the worldview and social structure, the personality system was undifferentiated and it was undifferentiated from the culture and the society (Parsons 1951).

[35.] Working from Parson's argument that the degree of conformity to social norms directly relates to the amount of positive sanctions one receives, Milner (1987: 1055) states, "the more in demand a particular kind of conformity, the more highly it will be rewarded." I argue that this relationship is reciprocal. Thus, the greater the reward, the more conformity is required to achieve it (see Hawdon 1996b: 189).

Pre-Modern Societies

Eventually humans stopped wandering as hunter and gatherers and began to settle in geographically fixed, horticultural societies (see, for example, Lenski 1966). As we began to live in and develop ties to geographically fixed settlements, we could no longer avoid social power. That is, we could not simply run away when someone exerted his or her will over us (see Mann 1986). Now, we were forced to stay and interact, despite our subordinate position. Thus, the densities of our settlements grew. Since we were forced to interact with others who we would have avoided during earlier times, "there is an exchange of movements between parts of the social mass which, until then, had no effect upon one another" (Durkheim [1915] 1964: 257; also see Mann 1986). With this increased social mobility and interaction, society began to differentiate. This differentiation resulted in changes in both the culture and social structure of society.

Culture

As the rationalization process continued, the belief in "spirits" emerged. It was no longer believed that the concrete object possessed extraordinary power. Instead, "certain beings (were) concealed 'behind' and responsible for the activity of the charismatically (magically) endowed natural objects, artifacts, animals, or persons" (Weber [1922] 1964: 3). During these early stages of abstract belief, the spirits were not personalized and were not necessarily enduring. They did not have personal names but were instead named by the process or processes they controlled. Still, spirits began to be regarded as "invisible essences that follow their own laws, and (were) merely 'symbolized by' concrete objects" (Weber [1922] 1964: 3). Now, concrete objects and actual events were not the only forces acting on humans, as was previously believed; "now certain experiences, of a different order in that they only signify something, also play a role in life. Magic is transformed from a direct manipulation of forces into a symbolic activity" (Weber [1922] 1964: 6). Thus, there was an eventual move away from naturalism toward an abstract, symbolic understanding of the world. As this process occurred, there was greater differentiation between the natural and supernatural worlds. Gods became increasingly personified and connected to a community "for which he has special significance as the enduring god" (Weber [1922] 1964: 10). With time, a pantheon typically emerged "once systematic thinking concerning religious practice has taken place and a certain . . . level of rationalization of life generally has been attained" (Weber [1922] 1964: 10).

With the further differentiation between the world of society and world of nature, humans gained greater knowledge of the natural world. They therefore gained knowledge of how to manipulate the world through rational calculation (see Habermas 1979). Yet this manipulation could not happen until the worlds of the sacred and profane began to separate so that fewer objects were considered sacred. Only appropriately reverential actions can manipulate sacred objects. As more objects fell to the world of the profane, however, the rational manipulation of them became legitimate. We begin to see a very early form of *instrumentally rational action*[36]

Structure

As the density of our settlements grew, the structure of society began to differentiate. Our social relations became more numerous and a division of labor began to develop (Durkheim [1915] 1964: 257). A more complex division of labor allowed for specialization, which, in turn, increased efficiency and surplus. At this point, societies became stratified instead of segmented (see Parsons 1977). When there was virtually no surplus, such as in hunting and gathering societies, goods were divided based on need. However, as surplus increased, stratification became possible. The greater surplus allowed the elites to procure and hoard

---

[36.] Weber ([1922] 1978: 24) defines instrumentally rational action as being "determined by expectations as to the behavior of objects in the environment and of other human beings; these expectations are used as 'conditions' or 'means' for the attainment of the actor's own rationally pursued and calculated ends."

commodities after members of the lower social classes, or at least most of them, consumed enough to survive (see Lenski 1966). Yet, stratification was still based largely on ascribed characteristics, thus there were still limited chances for social mobility (see Lenski 1966).

With the advancing complexity of the division of labor, there was an increase in inter-group contact. While the preponderance of interaction still occurred among persons who occupied similar structural positions (see Blau 1977; Laumann 1973), the still-limited-but-increasing rates of social mobility fostered inter-group contact by reducing the social or physical barriers that previously limited such interactions. The formerly diverse groups began to integrate into the society as a whole, thereby increasing the heterogeneity of society. The group members, although now belonging in one group instead of numerous smaller groups, were no longer as similar in their attitudes, beliefs or behaviors. With this greater heterogeneity, the normative consensus, while still strong, began to weaken. With the weakening of the collective conscience, the group's ability to exercise social control was also reduced. Individuals were beginning to gain independence from their collectives.

Advanced Pre-Modern Societies: *Culture*

Jumping ahead significantly in the rationalization and modernization process, the proceeding developments eventually resulted in the further differentiation between the natural and social worlds, as well as the economic, political and religious subsystems (see Parsons 1977).[37] Universal cosmologies and the "higher religions" replaced myth and narratives as means of explaining

---

[37.] It should be noted that when I speak of institutional or structural differentiation I do not mean to imply that the process is inevitably cumulative and irreversible. This is especially true with respect to the modern state and regional intergovernmental organizations (see Hawdon 1996c). In fact, as Poggi (1978: 14) argues, the distinction between the state and society have become somewhat de-differentiated. However, it is undeniably true that modern societies are more differentiated than pre-modern societies. Therefore, I use the term and rely on the basic theory of institutional differentiation, especially as espoused by Parsons.

and justifying traditions (Habermas 1979; also see Weber [1922] 1964; Weber [1922] 1978; Parsons 1977).[38] There was the rise of a professional priesthood who maintained and transmitted knowledge that had been codified around abstract concepts and moral principles. A universalistic worldview developed that explained the world in a unified, orderly fashion based on "universal truths."[39] Otherworldly deities dominated religion and the gods were no longer completely enmeshed in the social fabric. They were "great lords" who could no longer be manipulated through magic (Weber [1922] 1964: 27). The gods derived their legitimacy not by possessing extraordinary powers, per se, but from a transcendent source. Like the worldview in general, the appeal of the gods became universal. They were "regarded as either subject to some social and moral order or as the creators of such an order" (Weber [1922] 1964: 36). Thus, universal norms began to place increased ethical demands on humans and their gods (Weber [1922] 1964).[40]

While religion was still a significant source of authority and social control that promoted stability by directing believers to otherworldly goals such as salvation, the rise of abstract moral principles and the further differentiation between the state, economy and church allowed complex systems of law to emerge. With this further rationalization, the secular, yet universal, values became increasingly acceptable and religious views became less dominant.

[38.] By "higher religions," Weber primarily meant the world religions of Hinduism, Buddhism, Judaism, Christianity and Islam. The emergence of higher religions is associated with the stage of development that Habermas calls "developed societies." This is roughly equivalent to Parson's "intermediate societies."

[39.] Examples of such universalizing ideologies are Roman law and the Hebrew and Greek moral orders (see Parsons 1977: chapters 4 and 5).

[40.] This is one way in which occidental religions and eastern religions differed. Hinduism, for example, did not create universal ethical norms. Instead, it demanded that believers follow the exemplary path to salvation of the Brahmans. Weber ([1922] 1978: 447 - 448) classifies Hinduism as an "exemplary" rather than an "ethical" prophecy. This difference is one of the reasons that occidental rationalism developed in such a "specific and peculiar" fashion.

Religious authority weakened as ethical and pragmatic principles explained and justified that which once was interpreted solely through religion (Parsons 1977; also see Parsons [1963] 1967). There was an "increasing scope of a rational comprehension of an external, enduring, and orderly cosmos" (Weber [1922] 1964: 35). Having been removed from the realm of the purely sacred, nature became increasingly open for systematic evaluation. The cultural world also came under systematic scrutiny as increasingly complex social relationships became regulated by "conventional rules" and humans became increasingly dependent on observing those rules when interacting with each other. Finally, as the need for "the reliability of the given world increased, the principles of efficiency and calculation (were) increasingly applied to economics and politics" (Weber [1922] 1964: 35-36; also see Weber [1922] 1978; Weber [1904-05] 1958).

Structure

As modernization continued, the various subsystems of the larger system became increasingly specialized as they assumed primary responsibility for meeting various social functions. Thus, the economy, the state, religion, education, the family and other social institutions could clearly be distinguished. While the subsystems were integrated into a whole, they operated relatively independently of each other (see Parsons 1951; Parsons 1977). As the subsystems differentiated from each other, they also became increasingly differentiated internally. In the economic subsystem, labor became increasingly divided as people began performing more specialized economic tasks. As the division of labor advanced, work, which was once divided primarily on the basis of age and gender, began to be divided on dimensions of ability, interests and power as well. Differing classes of workers such as farmers, priests and magicians, politicians, artisans, warriors and others began to form (Lenski 1966).

While the occupational structure of advanced pre-modern societies were more complex, they still required most workers to toil in similar occupations and live at similar levels of subsistence (Lenski 1966). Therefore, the occupational structure was relatively fixed, and there were very few possibilities for individuals to advance their positions. Society became more stratified as increased productivity led to more surplus. In fact, advanced pre-modern, or advanced agricultural, societies are the most stratified type of society (see Lenski 1966). As society differentiated, the division of labor became more complex and stratification reached its apex, the group members became increasingly dissimilar. The solidarity that had been based on the members' similarity began to break down. As a result, the collective conscience began to erode and the once widely held normative consensus became more difficult to maintain. Increasingly, social life became regulated through contracts as complex systems of law emerged (see Durkheim [1915] 1964).

Modern Societies: *Culture*

The onset of the modern age witnesses even more dramatic changes in the culture and structure of society. The general process of rationalization continued as instrumental rationality began to dominate. Nature and society, the natural and supernatural, and, therefore, the sacred and the profane were now clearly separated. As this occurred, world views, such as those of the universal religions, that once monopolized truth "retreat(ed) to private conviction" because the continued systematic analysis of nature and culture furthered our objective understandings of these realms (Weber [1922] 1964). Non-sacerdotal ideology began to challenge the validity of an orderly cosmos with universal meaning. The increasing differentiation of the social system and worldview created tension between the unified religious worldview and profane knowledge of an increasingly secular world (Habermas 1984b: 88 - 89). Even religion itself became more self-conscious and reflective (Habermas 1979), and "the conflict between empirical reality and this conception of the world as a meaningful totality . . . produce(d) the strongest tensions in man's inner life as well as in his external relationship to the world" (Weber [1922] 1964: 59). Eventually "the tension between the value spheres of 'science' and the sphere of 'the holy' (was) unbridgeable" (Weber [1919] 1946: 154), and the widespread acceptance of the universal cosmologies crumbled under the weight of empirical evidence. The greater objective understanding of the natural and social worlds replaced the once unquestionable religious understandings of the world.

This greater "objective understanding" of the natural and social world is now firmly located in the province of science. The rise to dominance of science as a means of explaining the natural and social world marks the on-set of the modern age. While the modern, scientific worldview continues to insist on an ordered, coherent world, this unity is not simply accepted based on absolute laws concerning God or nature. Instead, unity, order and coherence are found only in the nature of reason itself (Habermas 1979). Moreover, science, like religion, aims at an all-encompassing system of thought. However, this system aims to explain *causes* rather than otherworldly *meaning* (Weber [1922] 1964; Weber [1922] 1978). Since the realm of the sacred has diminished in importance and science dominates the contemporary worldview, instrumentally rational action dominates the modern, rationalized world. This results in the principles of efficiency and calculation being applied to an ever-increasing range of behaviors and social contexts. The rational style of thought that dominates the modern world even enters familial life (see Habermas 1970).

Structure

As with the culture of society, the onset of modernization marks dramatic changes in the social structure of society. In the modern age, the social system becomes fully differentiated (see, for example, Parsons 1977; Parsons [1964] 1967). The division of labor becomes increasingly complex as greater differentiation and specialization occurs. The means of production are fully transformed from relying on human and animal labor to using mechanical labor. Because of the greater specialization and use of mechanical labor, so much is produced that elites can possess riches beyond the dreams of even the wealthiest ruler of agrarian societies and still make concessions to the laboring classes. Thus, surplus reaches a level where so much is produced, there is an excess of goods that can be redistributed (see Lenski 1966). Eventually, levels of stratification begin to reduce, although they never reach the near-egalitarian levels of hunting and gathering societies. As Lenski (1966: 308) states, "the appearance of mature industrial societies marks the first significant reversal in the age-old evolutionary trend toward ever increasing inequality."

Next, an advanced division of labor and a decrease in inequality associated with the emergence of modern societies also corresponds to a "decline in the importance of ascribed factors in the distributive process" (Lenski 1966: 410; also see Parsons 1951). While ascribed characteristics such as race, ethnicity, religion, kinship and gender typically fixed one's social position in earlier times, achieved characteristics, such as social class and education, increasingly become the major determinants of social stratification in industrial societies. Lenski (1966: 410) argues that in modern societies,

> Ancient hereditary distinctions between nobles, freemen, and slaves have been all but eliminated. The advantages and disadvantages associated with the ascribed, or largely ascribed, qualities of race, ethnicity, and religion have also declined in importance.

While ascribed statuses remain important sources of stratification in modern societies, their power to define one's position weakens. Thus, as societies industrialize, the form of social differentiation begins to change as social space is divided vertically more than horizontally. That is, stratification, which separates groups on some graduated parameter such as income, becomes increasingly important while heterogeneity, which divides groups on some nominal parameter such as gender or ethnicity, becomes less relevant.[41]

[41.] For a more detailed discussion of the difference between stratification and heterogeneity, see Blau (1977: 6 - 11).

As the occupational structure becomes more differentiated and ascribed statuses have less of an effect on limiting one's advancement, "the potential for movement is much greater" (Lenski 1966: 411). In general, people are able to change their structural positions more in modern societies than in pre-modern societies. For example, there is more occupational mobility in modern societies than in pre-modern societies.[42] Similarly, access to educational opportunities increases for most members of modern societies and a greater percentage of the population change their educational status from "uneducated" to "educated." Residential mobility is also more common in modern societies. In short, with a decrease in the importance of ascribed status in determining one's social standing, individuals are able to achieve greater movement between structural positions; therefore, social mobility increases. While birth still determines in part how likely one is to succeed in life, it is easier to overcome the accidents of birth in modern societies than in pre-modern societies. By emphasizing achievement over ascription (see Parsons 1951), modern societies create opportunities to move between status positions that simply did not exist in pre-modern societies.

With greater social mobility, the saliency of the structural parameters people use in determining with whom they will interact relaxes. Groups that were separated through customs, prejudice or distance, begin to interact more frequently. As a result, the relative heterogeneity of the social whole increases. That is, "the intermingling of many distinct but relatively homogeneous groups results in the formation of one heterogeneous group" (Hawdon 1996b: 186). This greater heterogeneity of modern societies erodes the widespread acceptance of the traditional worldview as individuals are exposed to new beliefs and behaviors. As these new ways of doing and thinking about things present themselves to

---

[42.] It should be remembered that, as Lenski (1966) notes, there was considerable occupational mobility in agrarian societies. This mobility tended to be downward, not upward. Still, the direction of the mobility aside, I recognize there was considerable mobility in many pre-modern societies. However, when one considers the movement of people between all differentiated positions, there is undoubtedly more mobility in modern societies than in any type of pre-modern society.

individuals, their once taken-for-granted reality is brought into question. As they are presented with new alternatives, old norms and customs, once so well defined and firmly established in the collective conscience, become increasingly blurred. The spontaneous normative consensus found in pre-modern societies erodes (see Durkheim [1915] 1964; Habermas 1984b; Parsons 1977; Luhmann 1982), and it becomes increasingly difficult to determine what is "proper" and "improper" behavior.

Moreover, the modern era becomes dominated by contract law as increasingly specialized relationships must be regulated. Contract law defines both rights and responsibilities that each involved party are bound to and enjoy. They therefore apply to the parties involved, not necessarily to all members of the group. As the functions of society become increasingly specialized, "the more marginal they (become) to the common conscience" (Durkheim [1915] 1964: 127). The rules that regulate these specialized functions and the contracts that govern them, since they are not applicable to everyone, lack the universal acceptance repressive laws typically have. They do not have "the superior force, the transcendent authority which, when offended, demands expiation" (Durkheim [1915] 1964: 127). Since cooperative laws assure that the functions they regulate occur in a regular manner, punishment is unneeded. Instead, if the regularity of such functions is disrupted, all that is needed is for the regularity to be restored. Thus, when compared to violations of repressive laws, contract law infractions do not produce a vehement collective response. Therefore, as contract law increasingly dominates a society, the collective conscience is weakened, and group members no longer share the normative standards that were once widely held (Durkheim [1915] 1964). There is no longer a value consensus; instead, people have competing value systems from which to choose. As Luhmann (1982) notes, dissent and a plurality of viewpoints are essential characteristics of the modern social system. Thus, due to their heterogeneity and relative lack of normative consensus, modern societies tend to be more pluralistic than pre-modern societies.

In addition, as society becomes more heterogeneous and social mobility increases, the types of people with whom one may interact and the number of groups to which one may belong increases (see Blau 1977). In the modern era, one is no longer limited to interactions with those like them. Instead, associations are based on any number of factors, including kinship, race, ethnicity, gender, social class, occupation, religious affiliation, religiosity, political interests, leisure interests, location, educational level, etc. As the number of possible associations increase, the exit costs associated with leaving any one particular sub-group are reduced since the likelihood of finding someone else to associate with is increased (see Hechter 1987). As the exit costs associated with leaving the group are reduced, the group's ability to demand conformity from the individual is undermined and sources of informal control are weakened. Thus, with an increased number of associations available for the individual to select from, he or she is freed "from the narrow confines of earlier circumstances" (Simmel 1955: 163; also see Simmel [1908] 1971; Simmel [1908] 1971b). As the group's ability to control the individual wanes, the individual becomes free to engage in a wider variety of behaviors and individual freedom increases.[43]

Some Consequences of Rationalization and Modernization

In general, the rationalization of society is a secularizing process. It is the extent to which the systematic ideas of science supplant the magical elements of thought. It is a process through which religious matters of the supernatural have gradually become differentiated from the social and natural world. They are removed from the total fabric of social life and brought under the control of secularized subsystems (Parsons [1964] 1967). This differentiation is instigated

---

[43]. Increased individual freedom is one of the most commonly cited consequences of modernization (see, for example, Simmel [1908] 1971; Simmel [1908] 1971b; Simmel 1955; Durkheim [1915] 1966; Weber [1919] 1946; Parsons 1951; Parsons 1977; Habermas 1984; Habermas 1984b; Luhmann 1982; Berger, Berger and Kellner 1973). This effect of modernization simply cannot be denied.

by the rise of universalistic belief systems and the loss of traditional worldviews, and it continues under the instrumental rational action of capitalism, bureaucracies, and science. The rationalization process increases the "steering capacity" of society and allows for greater adaptability (Parsons 1977). It allows increased ability to cope with natural and man-made disasters. It allows for the greater accumulation of wealth and material comforts. And, it is critical for the emergence of democratic political institutions and results in increased individual freedoms.

Yet, while instrumental rationality increases, it does so at the expense of substantive rationality (Weber [1922] 1978; Weber 1947). That is, just as the system and worldview have differentiated, there is greater differentiation between statements referring to nature and truth and the norms used to derive these statements and validate them (Habermas 1979). Science is concerned with causes of natural-occurring phenomena. It is concerned with efficient manipulation and tangible results. It does not concern itself with values. Because science dominates the contemporary worldview and excludes values from consideration, there is an emphasis on technical skills at the expense of values that define moral obligations. Thus, science harnesses the destructive and constructive power of atomic energy, irrespective of the collective values concerning human life or the environment. It produces increasingly powerful psychoactive substances without consideration of the morality of doing so. In traditional societies, the dominant institutions were oriented toward defining and meeting moral obligations, not the instrumental manipulation of nature. Now, however, science has substituted the manipulative rules of context-free knowledge for norms of solidarity and reciprocity (Habermas 1970).

Moreover, the domination of science results in a state where cultural traditions are constantly criticized and renewed (Habermas 1984b). Competing validity claims are continuously made, yet no universally accepted normative guidelines exist to judge these claims. Moral traditions become relative, not absolute. The unified worldviews that once secured the cohesion of social life are weakened and, with them, normative consensus becomes more difficult to achieve (see Luhmann 1982). The integrative functions that were once met by ritual practice are now forged through rational argumentation as "the authority of the holy is gradually replaced by the authority of an achieved consensus" (Habermas 1984b: 77). However, this consensus can be difficult to achieve because it is no longer given and taken-for-granted; instead, "consensus formation rests in the end on the authority of the better argument" (Habermas 1984b: 145). Thus, societies become increasingly difficult to integrate and the integrative systems become increasingly complex as specialized subsystems develop to integrate the systems (Parsons 1977). With competing moral traditions and an ever-increasing emphasis on instrumental behavior, orientations become concerned with the satisfaction of immediate, everyday needs (see Weber [1919] 1946). The rationalization process results in "the substitution for unthinking acceptance of ancient customs of deliberate adaptation to situations in terms of self-interests" (Weber [1922] 1978:30). Personalities become more autonomous as they are increasingly differentiated from culture and society (Habermas 1984b), and interests are privately, as opposed to collectively, defined (Parsons 1951). As individuals are freed from the cultural traditions and universalistic moral principles, they gain freedom in their choices and behaviors, but they also lose the security and definiteness of earlier times.

The result of this process of rationalization, at least according to Weber, is disenchantment. Mystery is removed from the world and all becomes knowable, at least theoretically. The increasing intellectualization and rationalization of the modern world

> means that principally there are no mysterious incalculable forces that come into play, but rather that one can, in principle, master all things by calculation. This means that the world is disenchanted. One need no longer have recourse to magical means in order to master or implore the spirits, as did the savage, for whom such mysterious powers existed. Technical means and calculations perform this service (Weber [1919] 1946).

Thus, the process of rationalization has led to the domination of scientific thought over religious thought. It has secularized the world. It has removed sacred mystery and substituted profane understanding. While we can better control the natural world around us, we are left disenchanted.

Similarly, modernization is the process of continued structural differentiation and advancement of technology. It results in greater surplus, less stratification, more social mobility, greater pluralism, more tolerance and extensive individual freedom. However, the greater freedom enjoyed at the individual level comes at a cost. Given the plurality of worldviews and normative standards and the resulting disintegration of the collective conscience, "behaviors that would once provoke a societal response and earn the practitioner a deviant status are no longer considered foreign or heinous" (Hawdon 1996b: 189). With greater individual freedom and a relative lack of consensus concerning what behaviors are acceptable and unacceptable, the variability in behaviors will undoubtedly increase. Behaviors that were once unthinkable now become possible. Some of these behaviors will likely upset those that hold values that are more traditional. With a greater variety of behaviors being engaged in, the likelihood of someone objecting to some of these behaviors would increase. If the objecting party is in a position to persuade enough people to also object to the behavior, the behavior will likely be defined as "deviant" (see, for example, Becker 1963; Hawdon 1996b). *Therefore, there will be more "deviant" behavior in modern societies than in pre-modern societies.*[44]

As "deviance" increases, there are attempts to control it. While the emergence of restitutive law is undoubtedly a function of modernization, repressive law does not decrease with modernization. Instead, there is simply

---

[44] It is critically important here to note that I use "deviant behavior" in a constructionist manner. The constructionist perspective emphasizes the subjective, social and political processes by which phenomena are defined as social problems (see Becker 1963). See Beckett (1994) or Hawdon (2001) for a discussion of these perspectives.

more law. While restitutive law emerges to regulate contractual relations, repressive law remains to attempt to forge the formerly held, but now lost, unified moral system. Although it may not be as successful as it had been in pre-modern societies, repressive law does exist in modern societies. The increase in law in general is due to the decrease in other forms of social control. As Black (1976: 107) notes, "law varies inversely with other social control." Finally, with more law and a greater variety of behaviors, *modern societies will have more crime than pre-modern societies.* The increase in crime is, in part, simply due to the increase in law. Since more behaviors are defined as criminal, more crime is likely to occur. However, the increase in crime is also due to the plurality of worldviews and the relative normative ambiguity found in modernized societies. This was a central point of the Chicago School of criminology. For example, Shaw and McKay ([1942] 1972: 170) state that,

> . . . in the areas of low rates of delinquents there is more or less uniformity, consistency, and universality of conventional values and attitudes with respect to child care, conformity to law, and related matters; whereas in the high-rate areas systems of competing and conflicting moral values have developed. Even though in the latter situation conventional traditions and institutions are dominant, delinquency has developed as a powerful competing way of life.

Thus, crime is inversely related to the extent to which residents hold similar attitudes and values. Moreover, crime is inversely related to the collective's ability to enact informal social control (see, for example, Shaw and Mckay [1942] 1972; Bursik and Grasmick 1993; Bursik and Grasmick 1995; Sampson, Raudenbusch and Earls 1997; Sampson 2001; Rosenfeld, Messner and Baumer 2001; Snell 2001; Hawdon and Ryan 2003). The modern age, an epoch marked by pluralism and decreased collective control, is, therefore, an epoch of high rates of crime. Thus, the highly interrelated processes of rationalization and modernization result in a world that is more physically comfortable, relatively more egalitarian, and more tolerant and free. Yet it is a world that is largely void of sacred meaning. Norms are not only secularized, they are relativized. Beauty is not the only thing that is in the eye of the beholder; now, so is good and evil.

Without clear guidance concerning how to behave, concerning what is wrong and right, concerning what is moral and immoral, people are freed from their moral conscience to engage in behaviors few of our ancestors would have thought possible. With the group's ability to control them weakened, people are free to engage in behaviors few of our ancestors would have dared.

While individuals living in modernized societies undoubtedly enjoy their greater freedom, this freedom comes at a cost. Although I enjoy my freedom, everyone else does too. Since people no longer necessarily agree on what they should and should not do, and since they are free to engage in a wide variety of behaviors, it is likely that one person's freedom will, eventually, conflict with another person's freedom. As our interests are increasingly defined privately, each of us is likely to conflict with those whose similarly defined interests do not suit our needs, desires or wants. Despite our advances in technology and ability to manipulate the physical world, society remains in a Hobbesean state where individual freedom is dependent upon granting power to a higher authority. Today, however, this higher authority is not found in God or the gods. It is no longer found in universal truths. Now, it is found in the political promulgations of the state, which, like our collective morality, are, to some extent, arbitrary. We are now directed by the arbitrary laws of men instead of the absolute laws of God.

While the modernization and rationalization processes remove the shackles of our ascribed statuses and free us to achieve great accomplishments, they also hold us more responsible for our actions and moral choices. While we are "free" to "earn" the status of CEO, lawyer, physician, or president, we are also "free" to "earn" the status of "doper," "addict" and "criminal."[45] While we are largely freed from the confines of our group, we are also deprived of the many supports communal life offers. We face choices that were once answered for us

---

[45.] Or, more accurately, we are told we are "free" to "earn" these statuses.

by the group, and, since we have the "choice," the consequences of our decisions become more pronounced. We no longer have the luxury of having the group dictate what we do. Instead, we are told to "make up our own minds," "do our own thing," and "just do it." Yet, we are not always sure what it is we are to "just do." As Simmel (1955: 141) states, "the security and lack of ambiguity in his former position gives way to uncertainty in the conditions of his life." We are free, but lack clear direction. We are free, but lack clear purpose. We are free, but disenchanted.

RATIONALIZATION, MODERNIZATION AND DRUG CONSUMPTION

The processes of rationalization and modernization have influenced every major aspect of modern life. They changed religion, economics, politics, law, education, and medical practices. They have even altered familial and personal relationships. They change how we work, recreate, worship, select our mate, educate our youth, govern our societies and enforce social norms. They affect how much we travel and how we communicate with each other. They alter interaction and the normative structure of society. There is little reason, therefore, to expect that modernization and rationalization would not influence the use of drugs. Let us now consider how the above changes associated with the modernization and rationalization processes would apply to the consumption of drugs.

First, religion typically dominates undifferentiated societies. Given that pre-modern societies are relatively undifferentiated with a simple division of labor, we would therefore anticipate that most drug use would be sacred in premodern societies and would therefore fulfill religious function.

Second, given the sacred nature of use, we would anticipate relatively low levels of use. Since rites govern sacred use, there will be clearly defined controls over use. Moreover, the solidarity of pre-modern societies is based on homogeneity. This homogeneity is evident in not only physical appearance and behavior, but also in the normative structure of the society. That is, there is a general normative consensus in pre-modern societies. This consensus is maintained through high levels of informal social control. And, on the rare occasion the consensus is violated, repressive law would be invoked to strictly punish the transgressor. Therefore, rates of drug use, like everything else in pre-modern societies, would be tightly controlled.

Third, ascribed statuses will most likely determine who uses drugs in pre-modern societies. Although not always, drug use is likely to be limited to adult males. Other societies, such as those who limit shamanism to females, will restrict use to adult females. Even in more advanced pre-modern societies, use is often restricted to the ruling elites. Such was the case in the Inca Empire (see Phillips and Wynne 1980). Yet, as Lenski (1966) points out, ruling class status was a function of heredity in pre-modern societies and thus it too was an ascribed status.

Fourth, drug use is unlikely to be defined as "deviant behavior" or criminal in pre-modern societies. Since there is little individuation in pre-modern societies, there will be little variation in what drug is used, when it is used, and by whom. Drug use, like most behaviors, will be communal and strictly governed by the group. The group will determine what drug is used, when, how it is used, and by whom. Drug use that does occur outside of these clearly defined normative boundaries will be punished repressively. Given the widespread normative consensus, restricted amounts of individual freedom, high levels of informal social control, and severe punishment for those who violate the normative code, "deviant" drug use will be unlikely.

Finally, the low rates of social mobility will mean that these patterns will be maintained for long periods. That is, social change will be relatively slow.

We would anticipate very different patterns of use in modernized societies. Given the highly differentiated nature of modern societies, drug use will occur in numerous settings, including religious, medicinal, economic and recreational. Contrary to pre-modern societies, the religious use of drugs will be relatively infrequent. Drug use, generally speaking, will be profane in modern

160

societies. Because of this profaneness, rites will no longer govern use. Therefore, use, when it occurs, will be less regulated in modern societies than in pre-modern societies, and those regulations that do exist will tend to be informal and generally permissive.[46] Since use will be less regulated, a greater variety of drugs will be used. Fewer drugs will be deemed sacred and thus any drug becomes eligible for use. Moreover, the relative lack of a normative consensus, low levels of informal social control and high levels of individuation associated with modern societies will free individuals to "experiment" with new behaviors, including drug use. Therefore, a greater variety of drugs will be used and more drugs will be used in combination. Plus, the method in which the drug is used will become more variable since neither religious interdictions nor a strongly held normative system limits the manner of use. With fewer restrictions over use, greater freedom from group constraints, a wider range of functions fulfilled by use, and greater individuation, we would expect relatively high rates of drug use. Therefore, all else being equal, *drug use will vary directly with modernization.*

Next, drugs will no longer be limited to the class of the "holy." Instead, anyone who wants to use the drug and can obtain a supply of the drug will likely use it. Consequently, we are likely to observe greater variation in who uses the drug. However, achieved characteristics rather than ascribed will determine who consumes drugs in modern societies. Now, we begin to see differences in use patterns based on factors such as social class instead of gender. Although ascribed statuses still play a role in determining who will use and who will not use drugs, these characteristics will have less of an influence relative to achieved statuses in modernized societies. Moreover, the more modernized a society becomes, the less stratification in drug use we would expect to see. That is, gender differences in drug use will likely converge. Similarly, there will be less

---

[46.] I emphasize again that we are making a distinction between the negative regulations of law and the negative and positive regulations associated with rites. While profane use is highly regulated in terms of what is and is not legal to use, these distinctions set only broad parameters on drug use. Once use of a profane and illicit drug occurs, it will be relatively unregulated.

difference with respect to drug use among various ethnic groups within the society. Even class differences will likely converge relative to those that existed in advanced pre-modern societies.

Yet, while inequality will decrease and drug patterns will converge, drug use itself will become a source of stratification. That is, the use of a drug will not only symbolize one's position in society, use can, in-and-of-itself, determine one's position in society. Once drug use is criminalized, use can affect the allocation of social rewards and punishments just as income or education does. Thus, drug use becomes a dimension of stratification. This situation differs dramatically from using drugs to symbolize group membership. For example, while the use of drugs was a common practice among shamans from numerous cultures, it was not the drug use, per se, that made them shamans. They were shamans and, as such, allowed to use drugs. They therefore used them and this use *symbolized their standing*. Similarly, the use of wine religiously does not make one Catholic. Catholics use wine ceremoniously *because they are Catholic*. And, even though it is profane drug use, it is not the use of marijuana that makes a Jamaican working-class. His membership in that group determines his use of the drug. With deviant drug subcultures, however, drug use not only symbolizes what group one belongs to, it determines the group. Illicit drug users are deviant and criminal *because* they use drugs. *Their use does not only symbolize their deviant standing in society, it determines it.*

In addition, drug use will be less integrative and more divisive in modern societies relative to pre-modern societies. As discussed above, drugs used religiously, as they often were in pre-modern societies, will reinforce the worldview of those who believe in the sacredness of the drug. The rituals reaffirm group membership and the rites governing those rituals protect the group from profane outsiders. Conversely, drugs used in a profane setting are more likely to promote solidarity among a sub-group of the larger population. Moreover, given the fractured collective conscience, fewer people will deem the use as sacred or necessary. Given the greater individuation, use will be defined as a "choice." Given the lack of clear moral guidelines, any choice, including the one to use or not use drugs, becomes open to scrutiny. Use will therefore be increasingly labeled as "deviant." Once use is defined as a "choice" and that choice is defined as "deviant," it becomes acceptable to punish the user (see Hawdon 2001). Thus, drug use now becomes a determinant of social status instead of simply a marker of it, and using the drug now confers the status of "deviant." The status of deviant divides the population into "us" and "them," into conforming members and deviants, into "insiders" and "outsiders" (see Becker 1963). Consequently, there will be more debate concerning the merits of drug use -- or at least the merits of using some drugs under some circumstances -- and drug use will become a divisive issue.

Related to the increased deviance of use, there will be more "drug abuse" in modern societies than there was in most pre-modern societies. First, as mentioned earlier, the less stringent control of use will result in more use. The greater the use of any drug, the more likely that use is to be "abusive" in an objective sense -- that is, causing empirical harm such as coma or death -- and defined as abusive subjectively. Moreover, the individualistic orientation of modern societies will lead to occasional incongruencies between individual interests and collective interests. When such problems arise and the cause of the dissonance is deemed to be drug use, it is likely that the collective will define the use as "abuse." Finally, the term "drug abuse" implies deviant behavior, and, as just mentioned, drug use is more likely to be defined as "deviant" in modern societies than in pre-modern societies.

Finally, the high rates of social mobility found in modern societies will mean that patterns of use will change relatively quickly. That is, the fads and fashions of drug use will come and go rapidly. So, for example, what drug is used will change in a matter of years instead of decades or centuries. Similarly, which ethnic or status groups are most likely to use drugs will change more hastily in modern societies. Lastly, how frequently drugs are used will vary over time more in modernized societies than in pre-modern societies.

Consequently, the rationalization and modernization processes, by transforming drugs from sacred to profane objects, changes the "who", "what," "how," "how often," "how much," and "what with" of drugs. In all of these dimensions of use, we would predict greater variation during the "modern age" than at any other time in history. While these processes have altered social relations in every country of the world and among most people in most countries, there is little doubt that the western societies have led the way. As Weber argued, these processes helped fuel the west's ascent to world hegemony. In fact, these processes have been most pronounced in the Protestant west, although this is not as true today as it had been when Weber was writing. It therefore fits the data at hand that nations with a western-orientation would lead the world in capitalistic production *and* drug consumption. While all nations have undergone the rationalization and modernization process to some extent, westernized nations are the most rationalized, at least in terms of the functional rationality of which Weber spoke, and the most modernized. It therefore appears that this general theoretic perspective may be useful.

CHAPTER FOUR SUMMARY

This chapter has developed the basis of a sociological theory of drug consumption. By noting the difference between sacred and profane drug use, we derived several general propositions concerning use. In general, sacred use, regulated through rites, is more tightly controlled than profane use. In the profane world, however, use is often unregulated or regulated permissively. For those profane drugs that are regulated tightly by secular laws, the negative nature of law (that it forbids use in total) renders the law mute once use occurs. As a result, we will see more variation in who uses drugs, what drugs are used, what combination of drugs are used, when drugs are used, and why they are used if drugs are classified as profane objects instead of sacred objects. Next, we anticipate that the variation in drug use has increased over time as the processes of rationalization and modernization have moved the general class of objects we call drugs from the realm of the sacred to that of the profane.

A casual consideration of the patterns of drug consumption presented in Chapters Two and Three suggests that the general theoretical framework developed in this chapter may have utility. As seen in the earlier chapters, western nations tend to consume the most drugs. These nations have also rationalized and modernized most extensively. The theory would therefore predict the relatively high rates of use in those nations. Indeed, the extent to which cultures have experienced the processes of rationalization and modernization is more likely to account for variations in cross-cultural patterns of use than individualistic variables such as self-esteem or self-control. As argued at the beginning of the chapter, the critical factors are more likely to be social-level phenomena than individual-level phenomena. Indeed, it is likely that rationalization and modernization have influenced the commonly cited social-psychological variables that are frequently offered as explanations for drug use. In fact, it is highly probable that this has happened.

While casual observation suggests the theory may be on the right track, it clearly remains underdeveloped. Although differences in the extent to which cultures have rationalized and modernized can account for *between culture variations in use*, it cannot, at least as developed thus far, account for *historical fluctuations in use within a culture*. While the general tendency to rationalize could explain monotonic increases in drug use over the course of a nation's history, such an explanation would not accurately fit the cyclical nature of drug use that has occurred in most western nations since the early 20th century. Linear trends cannot explain curvilinear trends. Somehow, it must be shown that the rationalization and modernization processes, while generally proceeding monotonically, can create forces that ebb and flow. This explanation will come later. We will first consider how rationalization would likely influence the reasons people use drugs. We now turn to "why people use drugs" and the social functions of drug use.

CHAPTER FIVE
THE "WHY" OF DRUG USE

As argued in the preceding chapter, the basis for understanding the empirical patterns of drug consumption lies in the distinction between sacred and profane drug use. Sacred use, generally speaking, is more strictly regulated and controlled than profane use. Next, the process of rationalization, particularly as it occurred in the west, changed the classifications of drugs from sacred to profane objects. The following chapters aim to further expand this general perspective and offer empirical evidence supporting the theory.

In this chapter, we turn to a central question of drug use. Namely, why do people use drugs? Why have so many cultures used drugs for such a long time? To answer these questions, we will consider the *social reasons* people use drugs. By social reasons, I mean the reasons people use drugs that serve some need, desire or function of the social group in which they live, not the host of idiosyncratic reasons individuals use drugs. I mean those reasons that have led all societies known throughout history to allow their citizens some means of altering their consciousness. The rationalization and modernization processes changed the relative mix of these *social reasons*, and these variations across time and space in why people use drugs can help account for consumption patterns in the contemporary world. So, let us consider the social functions of drug use.

The Social Functions of Drug Use

From historical and anthropological accounts of drug use, we can identify several social functions of drug use. The functions discussed below may not be exhaustive, and they are certainly interrelated. Nevertheless, the list provides an example of the varied social reasons people use drugs. The social functions drugs fulfill include a host of religious functions and sub-functions. Obviously, use for these religious reasons is always sacred in nature and will therefore be somewhat unique. However, religious uses of drugs are not the only social reason people use drugs. The other social functions fulfilled by drug use include to mark a rite-of-passage, to mark social boundaries between groups, to promote social solidarity, to celebrate and bind a social contract, to promote economic production, to promote effective warfare, to serve as a social lubricant, as a means of recreation, and for medicinal purposes. All of these functions can occur in either a religious or a more secular setting, and, as discussed before, this broader social context will influence the patterns of use. With that condition in mind, let us now consider each of these functions in more detail.

Drugs and Religion

Drugs can fulfill religious functions. In fact, religious use is likely one of the oldest, if not the oldest, social functions of use. Evidence of the use of drugs in religious settings date back to the Neanderthals of Europe and Asia more than 50,000 years ago (Furst 1976; Li 1975; Furst 1990) and can be found in Neolithic sites from seven thousand years ago (Gimbutas 1982; Vogal 2003). Amerindians have been using hallucinogens in religious ceremonies since at least the time of the pre-Columbian Aztecs (Schultes and Hofmann 1979). As Benet (1975: 44) says, "almost all ancient peoples considered narcotic and medicinal plants sacred and incorporated them into their religious or magical beliefs and practices." Written history also confirms the widespread use of drugs in religious ceremonies. For example, according to Herodotus (5th century BC), the Indo-European Scythians used hemp seeds during funeral rites to induce a trance among those in attendance. One of the major Hindu texts, the *Rig Veda,* devotes an entire book to the hallucinogen soma. The original Hebrew Old Testament refers to hemp both as incense -- which was integral to religious celebrations -- and as an intoxicant. The Bible mentions wine and "strong drink" over 200 times.[47]

Although used to some extent in the major world religions of Hinduism, Judaism, and Christianity, drugs tend to assume a more prominent role in shamanistic religions. In fact, according to Langdon (1992), the use of hallucinogens constitutes the most common form of acquiring shamanic knowledge. Shamans from a variety of cultures smoked, ate and sniffed a variety of drugs including tobacco, alcohol, cannabis, and a host of hallucinogenic drugs such as yage, peyote and the fly agaric mushroom. Shamans also consumed drugs in more exotic fashions such as via enemas and by drinking the urine of an intoxicated elders or reindeer (Furst 1976; Thompson 1970; Hirschfelder and Molin 1992). Shamans used drugs for numerous religious purposes. As Langdon (1992b: 41) states, "shamans use hallucinogens not only to transform into jaguars, but also divine the future, to adjudicate quarrels, to perform sorcery, to identify hunting and fishing areas, and to heal illness."

In addition to using drugs for religious purposes, some cultures deified particular intoxicating substances. The Romans, for example, had a god and goddess of wine, Bacchus and Bellona. Several Amerindian groups in ancient Middle America personified and deified tobacco (see, for example, Thompson 1970). Further to the south, other Amerindian groups deified peyote (see French 2000; Crow Dog 1984). The Cahuilla of southern California regarded the hallucinogenic plant datura as a great shaman with whom they could communicate in the course of their ceremonies (French 2000). The Aryans, as well as the ancient Zoroastrians, deified haoma (or soma) and considered it "the god Haoma on earth" (Parrinder 1983: 181).

Drugs fulfill several sub-functions under the general rubric of "religious functions." The range of functions drugs fulfill run from the extremely personal,

---

[47.] The number of times wine is mentioned depends on which bible is being used. For example, the King James Bible mentions wine 212 times, the American Standard mentions wine 210, and the Third Millennium Bible refers to wine 229 times.

such as healing a specific individual, to the extremely communal, such as to help ensure good harvests. Some are very practical, such as to identify sources of food or to provide an advantage to the group's soldiers, while others are highly symbolic. We will discuss these in more detail later in the chapter.

Drugs as a Source of Magic

Drugs also serve as a source of magic.[48] While the curing shaman's tobacco smoke was therapeutic for the Warao of Venezuela, shaman intent on causing harm through magic could also speed projectiles of sickness and death to their victims with blasts from their cigars (French 2000). Similarly, Huichol shamans "with a bad heart" would use tobacco to speed "arrows of sickness" to their victims. Huichol shaman "with a bad heart" are differentiated from other shaman and known as sorcerers (Furst 1976; French 2000).

Drugs mark Rite of Passage

Drugs can also mark a rite-of-passage. The Huichol Indians of Mexico use peyote as a rite-of-passage for adult males. Those who successfully complete the test obtain the critical dimension of wisdom and become respected elders (French 2000). Allegedly, Huichol shaman will tell initiates to "eat peyote so that you will learn what it is to be Huichol" (Vitebsky 2001: 85). Similarly, those among the Guajiro in Venezuela and Colombia who aspire to be a shaman drink tobacco juice to see if they will be a "true shaman." If the person vomits, it indicates that her "spirits are bad" and that she will experience difficulties as a shaman. If she tolerates the tobacco, she may become a true shaman (Perrin 1992). The use of drugs to mark a rite-of-passage does not only occur in a

---

[48.] Magic is similar to religion in that it too is made of up rites and beliefs. It also has ceremonies, prayers, chants and other religious-like behaviors, and it often calls on the same supernatural forces to achieve its goal. However, magic and religion differ in that religion "has a definite group as its foundation," whereas magic "does not result in binding together those who adhere to it, nor in uniting them into a group leading a common life." (Durkheim [1915] 1968: 59 - 60).

religious context. For example, The Yanomamo of Brazil and Venezuela use the powerful hallucinogen ebene as "a form of initiation" into man-hood (Vitebsky 2001: 85). Similarly, among the Jamaican working-class, "the smoking of ganja is considered by the young almost as a *rite de passage*, an audacious act signifying transition from adolescence to maturity" (Comitas. 1975: 129, emphasis in the original; also see Dreher 1982). Rubin (1975) discusses the vision of the "dancing lady" that young Jamaicans have that symbolize the initiate's transition into the ganja subculture. She states that, "it is the rare working-class Jamaican male who has never had an initial experience of smoking cannabis as an informal rite de passage to 'manship'" (Rubin 1975: 261; also see Dreher 1982).

Using drugs to mark a significant life transition is also common in the contemporary United States. American youth can work and drive at age 16. They can vote, serve in the military and view pornographic materials at age 18. However, full adult status does not come until they can legally consume alcohol at age 21. The use of alcohol to mark their ascendancy to adulthood on their 21st birthday is common among young adults in the United States.

Drugs Mark Social Boundaries

At times intricately linked to a rite-of-passage, drugs can mark social boundaries between groups. Legally being able to consume alcohol distinguishes youth from adults in the contemporary United States. Yet drugs can mark social boundaries without marking a life transition. They can reflect social distinctions based on wealth or other forms of status. For example, the Incan ruling class restricted coca use among the common people. Its use "was to be an exclusive privilege and right of the rulers and religious leaders and a symbol of their authority and control over the populace" (Phillips and Wynne 1980: 7). The elite's drug use was a visible marker of their social standing and clearly separated them from the commoners. Another example of how drugs can mark social groupings is the use of marijuana in Rwanda. At least at the time of her writing,

Helen Codere asserted that marijuana use distinguished the Twa from the Tutsi and Hutu. According to Codere (1975: 225), marijuana use is

> linked with all that sets the Twa apart as a social group, and the Twa obligingly see cannabis in ways that could have little appeal to other Rwandans, who, whatever their other dissatisfactions, remain quite content with their beer and with not being Twa.

Even within ethnic groups, drugs can signify differing status. As referenced earlier, marijuana smoking is highly associated with the working-class in Jamaica and middle-class Jamaicans forego the practice, or at least conceal it as much as possible. As Dreher (1982: 128; 130) says,

> the smoking of ganja is one of the institutions that distinguish the various segments of Jamaican society . . . (the debate over ganja) is the expression of a sharply stratified society in which ideological opposition serves to distinguish the sections and identify one's position therein.

Not only does the use of a particular drug signify a group member, the refusal to use it when it is popular among the dominant group or using it in a distinctive manner can mark social boundaries. Palgi (1975), for example, discusses how Jews living in Morocco rarely used hashish, despite the popularity of the drug among the Arab residents there. Their abstinence was a means of maintaining their separate identities. Palgi (1975: 211 - 212) says,

> At no time was the drug, per se, presented as a danger because of its particular chemic properties. Its rejection was always discussed within the framework of Jewish–Muslim relations in Morocco. . . . The smoking of *kif* (hashish) was thus regarded as a dangerous bluffing of borders between the Jewish and Muslim communities.

After the French occupation of Morocco in 1912, some young Jewish males began using hashish. However, they smoked hashish in a manner that was "not

Arab in style" (Palgi 1975: 213). A similar phenomenon is seen with the use of alcohol by Jews in various nations around the world. It has been suggested that Jews invert the dominant norm concerning alcohol consumption. In Europe, where drinking is prevalent, Jews tend to drink less than non-Jews do. In Muslim nations, where strong norms against alcohol consumption limit its use, Jews tend to be more permissive about drinking (Toch 1990). This inversion of local norms to distinguish between Jew and non-Jew was pronounced in Poland and Russia. According to Toch (1990: 489), "drinking was especially abhorred by the Jews of this region, and this led them to invert the norms of the dominant heavy-drinking majority by despising drunkenness."

Although not as institutionalized, tobacco use appears to be becoming a social marker in many cultures. In the not-too-distant past, tobacco use was common among all socioeconomic groups in the United States. Today, however, smoking is quickly becoming associated with low status groups. If the correlation between socioeconomic status and cigarette smoking continues to strengthen -- and there is little reason to anticipate it doing otherwise -- tobacco use may become as polarizing an issue in the United States as ganja is in Jamaica. In other countries, however, this trend is reversed. Tobacco use reflects high, not low, status in many Asian nations. As Shenon (1998: 57) notes, "nothing confers greater status than a pack of American or European brand cigarettes . . . In China, the choice is Marlboro. Among the gentry of Thailand, it is Dunhill." Regardless of whether the drug is being used by the upper or lower social classes, whenever such a "class gap" emerges in the use of a particular drug, use indicates to astute observers the user's social standing in society.

Drugs to Promote Solidarity

Sharing drugs can also increase in-group solidarity. For example, anthropologists have noted the solidifying effects of sharing drugs in numerous cultures. In Egypt, for example, sharing hashish unites people from different educational levels and occupational statuses who "meet in an atmosphere of brotherhood and equality" (e.g., Khalifa 1975: 203). Sharing marijuana is also a unifying event in Jamaica and "a sign of friendship and trustworthiness" (Comitas 1975; 129; also see Rubin 1975; Dreher 1982). Dreher (1982: 68), when discussing marijuana use among working-class Jamaican males, states,

> a refusal to accept the offer of a draw with either old or new acquaintances, provided the situation is appropriate, is regarded as a refusal of friendship and social intercourse. Among ganja smokers, smoking with companions symbolizes comradeship, equality, and belonging; it is a sign of friendship and trustworthiness.

Marijuana is such a powerful symbol of camaraderie that working-class men who do not smoke ganja, except for the elderly or those that decline because of religious or medical reasons, are viewed suspiciously, at least initially (Dreher 1982).

In other cultures, other drugs play the role of solidarity-builder. Chewing qat serves this function in Yemen. As Weir (1985: 126) says, "at qat parties individuals create and strengthen dyadic relationships, and are assimilated into a group and have their group affiliation affirmed and publicised." Ecstasy is also known to stimulate intense feelings of companionship (see, for example, Millman and Beeder 1998; Ter Bogt et al. 2002). Ter Bogt and his associates (2002: 161), for example, claim that English and Dutch ecstasy users were "fused together by a sense of warmth, friendship and solidarity, which was induced by XTC and dance." Alcohol assumes the role of solidarity builder in many places, such as the United States, Europe and Australia. For example, "in Australia, alcohol has traditionally been used convivially as part of the 'mateship' philosophy and in order to support the solidarity of the male adult group" (Sargent 1990: 510). Many English and Americans find fellowship in beer drinking. As Szasz (1974: 56) argues,

> the principal places where people commune or congregate to articulate, affirm, and experience a feeling of conviviality, of belonging to a group of

like minded persons -- where each person's habit validates that of every other -- are the bar, pub, or tavern. In America and in England, there are far more places for drinking than for worshiping.

Ethnographic accounts of drug-using subcultures in the United States and Western Europe also support the notion that drug use can promote group solidarity (e.g., Partridge 1973; Pope 1971; Millman and Beeder 1998; Palacios and Fenwick 2003; Ter Bogt et al. 2002). For example, when discussing the drug subculture, Harrison Pope Jr. (1971: 74) notes that, "sharing experiences that most people have not known, they (drug users) develop an *espirt de corps* which increases with deepening involvement in drugs." For its members, the subculture provides "a powerful fraternity . . . to many it is more attractive even than the effects of the drugs themselves" (Pope 1971: 77; also see Partridge 1973). A former student of mine once told me, "I'm a lot closer to my friends who use drugs than the ones who don't use. There's just something about sharing a bong that brings you together." The fraternity provided by common drug use can explain why adolescent drug users typically have drug users as friends (for evidence of this relationship, see, for example, Krohn et al 1996; Thornberry 1987; Thornberry 1996; Thornberry et al 1994; Inciardi, Horowitz and Pottieger 1993; Hawdon 1996; Hawdon 1999).

Drugs to Consummate Social Contracts

Drugs have marked treaties and closed business transactions for centuries. Probably the most well known example of this is the use of tobacco by Amerindians. Numerous North American tribes commonly used the "peace pipe." Mooney (1891: 423), noted this use of tobacco by saying,

> Tobacco was used as . . . the guarantee of a solemn oath in nearly every important function -- in binding the warrior to take up the hatchet against the enemy, in ratifying the treaty of peace, (or) in confirming sales or other engagements.

A similar practice was common among the African Bashilenge who lived on the northern borders of the Lundu. For the Bashilenge, the hemp pipe held symbolic significance, and "no holiday, no trade agreement, no peace treaty was transacted without it" (Benet 1975: 45).

We see a similar custom of drugs used to consummate a social contract in Yemen. According to Weir (1985: 126) so much business and "so many important matters are customarily settled in the context of qat parties that official meetings in stark modern offices can be considered less valid." Forging a deal while chewing qat makes it personal and less duty-bound. It therefore binds the parties on a personal and moral basis instead of solely a contractual basis.

To a lesser extent, the practice of toasting business deals serves a similar function. Alcohol use to help conduct business is quite old, especially in Europe. As Szasz (1974: 44) states, "in 1672, there was no business which could be done in England without pots of beer." Similarly, in contemporary Japan, "alcohol is indispensable at social occasions and business meetings" (Vaughn, Huang and Ramirez 1995: 503). While such toasts are not legally binding in modern westernized cultures (that would simply be too "irrational"), they symbolically reflect that a consensus has been reached and the deal has been successfully concluded.

Drugs to Promote Economic Production[49]

Drugs, especially stimulant drugs, can provide energy and promote economic production. Westerners are familiar with the use of caffeine for this purpose. Indeed, one can easily find a cup of coffee in almost any workplace in

---

[49.] Drugs have also been used as a reward for economic production, such as when Andean miners were paid in coca leaves. They can serve as a source of social control such as when American slaves were given alcohol or when Brazilian slaves working on the sugar plantations were allowed to smoke marijuana. In these cases, drugs were used to placate the slaves and maintain order. While these are social functions of drugs, they are not considered social functions of *drug use* since this is not the likely reason these people were *using the drugs*, it is the reason they were *permitted to use the drugs* and *provided with the drugs*.

Europe or North America. The mild stimulating effects of caffeine can increase concentration, reduce drowsiness, and, therefore, increase the amount of work its drinkers can accomplish. The compatibility of caffeinated beverages with work has been offered as an explanation for caffeine's widespread appeal in Europe during the 18th century. These substances were "compatible with the emergent capitalist order" (see Courtwright 2001: 59). The connection between caffeine and work can also explain Japan's fondness of caffeine (see Vaughn et al 1995).

Other drugs can also increase economic production. Andean miners, for example, chew coca leaves while working. It has been estimated that workers will chew up to 2 ounces of coca during an average day. The miners argue that coca gives them the strength and endurance needed to do the arduous work of mining at high altitudes (Phillips and Wynne 1980). Workers who need to stay awake for long periods such as long-distance truckers, physicians and students often use amphetamines. The compatibility of the Japanese work ethic and the stimulating effects of amphetamine accounts for amphetamine's popularity in Japan (Courtwright 2001). Indeed, during the World War II era, the Japanese civilian workforce was encouraged to use amphetamine to enhance production (Vaughn et al. 1995). Throughout eastern and southern Africa, leonotis leonurus, an intoxicant with similar effects of cannabis, is used to increase work production (Du Toit 1975). Despite being a relaxant and not a stimulant, field workers in Jamaica use marijuana to increase work capacity. According to the workers, marijuana, either drunk in teas or smoked, makes them work faster, harder and longer. Smoking, they argue, provides them with the strength, energy, and concentration needed to complete their workday. Marijuana is used "in the morning, during breaks in the work routine, or immediately before particularly onerous labor" (Comitas 1975: 129).[50] Similarly, the Efe, a Pygmy group of central Africa, smoke marijuana before working because they believe that

---

[50.] Studies indicate that the workers do expend more energy after smoking marijuana (Dreher 1982). It is questionable, however, if their efficiency increases.

smoking can give them the "power to kill elephants" (Du Toit 1975: 101). Agricultural workers in Réunion also use marijuana (Benoist 1975).

Drugs to Promote Efficient Warfare

Drugs have not only been used to enhance worker efficiency, they have also been used to promote efficient killing in warfare. In fact, the word "assassin" is allegedly derived from "hashishin" (hash-taker). Twelfth-century European crusaders attributed the renowned fighting ability of the Arabic hash-using cult to their use of cannabis (Palgi 1975). Likewise, Swazi warriors used marijuana before battle (Du Toit 1975). Evidence suggests that numerous governments have either provided stimulant drugs -- or failed to restrict their use -- to their soldiers during times of war. Soldiers on both sides of the American Revolutionary War routinely received rum as part of their rations. During the American Civil War, soldiers' rations included a mixture of opium and alcohol known as Hosteller's Bitters to help control dysentery and relax the troops before a battle (Helmer 1975). Prussian soldiers were given cocaine during the Wars of German Unification. English soldiers during World War I also selected cocaine as their stimulant of choice. During World War II, The Japanese government contracted with pharmaceutical companies to produce methamphetamine for their soldiers (Vaughn et al 1995). Americans soldiers also used amphetamine during the war effort. According to Courtwright (2001: 78), the U.S. military "issued upwards of 180 million (amphetamine) tablets and pills to bomber crews and jungle fighters during World War II." This practice resulted in an "amphetamine epidemic" immediately after the war in both countries (see Morgan 1981; Vaughn et al 1995; Greberman and Wada.1994). More recently, American pilots during the 2003 war in Iraq claimed amphetamine use and long hours were the cause of their accidental bombing of allied forces. While not officially condoned by the U.S. military, soldiers nevertheless used the drug to help them through combat. Whether it is to prepare for battle, be more effective during battle or to relax and get much-needed rest after battle, warriors have used drugs for centuries.

Drugs as a Social Lubricant

Another social function of drug use is to encourage social interaction and discussion. That is, drugs can serve as social lubricants. Some drugs, such as alcohol, are noted for reducing inhibitions and therefore can encourage social exchanges among persons who would not otherwise interact. The drug acts as an "ice breaker." People often serve alcohol at parties for precisely this reason. The lubricating effects of alcohol are a popular reason for its use, especially among the young. For example, in a study of British teens, 58.3% reported they used alcohol to socialize and an additional 13.1% said they used it because it reduced inhibitions (Parker, Aldridge and Measham 1998). Other drugs such as marijuana and LSD are also used to enhance social conversations. The infamous "rap sessions" of American hippies in the late 1960s are examples of social discussions promoted by the use of drugs (see Partridge 1973 for a detailed discussion of rap sessions). Similarly, Eastern Europeans have long known the socializing effects of marijuana. When discussing traditional uses of marijuana in Eastern Europe, Benet (1975: 43) notes that,

> the odor of European hemp is stimulating enough to produce euphoria and a desire for sociability and gaiety and harvesting of hemp has always been accompanied by social festivities, dancing, and sometimes even erotic playfulness.

Cocaine is also a social lubricant. Cocaine is a popular recreational drug in part because users consider it "a social drug which facilitate(s) social behavior" (Siegel 1995: 166). Club drugs, such as GHB and ecstasy, also lower inhibitions if used in moderate dosages. As such, these drugs have earned a reputation for stimulating conversation and easing the anxiety of social interaction. Ethnographic accounts of ecstasy use report that users will "sit for hours . . . talking with friends" (Rosenbaum, Morgan and Beck 1998: 202). As one respondents said when she was using ecstasy,

> I won't care what people think of me . . . I just don't care because it becomes about meeting people and just meeting different kinds of people (quoted in Palacios and Fenwick 2003: 278 - 279).

Indeed, the popularity of rave parties in England and the Netherlands is largely due to the lubricating effects of the drug.

> (Youth) experienced that XTC could turn a stranger into a friend. At least during the time they were high, they were liberated from doubts and fears, and in this sweaty, cuddly and ecstatic circle of kindred spirits they felt 'more in touch with themselves' than ever before (Ter Bogt et al 2002: 169).

Ecstasy earned the street name "the hug drug" because of this effect.

Another prime example of drugs serving as a social lubricant is the use of qat. It is widely held that qat stimulates discussion and social interaction. According to Weir (1985: 125), qat parties in Yemen are "the hub of the local communication system." At "everyday qat parties" held in individual's homes, local and national affairs are discussed, community issues are debated, and grievances are aired and settled. Qat parties even serve as a forum to discuss official business or political matters. Weir (1985: 126) states,

> Many political matters are hammered out in the context of qat parties. . . People can of course meet privately for legal or political purposes at any time of day or night, and do so, but it is in the context of the qat party that the public expression of views, discussions of problems and decision-making takes place.

Thus, qat parties in Yemen and throughout the Middle East serve a similar purpose as coffee houses, cafés, and 18th century taverns. They are places where people gather, use drugs, and discuss the topic of the moment.

Drugs for Recreation

Drugs are, have been, and will likely continue to serve recreational purposes. Numerous accounts of drugs being used at festivals and communal celebrations have been handed down over the centuries. Indeed the use of drugs to alter consciousness has been a feature of human life across all societies and in every known age. Humans have used drugs to celebrate major events for tens-of-thousands of years. It is likely that they have used them to divert their attention from the mundane everyday world for nearly that long. Much of the drug use discussed in Chapters Two and Three is recreational in nature. People use alcohol, marijuana, opiates, cocaine, hallucinogens, and a host of synthetic drugs for recreational purposes. Whether it is the Tohono O'odham of southern Arizona drinking suguaro cactus wine during the Wihgita festival (see French 2000) or British youth using ecstasy at dance clubs (see Parker et al 1998; Ter Bogt et al 2002) or young adults in Bangladesh using the codeine-based cough suppressant linctus phensedyl (see Emdad-ul 2000), millions of humans have and continue to define drug use as "fun." They gather in groups to use drugs straightforwardly or to engage in some other form of entertainment while using drugs to enhance, prolong, augment, supplement or otherwise alter the experience.

Drugs as Medicine

I list the medicinal use of drugs as a social function, but I do so cautiously. It is true that the Neolithic peoples of northeastern Asia used cannabis medicinally, and the *Atharvaveda* (*circa* 1400 BC) mentions hemp as a medicinal plant. It is also true that an old Germanic catalogue of medicinal plants lists hemp as a tranquilizer, and Medieval Arab doctors, who learned about hemp from the Greeks and Indians, treated burns and diseased joints with cannabis (Benet 1975; Courtwright 2001). It is also true that Neolithic peoples in central Europe used opium as a medicine (M. Cohen 1989; Courtwright 2001), and Amerindians applied ground peyote to their skin to treat joint pain (Furst 1976; French 2000). Similarly, millions of modern westerners consume antibiotics each year to recover from infections. While it is true that drugs have been used to treat illness for millennia, I consider this type of medicinal use a *personal reason* for using drugs, not a *social one*. The simple fact that they want to feel better or not die motivates individuals to use drugs to treat their personal illness. *Treating illnesses with drugs is therefore not included as a social function of use.*

However, the use of drugs to *prevent illness* is a social function of use, especially if this use promotes the social health of the collective. For example, drinking cannabisbased tea is common in Jamaica. In fact, it is the most prevalent means of cannabis consumption on the island and, unlike the practice of smoking it, all socioeconomic groups drink ganja tea. Ganja tea is recommended for infants and children to prevent disease and to maintain the user's good health (Comitas 1975). Similarly, the Cree and Ojibway were concerned about their health more than most other indigenous peoples of North America. While they took many precautions to promote their collective health, they also offered spirits "food and, more important, tobacco" (Hultkrantz 1992: 29). Similarly, the Tzotzil Maya of Chiapas Mexico believe that tobacco shields the individual and the community from the evil beings of the underworld and from death (Furst 1976). In contemporary western nations, the widespread use of drugs to inoculate against common illnesses reflects a similar attempt to prevent not only the specific individual from contracting the disease, but also to prevent an epidemic. Moreover, unlike using drugs to treat an illness, which is likely due to the culturally universal desire to feel healthy, the extent to which drugs are used to prevent disease varies across cultures and throughout history. The widespread use of preventive medicine to promote social health reflects the collective's attitudes toward health and illness and the role of drugs in promoting the one and preventing the other. While it is true that I take a fistful of anti-hypertensive drugs each morning for the very personal reason of preventing the stroke or massive heart attack that would surely ensue if I did not, the fact that I treat my hypertension with drugs instead of magic, meditation, prayer, aroma therapy, diet or a change of lifestyles reflects my faith in western medicine and my rejection of these more "primitive," "holistic," "organic," or "new age" practices. Because the aim of these practices, at least in part, is to promote the well-being of the collective, drug use for disease prevention represents a social function of drugs.

Rationalization and Drug Use

The above social functions that drugs fulfill are obviously interrelated. Rites-of-passages are often, but not always, marked by religious ceremonies. They can have sacred meanings (e.g., drinking wine during a Jewish wedding as a symbolic gesture) or be linked to purely secular life transitions (e.g., doing shots of tequila on one's 21st birthday). By definition, rites-of-passages separate groups. Any behavior that marks group boundaries can be ritualized, thereby promoting group solidarity. People use drugs recreationally, in part, because they are social lubricants and stimulate conversation. Most societies throughout history have likely used drugs in almost all, if not all, of these ways.

Given their interrelatedness, all of these functions likely emerged concomitantly, or nearly so. It is unlikely that these functions developed in stages. They probably did not "evolve" in the sense that one emerged first and was a necessary condition for the next one to develop. Yet the extent to which drug use fulfills each function varies cross-culturally and historically. It is this variation, the combination of the extent to which various functions are met through drug use, that can help explain why some cultures use more drugs than others do. How often drugs meet each of these functions, and therefore the relative combinations of each function, is likely a result of the rationalization process discussed earlier.

Applying the earlier insights from the discussions concerning sacred and profane drug use and the process of rationalization allows us to deduce that drugs, like most natural objects, have become increasingly profane over time. In the earliest stages of the rationalization process, drugs, as part of the undifferentiated natural world, would have been sacred objects. As such, they would have been treated reverently. The use of drugs would have been highly controlled by well-known and widely accepted rites. What drug was used, what was used with each drug, who used the drug, when it was used, and why it was used were highly regulated. As rationalization proceeds, some drugs would likely remain in the realm of the sacred and the use of these substances would remain highly controlled. Other drugs, however, would likely move to the realm of the profane. Once the drug became profane, sacred controls would not limit its use. As the modern age arrives complete with its scientific rationalism, the natural world is increasingly scrutinized and manipulated. Drugs, as part of that natural world, become, for the most part, profane objects. As profane objects, they come under the same scrutiny and manipulation as the rest of the natural world. Their use in the social world would also be analyzed, scrutinized and manipulated. The principles of efficiency and calculation are applied to drugs and drug use in the same manner as all other natural, profane objects. Since they are profane objects, the use of drugs no longer addresses otherworld meanings.

These changes brought about by the rationalization process would likely change the "why" of drug use. First, drugs will fulfill a religious function more in pre-modern societies than in modern societies. As rationalization and secularization occur, religion would occupy an increasingly narrow dimension of life. As rationalization removes the mystery from life and we replace our religious understanding of the world with a more scientific focus, we are more likely to contemplate "causes" in this world than "other-world meaning" (Weber [1922] 1964). That is, science focuses our attention on the here-and-now of this world instead of the distant and obscure otherworld. We therefore find fewer reasons to use drugs religiously and fewer restrictions on using them in other ways. Thus, *the more rationalized a society, the less likely drug use will serve a religious function.*

While this proposition seems obvious given the general secularization that occurs under the rationalization process, there are other, less obvious, changes in the functions served by drug use. As Weber notes ([1919] 1946), with rationalization, people become increasingly concerned with the satisfaction of immediate, everyday needs. Again, our focus becomes "the here-and-now" instead of the "hereafter." It would therefore follow that a similar phenomenon would occur with respect to drug use. Thus, we would predict that the use of drugs in a rationalized society would focus on practical, immediate, everyday concerns instead of more symbolic and abstract needs. The functional rationality associated with the "modern age" is concerned with practical, concrete outcomes, not symbolic, abstract meaning. Thus, while drugs will still be used to mark a rite-of-passage, promote solidarity, and mark social boundaries in modernized societies, use will occur less often for these reasons than it would among groups in societies that have not undergone the rationalization process as extensively, at least in a relative sense. The people living in a rationalized society, being focused on the immediate world around them, would likely use drugs for practical reasons such as to promote economic production or efficient warfare. They use drugs for everyday reasons like recreation and to promote social interaction. They would be more likely to use drugs as prophylactics as a means of manipulating disease than people living in "primitive" societies. Modern people are more likely to use drugs *to get something* than *to symbolize something*. They use drugs to do more work, produce more dead enemies, protect themselves and loved ones from illness, have a more relaxed conversation and therefore a better chance of getting a date, or to have fun. These are the functionally rational reasons for use. These will dominate modern users' motives for use. They will be relatively unlikely to "waste their time" by pursuing abstract, symbolic ends. Thus, *the more rationalized a society, the greater the use of drugs for earthly, concrete, and practical reasons.*

Therefore, we would anticipate the recreational use of drugs to be positively associated with levels of rationalization and modernization. Similarly, use to promote economic production will likely occur more often, in a relative sense, in highly rationalized societies than in less-rationalized societies. I anticipate the same positive correlation between the use of drugs to prevent disease and levels of rationalization. All of these functions are earthly, concrete and practical. Moreover, they are also likely to occur relatively frequently. Economic production and leisure from it are daily pursuits by most people in most places throughout most of time. How they produce and how they recreate vary. Since drugs have become increasingly profane in highly rationalized societies, their use becomes increasingly more likely to meet these social needs. Given the daily nature of these earthly concerns for which drugs are now used, it follows that, *the more rationalized a society, the greater the use of drugs on a daily basis.*

The relatively high rates of recreational drug use in contemporary England, United States, Ireland, Canada, Australia, New Zealand, Spain and France support these propositions. The high rates of pharmaceutical drug use in the United States, Japan, Germany, France, the United Kingdom and Canada also confirm these claims. The observations made concerning drug use in pre-modern societies by numerous anthropologists provide additional support for these propositions. As noted earlier, we are confident that the use of drugs in pre-modern societies typically occurred in a sacred setting. Recreational use was rare in these societies. Thus, for example, while the Cherokee and the Sioux used tobacco for sacred purposes, neither of these groups used tobacco casually. In fact, there is no indication that the indigenous populations of North or South America frequently used drugs for recreational purposes, despite the widespread use of drugs in religious settings (see Mooney 1891; Brown 1953; Furst 1976; French 2000). This was typically the case that the rapidly rationalizing European explorers of the 15th and 16th centuries found when they came across drug-using indigenous populations. While these peoples used drugs ceremoniously, they rarely used them for recreation (see, for example, Szasz 1974; Courtwright 2001). Although to a lesser extent, we see traditional Indian field workers using marijuana and tobacco for recreation on occasion; however, "only occasions of festivity and ceremonial functions are meant for using these drugs in most cases" (Hasan 1975: 245). Of course, we would expect there to be greater recreational use in late 20th century India than among the New World indigenous populations of the 19th century because of the differing levels of rationalization experienced by the two respective groups.

While the evidence does not provide a critical test of the propositions, it is, nevertheless, suggestive. The casual observer can note that contemporary drug use, at least that recorded by the U.N., is more likely to occur in modernized, advanced-capitalistic, western societies. The data reported in Chapter Three concerns, under most circumstances, recreational drug use since sacred use would not likely be defined by the authorities as "problematic" and governmental data collection officials are primarily concerned with illegal recreational drug use. So, as we have said before, western nations dominate the contemporary drug consumption markets. That is, drug use occurs in those societies most affected by the rationalization process. While there are exceptions, the general trend is undeniable.

However, if it is indeed the rationalization process instead of some other social, psychological, or biological factor that has led to this pattern, the general principle that rationalization will lead to the use of drugs for earthly, concrete and practical reasons should only hold for profane drug use. Yet, according to Weber, the rationalization process not only altered the secular world, religions have also rationalized. Assuming this is true and that rationalization influences drug use, it should change the nature of use in both the secular and profane worlds. However, in the realm of the sacred, we would expect an opposite pattern to emerge. That is, unlike the profane use of drugs, the religious use of drugs will become decreasingly this-worldly and increasingly other-worldly. To demonstrate this point, let us now consider the various functions that drugs can fulfill within a religious context.

The Religious Sub-functions of Drug Use

The religious use of drugs is probably the oldest social function that drugs fulfill. In fact, drugs can fulfill a variety of functions within a religious context. These religious sub-functions occur within a sacred context, but differ with respect to what they are to achieve. They range from the extremely personal to the extremely communal and from the practical to the abstract. The religious sub-functions of drugs include the following.

First, shamans use drugs religiously for *transcendental travel*. The use of drugs or even entering a trance is not always necessary for soul travel. Indeed, many shamanistic religions, such as the Sora of India or the Salish of North America, have soul voyages that are not drug induced (see Vitebsky 2001). Yet drugs are used to aid in transcendental travel in some cultures. For example, Shamans among the Cahuilla of Southern California used datura as a means of transcending ordinary reality and contacting spirits. They also travel to otherworlds to gain information, communicate with the dead, or retrieve lost souls (French 2000; Furst 1976; Bean and Saubel 1972). Similarly, Alaskan shamans "journey in altered states beyond ordinary boundaries for spiritual aid and help" (Hirschfelder and Molin 1992: 184). Yakut shamans of Siberia use a tobacco mixture for this purpose (Vitebsky 2001), while Siberian reindeer herdsmen shamans of pastoral Eurasia use fly agaric for their otherworldly travels (Dobkin De Rios 1975). Most transcendental travel, drug induced or not, involves arranging meetings with spirits or dead kinsmen who can provide the shaman with instructions. The Desana of the Amazon region also use large doses of hallucinogenic snuff for transcendental travel. However, the goal of the travel is different. During the shamans' initiation rites, the souls of those destined to become "true shaman," indicated by being able to tolerate the drug, travel to the Milky Way or roam through the jungle devouring their enemies (Vitebsky 2001).

Shamans also use drugs for *transformation*.[51] Most typically, the shaman transforms into an animal such as a mountain lion, jaguar, or eagle, depending on the culture (see Langdon 1992b). The use of drugs to transform into animal familiars seems to be more common among the shamans of the New World than

---

[51.] Again, transformation into animals is common in shamanistic religions and does not necessarily require the use of drugs. Whether or not a drug is involved in the ritual varies by group.

among those of the Old (Dobkin De Rios 1975; Vitebsky 2001), and the use of hallucinogenic drugs in shamanistic practices is more highly developed in the New World than it is in the Old (Vitebsky 2001). The Culina Indians of western Brazil, for example, use tobacco snuff for ritual transformation into a variety of animals (Pollock 1992). Shamans among the Matses, Desana and Guahibo, all of the Amazon region, use hallucinogens to become jaguars. The Matses blow large quantities of ebene snuff into each other's noses to bring about the transformation (Vitebsky 2001). The use of hallucinogens by shaman to transform into jaguars is common throughout the Amazonian region.[52]

Drugs can also induce *religious visions and facilitate communication with spirits*.[53] Visions, which are culturally patterned and institutionalized phenomena, may or may not be drug-induced. However, when a vision occurs, regardless of what induced it, "it is interpreted as communication from supernatural entities, and results in the recipient's acquisition of power, advice or ritual privileges" (Albers and Parkers 1971: 203). The vision revealed is that of "a true reality which in an ordinary state of consciousness remains hidden" (Vitebsky 2001: 84). While religious visions are not always drug-induced, they often are. For example, Amerindians living near the Gulf of Mexico use marijuana, or la santa rosa, to induce visions such as the Hill of Gold or "the place where God always goes" (Williams-Garcia 1975). Similarly, the ceremonies involving the African iboga plant, used by the Bwiti of Gabon and the Congo, foster communication with ancestral and deity images who provide advice (Batchelder 2001). Vision quests involving peyote are common among the Brule Sioux who seek to hear the voices

---

[52] Close identification between shamans and the jaguar is common in Amazonian shamanism. It is by far the most frequently revered animal in the region. In many languages, shaman and jaguar are variants of the same word (Vitebsky 2001: 46).

[53] For a discussion of what visions are and how they differ from idiosyncratic hallucinations see Albers and Parker (1971).

of the spirits and their divine ancestors who provide guidance (Furst 1976; Crow Dog 1984).

Shamans can also communicate directly with spirits without having visions and drug use can assist with this practice also. For example, before Desana shamans can approach any other spirit, they must use the hallucinogenic plant viho. Thus, the drug does not cause the vision; instead, it provides the shamans with access to the world of spirits (Vitebsky 2001: 84). The Siberian reindeer herdsmen use the fly agaric mushroom to communicate with malevolent beings, called *nimvits* (Dobkin De Rios 1975; Dobkin De Rios 1984). The Matsigenka of western Brazil and eastern Peru communicate with the benevolent "invisible ones" or "pure spirits" by drinking the hallucinogen ayahuasca (Baer 1992). Similarly, the Sioux sacred pipe sacrament, which is the most significant and common of the Sioux's seven sacred rituals, provides a form of communion and purification. The sacred tobacco smoke is the medium of spiritual communication with the Great Spirit *Wakan-Tanka* (French 2000; also see Brown 1953). The Warao of the Orinoco Delta in Venezuela smoked cigars and believed the smoke allowed them to communicate with the Great Spirit *Bahana* (French 2000). Similarly, Aztec priests ingested tobacco to contact deities, especially the goddess Cihuacoatl who was the mother of all Aztec gods and became manifest on earth as a tobacco plant (Winter 2000b). Drug induced spirit communication is also common among the Huichol of northern Mexico (Valadez 1992; Vogal 2003; Batchelder 2001; Winter 2000) and other indigenous peoples of North and South America (see French 2000; Lyon 1998).

Spirit communication can also take place through *possession*. For example, the Ifugaos of the Philippine island of Luzon are possessed by their ancestors during ritual celebrations. Their ancestors drink rice beer and speak to the assembled group through the mouths of the priest (Parrinder 1983). Throughout the Andean region, the Quechuan people chew coca together to gain communion with the earth, with the "sacred places," and with the ancestral dead (Allen 1988).

Related to communicating directly with spirits is the common practice of *divination*. Several cultures use drugs in their divinatory practices. For example, the use of drugs is one of three ways in which Zulus predict the future (Tedlock 2001). Similarly, Guajiros shamans in Venezuela and Colombia ingest tobacco "to see by other means than the eyes, to perceive the true nature of beings, and to recognize the various manifestations of the other world" (Perrin 1992: 111). Shamans living in the Andean region of South America used coca in many of their religious rites, including divination (Phillips and Wynne 1980). Using hallucinogens to predict the future is a common practice among many Amerindian people. For example, the Tamaulipecan, Tarahumari and Comanche all used peyote for divination (La Barre 1975). Langdon (1992b) contends that the main use of yage is as a divinatory aid in an attempt to understand the activities of the spirits and to bridge this world and the other realms. The Catimbó of the northern interior of Brazil use marijuana for divination (De Pinho 1975).

Drugs can also provide *healing powers*.[54] Drugs can either directly provide the power to heal or can provide insights into other cures. Many shamans in North America used tobacco to heal directly. The shaman would inhale tobacco smoke and fill their mouths with frothed saliva. The patient was healed by blowing smoky spray on his or her head (Lyon 1998). The Mazatec of Mexico use psilocybin mushrooms during ritualistic healing. Here, both the shaman and the patient use the drug (Vitebsky 2001). Aztec shamans, however, used drugs to heal in a more indirect fashion. The shaman would use piciétl in conjunction with chants to enlist the power of the creator gods to restore the patient's health (Furst 1976). The Siona, who live along the Putumayo River in the Amazonian lowlands, use yage for ritualistic healing in an even more indirect fashion. Yage does not have curing powers, per se. Instead, it empowers the shaman with

---

[54.] I emphasize that this function refers to the use of drugs by shamans to heal other people. Giving ill persons drugs to relieve symptoms or provide a cure for the illness is common in every known society. Here, the shaman either consumes the drugs to gain power or knowledge or as a means of administering a treatment to the ill.

knowledge which permits them control and influence over reality. It causes *dau*, the root of shaman power, to grow inside him, and *dau* imparts healing capacity to medicine. The Siona argue that all non-hallucinogenic medicines were discovered with yage knowledge (Langdon 1992b). Similarly, the Mestizo agricultural group in the northern coastal area of Peru use mescaline and datura plants to reveal possible cures. While under the influence of the drug, visions describe what remedies should be prescribed for the patients. Like the Siona, shamans do not use the drugs as cures, only to reveal them (Dobkin De Rois 1975; Dobkin De Rois 1984).

Drugs also serve in *purification rituals.* For example, the Siberian tribes of Pazaryk in the Altai region burnt hemp seeds to produce incense vapors during funeral purification ceremonies. This practice was also common among the Scythians (Benet 1975). The Cherokee's Black Drink, which produced mild hallucinogenic effects, induced vomiting which represented an internal cleansing and purification (French 2000; also see Mooney 1891). During the Cheyenne Buffalo Ceremony, participants used the sacred pipe to cleanse and purify themselves (Hirschfelder and Molin 1992). This practice is also a common function of the Sioux sacred pipe.

Drugs can also serve as *offerings to the gods.* Tobacco, used for a greater variety of sacred purposes than any other plant in the New World, served as a divine sustenance for the gods, mainly in the form of smoke. This function was nearly universal among the indigenous peoples of North America and very common among those of South America (Furst 1976; also see Thompson 1970). Tobacco smoke carried "the thoughts and words of religious practitioners to the Creator and spirit world for thanksgiving" (Lyons 1998: 299). The Warao of Venezuela provide an extreme example of the importance of tobacco as an offering. While the Warao used no other hallucinogen, tobacco smoke held together and sustained their metaphysical universe. According to French (2000), shamans smoked "incessantly" to fulfill their promise to the gods that they would provide abundant tobaccosmoke, which was the gods' proper and only food.

According to Winter (2000: 17), "the *manitous* (gods) of the Fox desire tobacco above all else." Drug offerings are also common in the Old World. Hindus devoted to Shiva offer him hemp on *Shivaratri,* the day Shiva was married (Hasan 1975; Szasz 1974). During *pūjā,* or worship, Hindus offer betel to the deity they are worshiping after they present him or her with food (Hopkins 1971). The Ifugaos offer their ancestral spirits rice beer during ritual celebrations (Parrinder 1983). In ancient Zoroastrianism, during the haoma ritual, the devout consumed the plant's intoxicating juice as a sacrifice (Parrinder 1983; also see Mehr 1991).[55] Numerous Greek and Roman rituals involved offering wine to the gods. Similarly, the Judaic-Christian Bible calls for an offering of "a quarter of a hin of wine" during rituals to consecrate a priest (Exodus 29:40).

Drugs can also assist in achieving a *higher state while praying or meditating.* Rastafarians, for example, smoke marijuana to achieve an altered state of consciousness. In this altered state, the devoted Rastafarian comes to realize the tenets of the faith. According to Barrett (1988: 254 - 255), "the herb is the key to new understanding of the self, the universe, and God. It is the vehicle to cosmic consciousness." Similarly, Hindu holy men visiting the Shivite shrines in Nepal use cannabis to aid meditation (Fisher 1975) and to "center their thoughts on the Eternal" (quoted in Szasz 1974: 43). The indigenous people of Mexico use marijuana to assist them in religious contemplation and as a "stimulant to thought" (Williams-Garcia 1975: 137). Peyote serves the function of aiding prayer in the peyote religions of the Carrizo, the Lipan Apache, the Mescalero Apache, the Tonkawa and the Caddo (Lyons 1998).

While drugs are often the centerpieces of a religious ritual, they sometimes merely supplement other rituals. Religious leaders can use drugs *to prepare for other rituals.* Holy medicine men among the Cree Indians would make tobacco offerings before starting healing rituals, for example (Hultkrantz 1992). Aztec priests would chew coca to prepare for their sacrificial rituals and healing rituals

---

[55]. In modern Zoroastrian faith, a non-intoxicating "haoma" is used during the sacrifice.

192

(Furst 1976). Mayan priests also consumed tobacco to prepare for other rituals. Similarly, the Huichol sacrificed tobacco to prepare for the peyote ritual and "all other ceremonies" (Furst 1976: 26; also see French 2000; Schaefer 2000; La Barre 1975). Related to such preparation, religious elites have used drugs to assist in fasts. For example, Szasz (1974: 43) notes this religious role of marijuana in Hinduism when he states, "by the help of bhang ascetics pass days without food or drink." During the late 19th century, French Catholic monks would use cocaine-laced elixirs to help them fast (Morgan 1981).

Finally, drugs use in religious rituals can *symbolically show devotion.* For example, the Inuit living on St. Lawrence Island in the Bering Sea sacrificed tobacco during whaling rituals. Unlike many other groups living further south, the Inuit did not use tobacco to enter trances, commune with spirits or in ritualistic healing, their religious use of tobacco was purely symbolic (see Winter 2000). Jews use alcohol symbolically during religious and family rituals (Palgi 1975). For example, in traditional Jewish weddings, the bride and groom share a glass of wine. The groom then smashes the glass to symbolize the destruction of the Temple. Similarly, wine symbolizes the blood of Christ during the Christian Eucharist (Parrinder 1983).[56] Members of the Native American Church use peyote in a similar symbolic ritual (Calabrese 1997). In shamanistic religions, tobacco often symbolizes spiritual power. For example, tobacco symbolizes *pülasü* power for the Guajiros of Venezuela and Columbia (Perrin 1992). Similarly, the Bashilenge of Africa "show (their) devotion by smoking (marijuana) as frequently as possible" (Benet 1975: 45).

The Irrationalization of Religious Drug Use

As can be seen, religious drug use can manipulate the gods or represent devotion to them. This direction, however, is not random. As Weber ([1922]

[56.] Among Catholics, the Doctrine of Transubstantiation holds that the wine is literally converted to the blood of Christ (Council of Trent 1552). Thus, drinking the wine technically results in possession.

1964) explains, the rationalization process not only leads to the secularization of society and forces religious understanding to give way to scientific understanding, the rationalization process also changes religion itself. From the earlier discussion concerning the rationalization process, one will recall that Weber argues that the most elementary forms of religious behavior are oriented to this world. Magic is used to manipulate the gods to bring food, fertility, or other forms of material wealth and comfort. As rationalization occurs, however, the gods become universalistic and god or the gods become "great lords" who can no longer be manipulated through magic (Weber [1922] 1964: 27). Thus, even religions are rationalized. Yet this rationalization, unlike that occurring in secular society, leads to *irrationalization*. Weber ([1922] 1964: 27 - 28) notes,

> Every aspect of religious phenomena that points beyond evils and advantages in this world is the work of a special evolutionary process, one characterized by distinctively dual aspects. One the one hand, there is an ever-broadening rational systematization of the god concept and of the thinking concerning the possible relationships of man to the divine. On the other hand, there ensues a characteristic recession of the original, practical and calculating rationalism. As such primitive rationalism recedes, the significance of distinctively religious behavior is sought less and less in the purely external advantages of everyday economic success. Thus, the goal of religious behavior is successively 'irrationalized' until finally otherworldly non-economic goals come to represent what is distinctive in religious behavior.

Applying Weber's general observation to the use of drugs, it follows that, unlike the profane use of drugs, religious drug use will become decreasingly this-worldly and increasingly otherworldly. Indeed, this is what happens.

In animistic or shamanistic religions, there is a blending of the natural and supernatural world. Just as the world of nature and the world of culture are undifferentiated, so are these two worlds. Before the natural and supernatural worlds a clearly distinct, there is also little distinction between human and gods. Indeed, humans can become gods, at least temporarily, through magical transformations. Not incidentally, some of the means by which humans become gods include the use of drugs. Moreover, while the two worlds of the natural and supernatural are close and intermingled, one can travel between them and return. Again, drugs are one way to transcend this world and enter the supernatural one. Similarly, there is little distinction made between the human world and the animal world. Humans are animals, animals are human, gods are human, humans and animals are gods. All are from and belong to the same mystical realm. Thus, humans can use drugs to transform themselves into animals. Of course, these animals are not merely jaguars, snakes or eagles; they are really gods. Moreover, since the gods and their worlds are close and penetrable, one can communicate with the gods directly, either through visions or through reading the signs of divination. With the worlds intermixed, there is little hierarchical distinction made between animal, human, and gods. Humans, or more accurately some special humans, are on near or equal grounds with the gods. Therefore, humans can bribe, chide, coerce, scare or force the gods to behave in a benevolent manner. Alternatively, they can be flattered or sweet-talked into giving. In either case, the gods can be manipulated, and, as Weber notes, this manipulation is always "oriented to *this* world" (Weber ([1922] 1964: 1, emphasis in the original). The gods are prodded and pleaded to provide food, shelter, health, fertility or some other good.

However, as rationalization leads to more universalistic gods, the distinction between human and gods becomes greater. While the Greek and Roman gods had many human qualities and flaws, they were still clearly more than human. At this point in the rationalization process, humans can no longer temporarily become a god or spirit. Unlike the Desana shamans who can transform into jaguars but become human again once the effect of the hallucinogenic snuff wears off, once the gods become universalistic, if a human were to transform into a god, they would remain a god. Moreover, as rationalization proceeds, the worlds of the natural and supernatural become increasingly differentiated. Now, one cannot travel to the "otherworld" and still return. Once you are there, you are there for good. This permanency reduces the incentive to leave "this world." It is true that the Greek and Roman gods came to this world and interacted with humans; however, humans did not routinely travel to their world. With rationalization, the functions of transformation and transcendental travel become obsolete. Humans can still talk to and bribe the gods with offerings, but they do so from earth, not in the gods' home.

As further rationalization leads to one universal God, the differentiation between human and God, between the natural and supernatural worlds, becomes complete. The hierarchical distance between God and human is immense and humans cannot transcend it. They cannot go to heaven and come back again. They cannot close the division between God and human because it is too vast. There is but one God, *and you are not it.* All monotheistic, or nearly monotheistic, religions clearly establish this hierarchical order of God over human. According to the book of Genesis, for example, God said to Adam before casting him from Eden for eating the forbidden fruit of knowledge,

> And the Lord God said, 'Behold, the man is become as one of us, to know good and evil: and now, lest he put forth his hand, and take also of the tree of life, and eat, and live for ever:' Therefore the Lord God sent him forth from the garden of Eden, to till the ground from whence he was taken (Genesis, 3: 22-23).

Having knowledge of good and evil made Adam and Eve god-like and therefore greater than other animals. However, God halted their ascent to the full status of god before they became immortal. The "tree of life" has been guarded from humans ever since to prevent them from becoming gods (see Genesis 3: 24). So, the hierarchy is complete. God is clearly more than human; humans are clearly not God.

With this great division, humans cannot become God, nor can they enter God's world and return to their own safely. Nor can humans bribe an omnipotent God since God would know the human motive. The religious use of drugs is therefore no longer oriented to this world. Instead, it becomes abstract and symbolic. Drugs are used to prepare for prayer or meditation, but these religious behaviors are typically directed to showing devotion, not gaining worldly benefits. Drugs can also be offered to God, as wine is offered to the Judaic-Christian God (see, for example, Exodus 29:40; Leviticus 23:13; Numbers 15:10), but this offering is made only in a symbolic way. One does not expect the hierarchically distant God to consume the offering literally. The Judaic-Christian God does not "eat the smoke" like the localized spirits of the shaman did (see, for example, French 2000; Furst 1976). Thus, drugs are relegated to the symbolic realm. As such, there are fewer reasons to use drugs and, therefore, drug use is less likely to occur.

Finally, we can understand how religious drug use is determined by the hierarchical distance between the gods and humans when we consider the atheistic religions. In atheistic religions such as the more traditional Theravāda Buddhism school, we find there are no deities (for a discussion of the atheistic aspect of traditional Buddhism see Prebish (1983) or Weber ([1922] 1978), and drugs are never used in a religious setting. According to traditional Buddhist thought, Buddha was, although enlightened, merely a man.[57] He was not a deity nor did he speak for a deity like the Jewish prophets, Christ or Muhammad. The Buddha simply provided the example for reaching nirvana. In Buddhist cosmology, there is no distinction between the natural and supernatural world. They are both natural, supernatural, neither and both. Neither really exists. Hence, there is no reason for transcendental travel because there is nowhere to go. Since there are no gods to manipulate or become, there is no drug use for these purposes. Nor is there anyone to communicate with or gain information from. Nor is there anyone or anything to offer drugs to, either literally or symbolically. There is no reason to show one's devotion to God, for there is no God. Hence, the fifth Buddhist precept of "do not take strong drink" is interpreted by traditional

---

[57.] The following comments do not necessarily apply to other, more liberal, schools of Buddhism. Some, such as the Pure Land School, not only have deities and semi-deities, they have a concept that is similar to the Judaic-Christian-Islamic concept of heaven (see Ehman 1983).

Buddhist to mean do not use anything that would tend "to cloud the mind" (see Parrinder 1983: 272; Prebish 1983).[58] In atheistic religions, there is no need to use drugs.

Thus, followers of many animistic, shamanistic and other naturalistic religions use drugs for a variety of reasons. Moreover, use tends to be oriented toward this world. The devout use drugs to communicate with the spirits, confront the spirits, implore the spirits, manipulate the spirits or personally gain the power of the spirits to bring about some concrete, earthly end. The use of the sacred drug will secure fertility, harvests, health or wealth. More "advanced" polytheistic religions also incorporate drugs into rituals; however, the more mythical uses, such as for transformation and transcendental travel, are not found. In these religions, worshipers use drugs as offerings, to prepare for rituals, or to show devotion. As we move to the monotheistic or nearly monotheistic religions, drug use, if it occurs at all, fills only symbolic functions. Drugs may symbolize some event, such as the destruction of the Temple in Judaism, or some sacred substance, such as the blood of Christ. When the religious offer drugs to God, these offerings are symbolic, not literal. Finally, followers of atheistic religions do not use drugs for religious purposes. Thus, we can state that *the greater the number of deities found in a religion, the more likely drugs used in that religion will be to promote earthly and practical reasons.* Conversely, *the fewer the number of deities in a religion, the less likely drugs will be used to promote earthly reasons and the more likely that religious drug use will serve only symbolic purposes.*

These propositions are useful in explaining why, for example, there is more sacred drug use in the vestiges of shamanistic religions still found in remote areas of North and South America, Asia, and Africa than we see in Hinduism. They also capture the fact that there is more sacred drug use in Hinduism than in

---

[58.] Interestingly, many less traditional Buddhist in the Mahāyāna school interpret the interdiction literally. That is, "strong drink" refers to alcohol, so other drugs can be used (Prebish, personal communication).

Zoroastrianism, Judaism or Christianity. They also apply to the absence of sacred drug use in traditional Buddhism or Confucianism. Yet these propositions also have ramifications for profane drug use.

Despite the rationalization of the modern world, religion is still a powerful socializing force in many people's lives. Although their society may be secularized, many people hold the doctrines of their faith sacred. Thus, religious prohibitions concerning *profane drug use* can limit the use of drugs. That is, if the sacred teachings hold that one "should not cloud the mind" with drugs or alcohol, the religiously devout will likely abstain from using these substances even in their daily, non-religious lives. However, when religions do not incorporate drugs into their rituals, the tendency is for them to avoid addressing their use in any manner. This tendency is seen in Judaism, Christianity and Islam.

For example, the Bible of Judaism and Christianity discusses the use of wine on several occasions and there are several warnings against its use. For example, in Proverbs (20:1) it says, "Wine is mocker, strong drink is ranging; and whosoever is deceived thereby is not wise." Similarly, Proverbs (21: 17) warns that, "He that loveth wine and oil shall not be rich" and "the drunkard and the glutton shall come to poverty" (Proverbs 23:21). Moreover, those who hath woe, sorrow and redness of eyes are "they that tarry long at the wine; they that go to seek mixed wine" (Proverbs 23:30). Priests are warned "Do not drink wine nor strong drink . . . when ye go into the tabernacle of the congregation, lest ye die" (Leviticus 10: 9). However, wine is also exalted. For example, Psalm (104: 14 - 15) says, "He causeth the grass to grow for the cattle, and herb for service of man; that he may bring forth food out of the earth. And wine that maketh glad the heart of man." Similarly, in the Christian New Testament, wine is Christ's blood (see Matthew 26: 28). Thus, alcohol is clearly revered. More importantly, the Bible never bans alcohol use. Thus, not only do Jews and Christians use alcohol sacredly, they fail to prohibit its use in daily life. Although there are norms of moderation expressed in the sacred texts, there are no outright restrictions placed on its use. There is also a noticeable lack of restriction on the use of other intoxicants. Although the Hebrews and early Christians knew of the use of marijuana as a medicine, fiber, incense, and intoxicant, the substance is not condemned in their holy scriptures. The same is true for opium. Although widely used at the time, God did not call for prohibitions against its use.

The same lack of restrictions on drug use is nearly as visible in the Qu'ran. While the Qu'ran is known for promoting abstinence, it does not explicitly forbid the use of alcohol. In fact, the word alcohol is never mentioned in the Qu'ran.[59] Wine is mentioned twice (both occur in Yusuf (12.36 and 12.41), but it is not condemned in these passages. "Intoxicants" are mentioned three times and being intoxicated is mentioned once.[60] Islamic leaders have derived the prohibition on alcohol from these four sections. In The Cow (2.219) it says, "They ask you about intoxicants and games of chance. Say: In both of them there is a great sin and means of profit for men, and their sin is greater than their profit." Similarly, in The Dinner Table (5.90 - 5.91) it says,

> O you who believe! Intoxicants and games of chance and (sacrificing to) stones set up and (dividing by) arrows are only an uncleanness, the Shaitan's work; shun it therefore that you may be successful. The Shaitan only desires to cause enmity and hatred to spring in your midst by means of intoxicants and games of chance, and to keep you off from the remembrance of Allah and from prayer. Will you then desist?

And, lastly, in The Women (4.43) it says "O you who believe! Do not go near prayer when you are intoxicated until you know (well) what you say." At least based on this translation, the Qu'ran never explicitly condemns alcohol use. Instead, intoxication is equated with the work of Satan. While interpretations of

---

[59.] I must admit that I do not speak Hebrew, Greek or Arabic; instead, I rely on the translations of others. For the Qu'ran, I use the 1983 Tahrike Tarsil Qur'an Inc, version which was translated by M.H. Shakir. I used the King James Bible.

[60.] Technically, being intoxicated from an intoxicant is mentioned only once. The word intoxicated appears three times in the Qu'ran.

the Qu'ran forbid the use of alcohol, they fail to explicitly forbid the use of other intoxicants. Like the Bible, the Qu'ran does not forbid the use of qat, marijuana, or opium, although these drugs were widely used throughout the Middle East during Muhammad's time.

Thus, by limiting or excluding the use of drugs in sacred rituals, the "higher religions" also removed the strict limitations on the *daily uses of drugs* found in the more "primitive" religions. In those religions that use drugs frequently in a sacred context, the sacredness of the drug prohibits its daily use since using the drug in such a manner would make it profane. The religious limitations therefore spill over into daily life, even that aspect of daily life that is not religious. In the "higher religions," however, restrictions are either limited to a specific substance, (such as the interpretations of the Qu'ran concerning alcohol or the restriction on alcohol use by the Brahmins in Hinduism), open for various interpretations, (such as the Buddhist precept concerning "strong drink"), or couched in norms of moderation (such as the Judaic-Christian warnings against the over indulgent use of alcohol). Failing to condemn intoxicating substances explicitly or specifically condemning only one allows the possibility that the use of other intoxicants is permissible. Indirectly, therefore, the major world religions condone drug use to an extent by not controlling it directly through sacred rites or neglecting to condemn it explicitly.

Again, we see that rationalization acts to lessen the restrictions on drug use. It does this by first moving drugs from the realm of sacred to that of the profane. Yet even in the religious realm, rationalization has an effect. As religious drug use moves from a this-worldly orientation to an otherworldly orientation and from the practical to the symbolic, religions tend to include drugs under their direct control less and less. Drugs become increasingly profane and are therefore uncontrolled through rites. The religious interdictions concerning drugs that remain typically either condemn the use of a specific substance or encourage abstinence through norms of moderation. These interdictions, however, tend to be more open for interpretation than the rites that directly include the use of a sacred drug. Therefore, in the "higher religions," religious authority has little control over drug use, especially as compared to those shamanistic societies that used drugs regularly in their religious rituals. The control that remains in modern religions is based on more general norms of moderation, "right living," or general morality. The control is thus derivative, not direct. While this control may remain strong among the extremely devout, control over the more secular sections of society is weakened, if not totally lost.

Summary

The re-defining of drugs as profane objects influences why people use drugs. In "primitive" societies, use was often highly symbolic. Members of these societies used drugs to meet religious ends or to symbolize a life transition, membership in a group, or a political agreement. While drug use still fulfills these functions, use is increasingly practical and serves every-day functions. Rationalization places a premium on results. Users now want to gain something more concrete than symbolic recognition. We therefore expect people in modern societies to use drugs for economic stimulation, the promotion of warfare and recreation. Because these "earthly concerns" are also "daily concerns," rationalization correlates positively with drug use.

The rationalization of drug use has also occurred in religions. However, the effect has been to "irrationalize" the functions of religious use. Instead of the functions becoming increasingly oriented toward this world, they have become oriented toward the otherworld. Instead of seeking to manipulate the gods to obtain some practical outcome, drugs are now only offerings or a symbolic sign of devotion. Religious use, at least in the "higher religions," has moved exclusively to the symbolic realm. This move was due indirectly to the rationalization process. It was a direct result of the growing distance between gods and humans that the rationalization process induced. Now, religious control over drug use is derived from other interdictions concerning general morality. The control is therefore no longer direct. This removal of religious controls has profound effects on daily drug-using habits.

Let us now return to contemporary patterns of drug use and determine if the propositions presented thus far withstand systematic empirical scrutiny.

CHAPTER SIX

EMPIRICAL TESTS OF A SOCIOLOGICAL THEORY OF

DRUG USE

The preceding chapter demonstrated how the rationalization of the social functions of drug use has changed the nature of use. We must now see if additional empirical evidence supports the theoretic propositions derived thus far. To do so, I will rely on both qualitative and quantitative analyses. I will also present contemporary and historical analyses. Let us consider each proposition and the empirical evidence that addresses it in turn.

WHAT IS USED:

VARIETY OF DRUGS USED

The first proposition is:

(1.0) The greater the level of modernization, the greater the variety of drugs used.

While most people in most places at most times had, or have, access to a relatively limited range of drugs, those in the most modernized societies have a well-stocked drugstore. The increased variety of drugs that are used is due, in part, to the fact that knowledge tends to be cumulative. That is, as we have discovered new drugs, or how to make synthetic drugs, additional substances become available for use. As Courtwright (2001: 65 -66) notes,

> The primary source of psychoactive novelty over the last hundred years has been and will continue to be the introduction of synthetic drugs by

> multinational pharmaceutical companies. . . . Inevitably, some of these products will find their way into the drug underworld.

However, while pharmacists introduce the public to new drugs, older drugs often remain popular. Thus, simply by adding drugs to those that are already available, modernization, and its related offshoot of science, increases the variety of drugs. Yet the use of an increasing array of drugs is also due to a worldview that considers the substances profane instead of sacred. As discussed in Chapter Four, the use of sacred drugs will be controlled by rites that will limit when, how, and with what the drug is used. Thus, the Huichol, for example, use only peyote and tobacco in the peyote ritual (see Furst 1976; French 2000; Schaefer 2000). It would be sacrilegious to use other substances during the ritual. Such controls do not exist on the use of profane drugs; therefore, in modern societies, only the availability of the drug, which is rarely a problem, and one's taste, morality and desires limit what drugs are used.

One can see the growth in the variety of drugs used in North America as modernization advanced there. While most indigenous groups limited their drug use to tobacco and corn beer (see French 2000), American colonists used tobacco, a variety of alcoholic beverages, caffeine and opium (see, for example, Morgan 1981). Modern Americans, however, use the common drugs of alcohol, tobacco, caffeine and a wide variety of pharmaceuticals. In addition, numerous illicit substances are available and used. Marijuana, cocaine, cocaine derivatives, LSD, ecstasy, inhalants, amphetamines, barbiturates, benzodiazepines, tranquilizers, natural hallucinogens, steroids, PCP, opium, morphine, heroin and other "exotic" drugs can all be found in the United States and are all used illegally or pseudo-legally (see Johnston, O'Malley and Bachman 2003). In a study of over 600 "serious delinquents"[61] in Miami, the sampled youth used marijuana, cocaine, and alcohol regularly. Moreover, over half the sample used depressants,

---

[61.] The sample of youth reported an average of 702 total offense and 29 index crimes even without including assault or larceny (Inciard et al, 1993: 56).

amphetamines, hallucinogens, heroin and inhalants on occasion. The researchers note that, "of the seven drug types studied, respondents used an average of 4.6 different types in the prior 90 days, and they used an average of 2.6 types regularly during this time" (Inciardi et al., 1993: 175).

Evidence from Western Europe also demonstrates the wide variety of drugs used in modernized nations. In a study of Dutch ecstasy users, for example, many respondents regularly used cannabis and amphetamine in addition to ecstasy. Cocaine, psilocybin, and LSD were less popular among the youth, but also used on occasion (Ter Bogt et al. 2002). Irish youth in Dublin were also found to use a wide variety of drugs. While cannabis was the most popular and frequently used drug by both "users" and "problem users," the majority of the drug-taking respondents used ecstasy, amphetamine, LSD, inhalants and tranquilizers extensively. Moreover, 81% of the "problem users" also reported the use of heroin and 88% reported they used cocaine. Not surprisingly, a greater proportion of "problem users" reported lifetime use of all the illegal substances (Mayock 2002). English youth also consume a wide variety of drugs (see Parker et al. 1998). The same is true for citizens of other modernized countries including France, Germany and Australia (see UNODC 1999).

UN data also reveal the extent to which modernized nations use greater varieties of drugs as compared to less-modernized nations. As discussed before, the UNODC collects data from national governments who report the extent of drug use in their nation to the UN. While there is a much higher response rate among developed nations than lesser-developed nations and more developed nations have better estimates of their citizens' use, it is telling if the percentages of nations that report use of a specific drug are compared. Classifying nations as core, semi-peripheral, and peripheral[62] and noting the percentage of each type of

---

[62.] These categories are associated with the world systems perspective. Core nations are the most developed and control the system. They control and dominate the international economy, political realm and even the cultural realm. Nations in the periphery are dependent on the core for manufactured goods and much of the primary commodities consumed there. Semi-peripheral nations have more diversified economies than peripheral nations, but they are still dependent on

nation that reports any drug problem reports a problem for a specific drug demonstrates that people in core nations use a wider variety of drugs. For example, while seventy peripheral nations reported the use of *any drug* to UNODC, only 4 (5.7%) peripheral nations reported the use of *ecstasy*. In contrast, 16 of the 18 core nations (88.9%) reported the use of ecstasy by their citizens. Similarly, only 24 peripheral nations (34.3% of those 70 that reported any drug use) reported the use of amphetamines, but all 18 (100%) core nations reported amphetamine use. Table 6.1 reports the percentage of reporting nations that reported the use of a specific drug by the nations' world-system position.

Table 6.1: Percent of Reporting Nations Reporting the Use of a Specific Drug by World-System Status

	Ecstasy	Amphetamine	Cannabis	Cocaine	Opiates
Peripheral (70 nations)	4 5.7%	24 34.3%	61 87.1%	29 41.4%	47 67.1%
Semi-peripheral (54 nations)	13 24.1%	45 83.3%	50 92.6%	39 72.2%	45 83.3%
Core (18 nations)	16 88.9%	18 100.0%	18 100.0%	17 94.4%	18 100.0%

As can be seen in Table 6.1, for every drug type reported by UNODC, almost all core nations report its use. The only exceptions are that Japan and Switzerland do not report the use of ecstasy and Japan does not report the use of

the core for most manufactured goods and finances. Those interested in more precise definitions of these nations should see Wallerstein (1974). Nations were classified using the classification schemes presented in Smith and White (1988) and Snyder and Kick (1979). I updated these schemes in one of my earlier works (Hawdon 1996c). Given the relative stability of the world-system (see Wallerstein 1974; Smith and White 1988), these schemes should still be valid for our purposes. It is clear that there has not been a radical change in the relative position of nations in the world-system.

cocaine. Conversely, among the reporting peripheral nations, only cannabis and opiates are reported by more than half of those countries that report the use of any illicit drug. The reporting rates for the semi-periphery are closer to the core's than the periphery's; however, there is still a substantial, and statistically significant, difference in the numbers of semi-peripheral and core nations reporting the use of ecstasy, amphetamine, cocaine and opiates.[63] While some of these observed differences in reporting could be due to differential abilities to determine and report the use of a drug, the magnitude of the differences suggests that the differences, even if somewhat inflated, are real. Moreover, available ethnographic and anecdotal evidence also supports the general pattern of a greater variety of drugs being used as levels of modernization increase. Finally, when one also remembers that residents of core nations are by far the largest consumers of varied types of pharmaceutical drugs, the difference in the variety of drugs used by level of modernization is even more pronounced.

DRUGS IN COMBINATION

A second proposition can be stated as:

(2.0) Modernization is directly related to the number of different types of drugs used in combination.

Just as the citizens of modernized nations use a greater variety of drugs than those of less-modernized nations, they will also use more drug types in combination. That is, there will be more polydrug users in modernized societies than in less-modernized societies. Not only are there a greater variety of drugs to select from and therefore use together in modernized societies, but the relative lack of social control over drug use will free users to experiment with a variety of

---

[63.] The difference in reporting the use of marijuana by semi-periphery and core nations is not statistically significant (χ^2 = 2.38; p = .123). All other differences between core and semi-peripheral nations are significant at the p < .05 level or better.

combinations of drugs. In most pre-modern societies and nations of the contemporary world-system's periphery, drug combinations are limited either by tradition, lack of availability or lack of funds. These limits usually do not pose a problem in the modernized societies of the core.

Again, we can see support for this proposition from ethnographic accounts. For example, from historical accounts we know that the various peoples of southern and eastern Africa have used marijuana for centuries. While many of these groups also used alcohol and, once introduced to the continent, tobacco, they rarely used these drugs simultaneously (see Du Toit 1975). Similar restraints over the simultaneous use of different drugs were observed in most Amerindian groups of North and South America (see French 2000; Furst 1976). Even in more contemporary times, in societies that are still removed from the full force of modernization, we see limited use of drugs-in-combination. In the relatively isolated provinces of China, for example, opium and tobacco are the only types of drugs used with any regularity, and only the elderly use these. Similarly, in Viet Nam, drug use was largely restricted to opium smoking, at least prior to 1975. Since then, heroin use has gained popularity (OGD 1997). Yet, these are similar drugs and they would not be used in combination. In Jamaica, where marijuana use is prevalent among the working classes, few users consume any drugs other than marijuana, with the possible exception of alcohol. Even among those marijuana users who drink alcohol, it is "abusive" to consume rum regularly (Dreher 1982). In remote areas of Armenia, opium is traditionally smoked, as it is almost everywhere in the Caucasus, but heroin and other synthetic drugs are still practically unknown (OGD 1997). In short, most pre-modern societies had a limited number of drugs that were used and, in most cases, very few combinations of drugs were used simultaneously. The same is true for less-modernized societies today.

As we move to "modernizing" nations, or those of the semi-periphery, we begin to see the emergence of more polydrug use. However, even here, the number of combined drugs is far less than is found in highly modernized societies. For example, heroin users in India are increasingly using licitly manufactured drugs with heroin. In particular, users often combine codeine-based cough syrups and benzodiazepines with heroin. Yet there is no mention of these users consuming drugs such as cocaine, hallucinogens or amphetamines as there is in more economically developed nations (see Drug Policy Alliance 2002). Hungarian opiate users are similarly developing a taste for synthetic drugs such as LSD, ecstasy and other amphetamine derivatives, but this is a recent development (OGD 1997). We can also see the influence of modernization by tracking the history of drug use in rapidly developing societies. In the Ukraine, for example, the *Sociological Research Centre of the National Scientific Institute on Youth Problems* investigated the consumption of drugs by the elite. Three generations of Ukrainian drug consumers were identified. The first group, which emerged in the 1960s, used cannabis derivatives and refused to use the traditional drug alcohol. Users in the 1970s, however, were more willing to explore the world of chemical stimuli and added drugs such as LSD to their list of preferred drugs. In the 1990s, cocaine and heroin have also become popular, especially among the nouveau riche (OGD 1997). A similar growth in the variety of drugs used in combination has been seen in the Russian Federation and Bosnia-Herzegovina since these areas have "westernized" (see OGD 1997; UNODC 2002). However, once again, the types of drugs available to use in combination are not as extensive in these semi-peripheral nations as is found in core nations.

Indeed, the situation is very different in more modernized, or core, societies. For example, in the study of serious Floridian delinquents mentioned above, James Inciardi and his associates (1993: 174) found that "polydrug use -- most often marijuana plus cocaine plus a depressant -- characterized the current drug use of these youths." Slightly over 82% of sampled youth used marijuana every day, 64% used some form of cocaine daily, and 91% used at least one coca product -- cocaine, crack or freebase -- three or more times a week. All of the respondents also used alcohol on a near-daily basis. Over 46% percent of the sample used depressants at least weekly. Over half the sample used amphetamines, hallucinogens, heroin and inhalants on occasion (Inciardi et al. 1993: 174 - 175). We can deduce from these numbers that a large percentage of these youth used marijuana and cocaine or some coca product daily. On many days, they used marijuana, cocaine, depressants and alcohol. In addition, on a limited number of days, they used all of these drugs plus others. Such polydrug use would rarely be seen in less modernized societies. In modern societies, however, polydrug use is almost the norm. We know, for example, that most users of illicit drugs other than marijuana also use marijuana (Goode 1999; UNODC 2000). In addition, polydrug use is common among American heroin users. As Goode (1999: 159, emphasis in original) states, after 1970, American "heroin addicts used heroin *in addition to* a wide range of other drugs; in short, heroin addicts and abusers became *polydrug users.*" In addition to cocaine, marijuana, alcohol, synthetic narcotics and barbiturates, heroin users commonly use benzodiazepines to augment the heroin's effect or to relieve anxiety when heroin is unavailable (Julien 2001). Similarly, the majority of American methamphetamine users are polydrug users (Greenwell and Brecht 2003). Polydrug use is also common among American adolescents who drink alcohol, at least among heavy drinkers. In a study of seventy-one adolescents in an alcohol treatment center, "the data suggest that polydrug use characterizes the large majority of adolescent alcohol abusers, and that such use is often quite extensive" (Martin et al. 1993).

While the above users may be extreme cases, other evidence indicates that polydrug use is common even among less immoderate American drug-using youth. We know, for example, that American marijuana users are also disproportionate users of tobacco and alcohol (see, for example, Kandel 1980; Javetz and Shuval 1990; Jessor 1987; Jessor, Chase and Donovan 1980; Donovan 1996; Hoffman et al. 2000). In a representative sample of New York State students in Grades 7-12, 21% had used marijuana and alcohol together at least once in the six months prior to being interviewed. Two percent had used alcohol and cocaine together in the previous six months (Hoffman et al. 2000). Similar patterns were found in a study of middle school students from a mid-size city in South Carolina (Hawdon 2003).

Polydrug use is also the norm among drug-using British youth. According to a recent study of youth in Britain, two-thirds of drug users were polydrug users. While alcohol in combination with another drug was the most common form of polydrug use, the surveyed youth also simultaneously used a host of other substances. In total, the respondents reported "fifty-two different combinations of drugs" (Parker et al 1998: 71). Similarly, a study of 135 regular drug users in Edinburgh found a "new preference for multiple drug use, found to be common among both injecting and non-injecting users" (Ruggiero and South 1995: 25). In addition to regularly using cannabis and alcohol, users frequently combined illicit drugs, such as heroin, with licit drugs such as temazepam and buprenorphine. A similar conclusion can be drawn from Ter Bogt's (2002) discussion of the emerging British ecstasy scene in the mid-1980s and more recent surveys of young British clubbers (see Measham et al 2001). Likewise, the Dutch youth who use ecstasy frequently combine it with cannabis or amphetamine (Ter Bogt 2002). Finally, based on the relatively high rates of use of marijuana, ecstasy, amphetamine, heroin and inhalants among Australian, German, Belgian, Spanish, and Canadian youth (see UNODC 1999), we can assume polydrug use is also relatively frequent in these core nations as well.

HOW MUCH IS USED

Modernization not only influences what drugs are used, but it also influences how frequently drugs are used. Proposition 3.0 states:

3.0 Drug use will vary directly with modernization

The greater individuation, freedom and physical comforts enjoyed by citizens of modernized societies permit them to use more drugs in more settings at more times than was possible earlier. While drugs may be an integral part of pre-

212

modern societies, higher levels of social control limited drug use to fewer people and fewer situations. In contrast, modern societies, with more individual freedom, are less restrictive. However, proposition 3.0 must be conditioned. While drug use will vary directly with modernization in a cross-sectional analysis, longitudinally, the relationship between drug use and modernization is curvilinear. That is, at any given time, more modernized societies will typically have higher rates of use, on average, than less-modernized societies. Thus, in the eighteenth century, the more modernized Europeans consumed more caffeine, alcohol and tobacco than did the less-modernized Americans (see Courtwright 2001). Today, the more modernized Americans consume slightly more drugs than the slightly less-modernized Canadians, and the more modernized British consume considerably more drugs than the considerably less-modernized Chinese (see UNODC 2002; UNODC 1999; WHO 2003b). However, the less-modernized Americans of the late 19th century consumed *more drugs* than the more-modernized Americans of the late 20th century (see Morgan 1981; Musto 1987; Hawdon 1996b). The same is true for most highly modernized nations. We therefore need to make a distinction between cross-sectional and longitudinal patterns when discussing rates of drug use. Let us consider the simpler cross-sectional case first.

MODERNIZATION AND DRUG USE IN THE CONTEMPORARY WORLD

In the contemporary global environment, drug use varies directly with modernization. This pattern is clear, even noting the problems with our data. Using available UN data (UNODC 1999; 2000; 2002; 2004) and data from the World Bank's *World Development Indicators* (World Bank 1998)[64] we see that a

---

[64]. I use 1998 data to establish temporal ordering clearly. That is, the proxy measures for modernization clearly precede the estimates of drug use. Thus, *if this relationship is causal*, it would have to be in the direction of modernization "causing" drug use and not in the opposite direction.

number of proxy measures for modernization correlate positively with drug use.[65] In addition, world system status, which can be considered a cumulative measure of modernization, is also positively correlated with drug use. From Table 6.2 we see that GDP per capita[66] is positively correlated with liters of alcohol consumed, percentage of the population who uses ecstasy, percentage of the population who uses cocaine, percentage of the population who uses marijuana, percentage of the youth population who use marijuana and total pharmaceutical sales. The percentage of roads that are paved is positively related to the use of tobacco, alcohol, opiates, pharmaceutical drugs, youthful marijuana, ecstasy and heroin use. Life expectancy at birth is positively related to the use of ecstasy, cocaine and opiates in the general population. It correlates positively with the youthful use of ecstasy and marijuana. Telephone mainlines per 100,000 population is also positively related to the use of alcohol, ecstasy, and youth marijuana use. World system status is positively correlated with alcohol, ecstasy, amphetamine, marijuana, cocaine, opiate and pharmaceutical use in the general population. World system status also co-varies positively with youthful use of marijuana and ecstasy. The only negative correlations in the table are not statistically significant.

[65.] I use non-parametric correlations due to the small samples in some instances and the poor quality of the data. As discussed earlier, the available data are probably not accurate enough to use parametric statistics that rely on interval-levels of measurement. However, Spearman's Rho, the correlational statistic used here, correlates relative rankings between the two variables in question, and therefore makes fewer assumptions about the distribution of the data than the typical Pearson's correlation coefficient. Although I believe it is relatively safe to assume that the data are accurate enough to use statistics based on rank orders, the statistics presented here should be viewed as tentative. With this caution in mind, however, the consistency of the correlations across several proxy measures of modernization and several types of drugs provides additional confidence that the *general pattern* that more developed nations tend to use more drugs is accurate.

[66.] GDP per capita is used as a proxy measure of rationalization and modernization since GDP is a function of modernization (see Estes 1988; Hawdon 1996c; Kennedy1987) and modernization is a related process to rationalization (see, for example, Martin 1999).

Table 6.2: Correlations between Drug Use and Measures of Modernization

	Per Capita GDP	Percentage of Roads that are paved	Life Expectancy	Telephone Lines per 100,000 population	World System Status
Per capita liters of Alcohol	.547 ** (147)	.344 ** (145)	.578 ** (171)	.634 ** (145)	.556** (172)
Percent of population ecstasy	.519 ** (30)	.292 (26)	.576 ** (30)	.482 ** (31)	.574** (33)
Percent of population amphetamine	.171 (81)	.100 (74)	127 (83)	-.208 (74)	.360** (87)
Percent of population cocaine	.337 ** (77)	-.129 (73)	.283 ** (79)	.118 (81)	.363** (85)
Percent of population opiates	.078 (100)	.303 ** (90)	.163 * (106)	.111 (107)	.175 * (110)
Percent of youth marijuana	.400 ** (48)	.388 ** (44)	.383 ** (49)	.421 ** (49)	.478** (49)
Percent of youth ecstasy	.366 * (26)	.380 * (24)	.528 ** (49)	.198 (27)	.423 * (27)
Percent of youth heroin	.201 (35)	.471 ** (22)	.165 (25)	.113 (25)	.234 (25)
Total pharmaceutical sells	.864 ** (11)	.584 * (10)	.491 (11)	.709 ** (11)	.775 * (11)

** p < .01   * p < .05 (one tail test)   Sample sizes reported in parentheses

The only drug for which data are available that were not correlated with at least one of the various modernization variables were the use of inhalants by youth. The correlations between this variable and the modernization variables were positive, however, they were not statistically significant. Generally speaking, then, drug use and modernization are positively related. This pattern holds for legal drugs, such as alcohol and pharmaceutical drugs. It also holds for illegal drugs, including marijuana, cocaine, amphetamine, ecstasy and opiates.[67]

From these data, we see that modernization, or at least factors highly reflective of it, and drug use are positively correlated. Of course, skeptics would argue that modernization is highly correlated with the region of the world and, therefore, the above correlations simply reflect the regional variations in drug consumption that are due, in part, to drug availability. That is, the relationship between modernization and drug consumption is spurious with respect to region. However, this argument is not supported. Modernization and drug use is positively correlated even while controlling for region. Table 6.3 reports the partial correlations between GDP per capita and the use of various drugs. Even while holding region constant, modernization, at least as measured by per capita GDP, is positively related to the use of alcohol, ecstasy, marijuana and cocaine.

Table 6.3: Correlations: per capita GDP and Drugs, controlling for region

	GDP per capita	N
Per capita liters of alcohol consumed	.469 ***	144
Percent of population who use ecstasy	.308 *	27
Percent of population who use marijuana	.142 *	109
Percent of population who use cocaine	.357 ***	74
Percent of youth population who use marijuana	.462 ***	45
Percent of youth population who use ecstasy	.232 *	23
Total pharmaceutical sales	.705 **	8

*** $p \leq .001$ ** $p < .01$ * $p < .05$ (one tail)

[67.] Other proxy measures of modernization, such as newspaper circulation, per capita number of hospital beds, radios per 100,000 population and literacy rates are also positively related to the use of drugs. However, most of these correlations are not statistically significant due to the small number of countries for which data are available for these variables. Moreover, these are not reported for sake of brevity. The general point that modernization, regardless of how one measures it, is positively related to drug use is supported by these data.

To further support the general proposition that drug use varies with modernization, a series of ANOVAs were conducted. In the ANOVA models, world system status (coded as core, semi-periphery, and periphery), Muslim nations, region and percent urban predict alcohol, marijuana, ecstasy, amphetamine and cocaine use.[68] World system status was a significant predictor of alcohol (p = .025), marijuana (p = .017), ecstasy (p = .002), cocaine (p < .001) and amphetamine (p = .034) use. "Muslim" was significant in predicting everything except ecstasy and cocaine use. Region significantly predicted everything except marijuana use.[69] Percent urban was a significant predictor of amphetamine use if a one-tail test of significance is used. All of the models were statistically significant and most accounted for substantial amounts of the variation in use. Table 6.4 reports the mean rates of use by world system status, adjusted for the other variables in the model. As can be seen from the table, core nations consistently have the highest rates of drug use. Moreover, with the exception of ecstasy, semi-peripheral nations have higher rates of drug use than peripheral nations. And, with the exception of marijuana and amphetamine use, the models account for a substantial amount of the variation in rates of use observed across nations. The influence of modernization on drug use remains even after controlling for the effects of region. Thus, the supply of drugs cannot account for all of the variation in use. Moreover, the influence of an anti-drug-using religion such as Islam does not eliminate the influence of modernization. Similarly, urbanization does not negate modernization's ability to predict drug use in the modern world system.

---

[68] For the variable "Muslim," a nation was coded as a "1" if more than 25% of the population were Muslim (data taken from the CIA's *World Fact Book*). The regions used were Africa, (including the Middle East), Asia (including Oceania), North America, South America, and Europe. None of the variables significantly predicted opiate use. The variation in opiate use is simply too small to model adequately with statistics.

[69] Region and Muslim were significant using a one-tailed test (p = .067 and .071, respectively).

Table 6.4: Mean Levels of Use by World System Status, controlling for Muslim nations, region and urban population

	Alcohol (Liters per capita)	Percent use marijuana	Percent use ecstasy	Percent use amphetamine	Percent use cocaine
Grand Mean	5.08	4.29	0.45	0.54	0.63
Periphery	4.00	2.41	0.28	0.12	0.21
Semi-periphery	4.38	3.08	0.23	0.47	0.54
Core	6.87	7.38	0.85	1.04	1.13
Model F	28.08	2.05	3.98	3.58	7.77
Eta^2	.579	.125	.559	.279	.467

* All models were significant at the $p < .05$ or better.

Finally, the correlation between modernization and rates of drug use can be seen at the regional level. Recalling the classification scheme presented in Chapter Three that categorized nations based on the relatively rankings of the various "major" illegal drugs, Asia and Africa were classified as "low use" regions, Eastern Europe and South America were "moderate use" regions, and Western Europe, North America and Oceania were "high use" regions. Correlating this classification with the region's average per capita GDP, we see a strong, positive significant correlation (Spearman's Rho = .794; $p = .033$). Thus, once again, development is positively correlated with drug use. Whether the analysis is conducted at the national or regional level does not matter; the more developed a nation or region, the higher the rate of drug use.

While there are obvious data limitations that weaken the confidence we have in these analyses, the evidence consistently points in the same direction. Modernization increases drug use. Coupled with the data from Chapter Two that indicates core nations also consume the most pharmaceutical drugs and caffeine, the proposition gains even more support. Thus, at least in a cross-sectional setting, proposition 3.0 is supported. As will be discussed shortly, historical data also supports the claim that modernization increases drug use, at least until modernization results in changes that curtail use. Let us now consider these changes and the longitudinal effect of modernization on drug use.

MODERNIZATION AND DRUG USE RATES OVER TIME

The general pattern has been for modernization to increase the use of drugs as societies transition from pre-modern to modern. The relatively slow increase in use that occurred in the early transition from pre-modern to modern societies hastened dramatically with the coming of the industrial revolution and the full transition to the "modern age." Courtwright (2001: 173) notes industrialization's profound effect on the use of drugs when he states that,

> If the single most important fact about the early modern world was the expansion of oceangoing commerce, the single most important fact about the following era was industrialization. During the nineteenth century psychoactive discoveries and innovations -- the isolation of alkaloids, the invention of hypodermic syringes and safety matches, and the creation of synthetic and semisynthetic drugs -- were married to new techniques of industrial production and distribution. Factories did for drugs what canning did for vegetables. They democratized them. It became easier, cheaper, and faster for the masses to saturate their brains with chemicals, making a lasting impression on their most primitive pleasure and motivational systems. We can see this pattern in numerous cases.

The relationship between industrialization and drug use can be seen in American history, for example. While Americans had been using opiates, tobacco and alcohol since the founding of the nation, drug use began to increase rather dramatically during the industrial revolution. Coffee consumption, for example, increased nearly threefold between 1830 and 1859 (Courtwright 2001: 21). Likewise, opiate addiction began "to grow in the 1870s, peaked in the late 1890s, and declined thereafter" (Morgan 1981: 30; also see Terry and Pellens [1928] 1970; Duster [1970] 1989). Opium smoking increased throughout the United States in the 1880s as imports of smoking opium increased by "more than two and a half times" between the 1870s and mid-1880s (Helmer 1975: 23). Helmer (1975: 28) estimates that the rate of opium smoking among Chinese in the United States nearly doubled between 1870 and 1890. Not only was smoking opium common among the young, single Chinese immigrants, but it also became "an important ritual in the white underworld and (laid) the groundwork for a criminal drug subculture" (Courtwright 2001: 36; also see Morgan 1981; Musto 1987; Morgan 1974). The use of medicinal opium also increased dramatically between 1870 and 1890 as "per capita imports of medicinal opiates doubled" (Courtwright 2001: 36). The use of opiates, in general, whether smoked, eaten or drunk for medicinal purposes or pleasure increased dramatically between 1870 and 1890. Duster ([1970] 1989: 32) estimated that narcotic use was "eight times more prevalent (at the turn of the century) than (in 1970)."

Similarly, cocaine, first introduced in the United States in the 1870s, quickly became popular (see Ashley 1975; Musto 1987; Morgan 1981; Hawdon 1996b). As Morgan (1981: 18) notes, "the number of vague, nervous complaints increased dramatically after the American Civil War" and doctors turned to cocaine to treat these "disorders." Morgan (1981: 18) continues,

> In short order, cocaine became the druggists' stock in trade, available to anyone who had read or heard of its virtues, to be swallowed, inhaled, or injected on demand . . . its overuse was as sudden as its notoriety.

Despite some popular concerns, cocaine use continued to increase throughout the late 19th century, fueled in part by the growing patent medicine industry. Cocaine imports and manufacturing increased rather steadily from the late 1880s through the early 1900s, reaching its peak at 9 tons in 1903 (Spillane 2000: 64).[70] With the increased supply, it is safe to assume that consumption also increased. According to Spillane (2000: 65), "consumption totals for the early 1890s increased by at least 500 percent by the following decade." It is estimated that between 56,000 and 100,000 Americans were chronic users of cocaine and approximately 1 million Americans used cocaine on occasion in the early 1900s (Hawdon 1996b). This figure is approximately 1/3 of the number of American cocaine users than in the early 2000s. Thus, while the number of cocaine users

---

[70.] Imports and manufacturing of cocaine decreased between 1892 and 1893, largely due to the economic "panic" that occurred. They were also relatively low in 1899. Generally speaking, however, the supply of cocaine increased dramatically between 1880 and 1903.

increased by 200 percent between 1900 and 2000, the population increased by over 277 percent. To put this in perspective, "Americans consumed as much coke in 1906 as they did in 1974, and there were less than half as many of them to do it" (Ashley 1975: 64).

America's use of other drugs also increased in the late 19th century. Marijuana, used medicinally in the United States for over 50 years, regained popularity in the 1890s (Morgan 1981; Hawdon 1996b). Once introduced, barbital also became popular in the United States around the turn of the century (Morgan 1981). In general, the use of all types of drugs increased after the American Civil War. This increase rate of use led H. H. Kane to note in 1881 that "a higher degree of civilization . . . seems to have caused the habitual use of narcotics, once a comparatively rare vice among Christian nations, to have become alarmingly common" (cited in Morgan 1981: 46). Thus, America's "first drug epidemic" began during the period of industrialization as Americans were witnessing the effects of modernization on a widespread basis for the first time, and "peaked around the turn of the century" (Musto 1987: 251; also see Morgan 1981; Hawdon 1996b; Jonnes 1996; Ashley 1975; Helmer 1975).

We see a similar pattern of increased use during the industrial revolution in Britain. The "drug boom" occurred in Britain before it did in the United States in part because the industrial revolution occurred in Britain before it did in the United States. It was in the mid-19th century that the English economy was transformed from being based on cotton textile production to the production and exportation of machinery, railroads, and steamships, and London's rise as the world's financial center (Hobsbawm 1968; Chase-Dunn 1998). And, it was in the mid-19th century that drug consumption in Britain increased at the fastest rate. While the per capita import of opium into Britain held relatively steady through the late 18th century and into the early 19th, it began to increase in the 1830s. Between 1831 and 1859, the per capita amount of opium imported into Britain for domestic consumption grew at a rate of just over 2% per year, and it continued to increase at approximately the 2% rate until at least the 1870s (Parssinen 1983:

10). Thus, between 1831-1835 and 1855-1859, the amount of opium imports retained for domestic consumption increased by 70%, once adjusted for the growth in population (see Parssinen 1983: 18). Again, opium was not the only drug that attracted an increasing number of users. Morphine and cocaine use also increased. In 1783, the English started to charge a stamp duty on all patent medicines of approximately 12% of the item's sale price. From 1827 until 1855, revenues increased a mere 40%. Between 1855 and 1905, however, revenues from patent medicine stamp duties increased from less than 50 thousand pounds in 1855 to nearly 325 thousand pounds in 1905. "From 1855 to 1905, the period of greatest growth, sales of patent medicines increased nearly tenfold, while the population just about doubled" (Parssinen 1983: 31). Tobacco consumption also increased in Britain during the middle of the 19th century (see Courtwright 2001), as did the consumption of alcohol (see Shaw 2001).

Thus, in both the United States and Britain, drug use increased rather dramatically once the industrial revolution began to hasten the pace of modernization. Other modernized nations, including Canada, France and Germany, saw similar trends in use. The history of opiate drugs is similar in many nations. For example, the Dutch, like the Americans and British, saw opiate use increase in the mid-19th century, especially among "doctors, chemists, and artists" and those who became addicted using it medicinally (van de Wijngaart 1990: 529). Cocaine use in Canada and continental European nations provides another supporting example. The use of cocaine increased rather dramatically in the early decades of the 20th century in Canada and Western Europe. In Germany, for example, the "cocaine problem" was noted when it was estimated that between 5,000 and 6,000 cocaine users lived in Berlin by the end of World War I (Phillips and Wynne 1980: 83). Use also increased sharply in France, Belgium and Austria (see Phillips and Wynne 1980; Ruggiero and South 1995). It is estimated that there were more habitual users of cocaine and morphine in Paris in 1924 than there were in the early 1990s (Ruggiero and South

19995). In all of these cases, drug use increased as the nations quickly modernized.

Thus, once again, we see modernization leading to increased drug use. Yet, based on available data, drug use was most widespread during the late 19th century and early 20th century. For example, the American "drug epidemic" began to wane in the early part of the 20th century and was fully over by 1920. Using available evidence, I estimate that narcotic drug use in the United States decreased by 52% between 1915 and 1922 (see Hawdon 1996b: 193).[71] Cocaine use also declined and was largely confined to a small segment of the population by 1920 (see Ashley 1975; Musto 1987; Morgan 1981). By 1939, the United States Treasury Department (1939: 14), which had the tendency to overestimate the amount of use, stated that, "the use of cocaine in the illicit traffic continues to be so small as to be without significance." Even the use of alcohol declined in the United States after the passage of national prohibition (see Goode 1999). Thus, alcohol, opiate and cocaine consumption was higher in 1880s and 1890s America than at anytime during the 20th century (see Morgan 1981; Musto 1987; Hawdon 1996b).

The same is true in Britain. As in the United States, England's "drug epidemic" subsided in the late 19th century and had ended by the early 1920s. For example, opium use was heaviest in Britain in the Fens where "by the middle of the nineteenth century, opium had become the all-pervasive drug of choice" (Parssinen 1983: 50). Opiate use continued to increase in the Fens throughout the 1850s, 1860s and 1870s. However, "by the 1890s, there were but a few opium-eaters remaining" (Parssinen 1983: 51). The decrease in manufactured morphia and heroin after 1916 provide additional evidence of declining rates of drug use. Between 1916 and 1921, manufactured opiates decreased by over 80% (see Parssinen 1983: 146). Similarly, the number of cases prosecuted under the

---

[71.] Data are taken from Morgan (1981), Musto (1987), Terry and Pellens ([1928] 1970) and Kolb and Dumez (1924). See Hawdon (1996b: 193) for a more detailed discussion.

Dangerous Drugs Acts during the 1920s reveals a declining drug-using population. While there were between 200 and 300 cases per year between 1921 and 1923, after 1923 there were never more than 100. And, all categories of drugs were affected, including opium, cocaine and morphia.

> One could argue that this decline reflects an addict population which did not change its rate of drug consumption, but which had learned, after a few years, how to avoid arrest. While that may be partially true, the decline is simply too substantial for that explanation to carry full weight. Surely this decline in DDA prosecutions reflects the reality of declining use of narcotic drugs after 1923 (Parssinen (1983: 167).

Even the use of alcohol, Britain's long-time favorite drug, follows a similar pattern. Based on data from the midland city of Lichfield, we can see drug use beginning to decrease in the late 19th century. Lichfield, a Cathedral and former garrison town, is, and has been, known for its pubs. In 1834, there were 72 pubs and approximately 5,000 people in Lichfield (Shaw 2001: 9). That is a ratio of 1 pub for every 69 people. By 1900, however, 52 pubs served 8,000 people, a ratio of 1 pub to 163 people. By 1936, only 46 pubs served the 8,500 residents, a ratio of 1 pub to 184 people. The declining pub-to-person ratio continued through the 20th century so that by the end of the century 30,000 Lichfieldians shared only 26 pubs. To achieve the 1834 "ratio of 1 to 69, Lichfield City area would need about 434 pubs" (Shaw 2001: 9). The history of person-to-pub ratio had a similar trajectory in many British towns, as British alcohol consumption declined rather drastically during the 1920s (Shaw 2001; Courtwright 2001).

Other modernized nations have similar patterns. Rather sharp decreases in use followed steep increases in Canada, France, the Netherlands, Germany and Austria. For example, opiate use decreased in the Netherlands around the turn of the century until "there was little demand for it in the Netherlands" (van de Wijngaart 1990: 529). The use of opiates in the Netherlands did not re-emerge on any large-scale basis until the early 1970s (van de Wijngaart 1990). Cocaine use, which increased in a number of European nations during the early part of the 20th century, decreased considerably beginning in the early 1920s (Phillip and Wynne 1980; Courtwright 2001). Ruggiero and South (1995: 45) note that in Austria, "as elsewhere in Europe, drug problems declined in the early decades of the twentieth century."

Thus, throughout the 19th and early 20th century, the general pattern, especially in those nations that were rapidly undergoing the modernizing effects of industrialization, was for drug use to "explode," slowly stabilize, and then decrease dramatically. The exact time of the "explosion" varied from nation to nation, as did the timing of the industrial revolution in these states. Yet, the period of dramatic decrease was similar, the 1920s. As Musto (1987: 251) claims when discussing this pattern in the United States, "gradually, at times imperceptibly, in the years after World War I, the use of narcotics . . . declined." This decline was evident in industrializing nations everywhere. In addition, none of the modern states have reached their pre-World War I levels of drug use to date.

While the "explosion" of drug use in these nations during and shortly after they experienced industrial revolution supports proposition 3.0, the subsequent declines in use do not. Yet, factors related to the modernization process can also account for this apparent contradiction.

MODERNIZATION AND INCREASED FORMAL RESTRICTIONS ON USE

As discussed in Chapter Four, modernization alters the nature of social control. While informal sources of social control weaken, more formal sources replace them. It was during the late 19th century that a number of forces, all brought about by the modernization process, converged and fanned the passage of anti-drug laws. While numerous pre-modern societies had laws restricting drug use,[72] or taxes on drugs that effectively limited use, the extent of this legislation

---

[72] For example, an Imperial edict outlawed the opium trade in China in 1729.

pale in comparison to the barrage of legal sanctions concerning drugs that were passed in the early 20th century (see, for example, Musto 1987; Duster 1989). While it may be tempting to attribute the emergence of drug laws in the late 19th and early 20th centuries to the increased rates of use witnessed in numerous societies at the time, as Savelsberg (1994) argues, the rise of an issues such as drugs as a political issue cannot be accounted for solely by increasing rates of use or increased public concern. Instead, the manner in which public and political knowledge is created shapes these issues. The relatively free exchange of knowledge between public, private, and academic sources that began to emerge as western nations modernized tend to create conservative macro-level punishment decisions (Savelsberg 1994, also see Iyengar 1991). This is exactly what occurred in the late 19th century. By that time, the modernization process reached the point where a growing anti-drug public sentiment was widely publicized and concerns of industrialists about economic productivity were firmly crystallized. Added to an increasingly complex international system, the forces of modernization created a wave of anti-drug legislation in many nations and at the international level.[73]

First, increased modernization promotes democratic institutions and, therefore, the common public's participation in political life (see, for example, Parsons 1977). In addition, due to improvements in literacy and mass printing techniques, the mass media in the form of the popular press was influencing the public's opinion. In the United States, Great Britain and elsewhere, the temperance movements, which had begun in the earlier decades of the 19th century, joined forces with anti-opium advocates, physicians and missionaries in a crusade against inebriation. Using the popular press and other media outlets, including works by literary greats such Charles Dickens, Arthur Conan Doyle and Oscar Wilde, the anti-drug message played on the populations' suspicions of Chinese coolies, other immigrants, or other relatively powerless groups. These

---

[73] Numerous authors discuss all of these "forces" in detail. I therefore provide only a brief summary of these. Interested readers should see the works of Helmer, Musto, Morgan, Courtwright, Parssinen, and Duster cited throughout the next few paragraphs.

campaigns equated drug use to crime, laziness, slothfulness, backwardness and other individual ills and failings. It was not coincidental that the first laws banning drug use in the United States, Britain, Australia and Japan outlawed smoking opium (see Helmer 1975; Parssinen 1983; Greberman et al. 1994; Musto 1987; Morgan 1981; Courtwright 2001). While opium drinkers and eaters were disproportionately Anglo, middle to upper class women, opium smokers were disproportionately Chinese, lower class males (see Morgan 1981; Helmer 1975). Other industrialized states of the time observed similar patterns. In Australia, where narcotic-laced patent medicines were consumed heavily, laws banning opium smoking were passed around the turn of the century. The difference between consuming opium through patent medicines and through smoking "was the association of opium smoking with coolies, consensus that it lacked therapeutic value, and fear that it would spread beyond the Chinese" (Courtwright 2001: 177).

Next, the industrial revolution elevated the relative power of the industrialist classes in every modernizing nation, and these wealthy industrialists were becoming increasingly concerned that workers who were drunk or using opiates were not producing at desired levels. Drug use threatened the industrial process itself, or so it was argued. Industrialists went to great lengths to try to establish a sober, diligent lifestyle and a drug-free workplace. For example, in Holumsund Sweden, the local sawmill company granted land, money and building materials to construct the local Good Templar Lodge so workers could have an alcohol-free recreation center. Similarly, Massachusetts manufactures banned drinking on their premises (Courtwright 2001). Other industrialists went further. Henry Ford sent his "Sociology Department" agents to workers homes to determine if alcohol abuse -- evidenced by a dirty home and unkempt children -- was occurring. Industrialists, and many labor leaders, believed that drug use and industrialization simply did not mix. As Courtwright (2001: 178 - 179) states,

As the social environment changed, becoming more rationalized, bureaucratized, and mechanized, the distribution of cheap intoxicants became more troublesome and divisive. A drunken field hand was one thing, a drunken railroad brakeman quite another. While the consumption of drugs might keep workers on a treadmill, over time and in certain industrial contexts it rendered their labor worse than useless. The growing cost of the abuse of manufactured drugs turned out to be a fundamental contradiction of capitalism itself.

Finally, in an attempt to improve their strained relations with China and promote its *Open Door Policy* there, the American government joined the Chinese in calling for international regulations on opium trafficking. At the request of the Americans, an international conference convened in Shanghai in 1909 where the major powers of the day, including the United States, Great Britain, France, Germany, Italy, Russia, China and Japan, met to discuss restricting the international trafficking in opiates and cocaine. The goal was to limit the trade and distribution of these drugs to medicinal purposes only. While nothing formal was accomplished at the 1909 convention, three additional international conferences were held at The Hague in 1911-1912, 1913 and 1914. In January of 1912, the *International Opium Convention* was signed and circulated for ratification, and by the spring of 1914, most major powers had ratified it. The international obligations outlined in that document strongly influenced the national drug laws of the powers who ratified the Convention. While World War I interrupted many national attempts to match domestic policies with the international treaty, Article 295 of the *Treaty of Versailles* committed its signatories to honor the *Opium Convention* (see, Helmer 1975; Morgan 1981; Musto 1987; Duster [1970] 1989; Parssinen 1983; Szasz 1974; Courtwright 2001).

The result of these forces -- public concern, industrialists' concerns, and international concerns -- was a series of drug laws in every major power of the time. In the United States, for example, the first major federal anti-drug legislation was approved on December 17th, 1914.[74] The Harrison Act of 1914, which governed all U.S. federal legislation until 1970, required those persons involved in the production and distribution of narcotic drugs -- which now included cocaine – to register with the federal government. It also require those persons buying or selling narcotics to pay a tax. Moreover, those who were not registered could purchase drugs only with a prescription and only for legitimate medical use. Similarly, England, the first European nation to pass legislation that controlled the use and distribution of drugs (Phillips and Wynne 1980), enacted Regulation 40B, the Defense of the Realm Act (DORA 40B) in May of 1916 (Parssinen 1983; Ruggiero and South 1995). Like the Harrison Act, DORA 40B required importers to be licensed and records of cocaine and opium sales to be maintained.[75] In addition, opium and cocaine were legally available only through non-repeatable prescriptions and prepared opium for smoking was totally prohibited. These wartime regulations were made permanent by the adoption of the Dangerous Drugs Act of 1920, and morphia and heroin were added to the controls originally outlined in DORA 40B. As the Harrison Act did for the Americans, The Dangerous Drug Act brought Britain's domestic policy in line

---

[74.] The Pure Food and Drug Act (1906) and the Native Races Act (1909) were actually the first federal laws that regulated drugs in the United States. The Pure Food and Drug act simply required all ingredients of products sold for human consumption to be labeled clearly. The Native Races Act outlawed opium smoking and the distribution of intoxicants to "primitive peoples." In addition, there were state laws that regulated drugs prior to the passage of the Harrison Act. While there was no significant legislation that regulated the distribution of drugs prior to 1887, Oregon passed the first state law, which regulated cocaine, in 1887. Soon after, numerous states followed Oregon's lead. Between 1887 and 1914, forty-six states passed legislation regulating the use, sale and distribution of cocaine (Phillips and Wynne 1980: 296). By 1931, every state had legislation that restricted the sale of cocaine and thirty-six prohibited the unauthorized possession of cocaine (McLaughlin 1973). Similarly, between 1887 and 1914, twenty-nine states passed laws in an attempt to control the distribution of opiate drugs. Most of these laws banned smoking opium (see Duster [1970] 1989; Musto 1987; Helmer 1975). New York state passed the first comprehensive piece of legislation that attempted to limit the distribution of narcotics, including medicinal narcotics, in 1904 (Duster [1970] 1989: 36).

[75.] The regulation covered all opium products. Products containing 0.1 percent cocaine or heroin were also included.

with its international obligations (see Parssinen 1983: 132 - 133). The other major powers, and most of the minor powers, adopted similar restrictions.

Therefore, both domestic and international concerns led to drug legislation in the United States, Britain, Germany, France, Australia, Italy and other world powers of the time. In addition, generally speaking, these laws were passed in the 1910s and 1920s. However, we must realize that the process of modernization united the forces that led to regulation. Modernization changed the manner in which public and political knowledge shaped our collective understanding of the issue. Thus, while the tactic of linking drug use to a relatively powerless minority group who could be easily demonized had been used before -- and would be repeated often during the history of anti-drug laws (see, for example, Helmer 1975) -- it was not until the modernization process increased media circulation and literacy that the message could reach a substantial number of citizens. And, it was not until the modernization process democratized the political realm that the public's opinion mattered. Similarly, modernization changed the nature and setting of work, thereby increasing the demands for a sober workforce. Finally, modernization "shrank" the international world and dramatically increased the linkage between domestic and international relations (see, for example, Hawdon 1996c). Morgan nicely summarizes how modernization led to attempts at regulating drug use in the United States. Morgan (1981: 88) states that,

> As the new century dawned, reformers set out to curb the power of business to affect individual lives and the larger society. They were equally determined to make politics more responsive to the popular will. And while they were as individualistic as their fathers or grandfathers, their definition of freedom changed along with the complex, interdependent society that competition ironically had produced. Personal actions that seemed to affect social institutions or stability faced the test of popular opinion. What people did as individuals now affected others in ways undreamed of only a generation before. A new powerful urge to 'purify' American life thus accompanied reforms in business and politics. This involved struggles against prostitution, alcohol, and drugs as much as it did efforts to improve the lot of women, children, and the poor.

Similarly, Courtwright (2001: 186) says that legislation, both at the national and international level, "was a manifestation of modernity itself." Stricter regulations seemed inevitable as the world became more industrialized and informed.

The laws passed in the United States, Britain and other nations helped reverse the general tendency for modernization to increase drug use.[76] Use decreased everywhere after these laws passed. Law, as a means of social control, does influence peoples' abilities and decisions to behave in certain manners. While it is true that laws, being negative in nature, do not restrict drug use as effectively as sacred rites do *once use occurs*, they do determine *if drug use occurs*. That is, while those who decide to ignore legal dictates that prohibit drugs are unregulated once they use the drug, the legal dictates tend to deter many people from ever using the drug in the first place. A cocaine user, for example, is obviously unconcerned with the laws that regulate cocaine use, but many people refrain from indulging in cocaine use simply because it is illegal. Simply put, most people obey the law most of the time. Therefore, legal drugs are more widely used than illegal drugs, all else being equal. World-wide and within nations, rates of marijuana, cocaine, LSD, heroin and other illicit substance use do not come close to those of alcohol, tobacco or caffeine, and this pattern is largely due to the respective legal status of these drugs. Thus, once modernization advanced sufficiently to lead to the widespread public concern and passage of anti-drug laws, use, which had increased considerably in the earlier stages of the modernization process, began to subside.

Consequently, while modernization increases drug use in a cross-sectional setting, the longitudinal relationship between modernization and drug use is curvilinear. As modernization increases, drug use increases, to a point. Once modernization transforms domestic and international politics, social control, and

---

[76] I emphasize that these laws were not the "cause" of the decrease in use. In most nations, use began to decrease *before* the passage of these laws. I will address the "causes" of fluctuations in use rates in Chapter Seven.

economics in a manner that is not overly conducive to excessive drug use, nations enact drug laws that curtail the use of those substances deemed illegal. Therefore, conditional statements must be added to proposition 4.0, and proposition 4.1 can be stated as,

(4.1) Drug use varies directly with modernization, at any given historical point.

Proposition 4.2 can be stated as,

(4.2) Over the time within a given nation, the relationship between drug use and modernization is curvilinear.

ILLEGAL USE

Related to the above discussion, we can also state proposition 4.3 as

(4.3) Illegal drug use varies directly with modernization

This change follows simply from some of the changes already discussed. First, modernization leads to increased use. Second, modernization leads to more laws regarding use. It therefore follows that in all likelihood modernization would lead to an increase in *illegal drug use*. While modernization increases the use of legal drugs also, most of the evidence presented above indicates a direct relationship between modernization and illicit drug use. Moreover, using available UN data (UNODC 1999), there are moderately strong, direct relationships between GDP per capita and the most common forms of illicit drug use among youth. For example, the correlation between GDP per capita and percent of the youth population[77] who uses marijuana is .414 ($p < .003$; $n = 49$). Similarly, the correlation between GDP per capita and percent of the youth

---

[77] "Youth" are persons less than 18 years of age in most reporting nations (UNODC 1999).

population who uses ecstasy is .423 (p = .028; n = 27). Thus, for both of these popular recreational drugs, as GDP per capita increases, youthful drug use increases. While I recognize that these correlations do not provide a "critical test" of the theoretic propositions, the evidence is, nevertheless, suggestive.[78] Moreover, case-study information also increases our confidence in the proposition that modernization increases illicit drug use. For example, in a study of Zimbabwean school children, the stronger the child's western cultural orientation, the greater her or his use of cannabis and inhalants (Eide and Acuda 1997). Similar results have been found in India and Bangladesh (see Emdad-ul 2000).

WHO USES

Modernization also alters the social characteristics that distinguish drug users from non-users. Proposition 5.0 can be stated as,

(5.0) The greater the level of modernization, drug use will be more equally distributed among the various sub-groups of a society.

As discussed previously, the onset of the modern age marked the reversal of the general tendency for levels of stratification to increase (see Lenski 1966). First, the greater surplus produced by modern means of production allows for a greater re-distribution of wealth and therefore less overall stratification (Lenski 1966). Thus, modern societies are less stratified than less-modernized societies. We therefore would expect drug use to be more evenly distributed across the various sub-groups of a society in modern societies than in lesser-developed

---

[78.] I realize that many nations do not report youthful drug use. The sample size, even for marijuana use, indicates that, in fact, most nations do not report. This finding, however, is still suggestive support for the proposition. First, there is a broad range of nations that do report (GDP per capita ranges from $1,162 to $32,211). Second, even relatively highly developed nations would vary with respect to rationalization as measured through GDP per capita. Thus, while the findings cannot be confidently generalized much past the existing sample, among those nations that do report, modernization, as measured herein, is positively related to the use of drugs to meet earthly, concrete, practical functions such as recreation.

societies. That is, while differences may still exist, the distribution of users across gender, age, ethnic and class lines will be less pronounced in modern societies than in pre-modern societies.

We know that drug use was highly stratified in many pre-modern societies, most often by gender and age. For example, only adults participated in the Cherokee's Black Drink ritual during the Bouncing Bush festival (Mooney 1891; French 2000). Similarly, the use of peyote by the Huichol Indians as rite of passage is limited to adult males (Furst 1976; French 2000). Adult males among the Siona, a western Tukanoan group living along the Putumayo River, are the primary users of yage. While women use yage on occasion, they are discouraged from drinking large amounts on a regular and intense basis (Langdon 1992b). In most shamanistic societies, drug use was limited to shamans and, on occasion, those participating in shamanistic rituals (see Langdon and Baer 1992; Lyon 1998; Pollock 1992). Shamans, the primary users of drugs in such societies, were typically either men or women. Very few shamanistic societies had similar numbers of male and female shamans (see, for example, Vitebsky 2001).

More recently, in lesser-developed, or peripheral, nations, drug use remains highly gendered and limited to adults. For example, at least until the early 1980s, "in Cambodia, it is essentially the males who smoke Kanhcha" (marijuana) (Martin 1975: 68). Similarly, marijuana use was "almost entirely confined to the male sex" in West Africa (Du Toit 1975: 101) and North Africa (see, for example, Joseph 1975). Until recently in Pakistan, "the use of cannabis (was) equally prevalent in both urban and rural areas and (was) virtually confined to adult males" (Khan, Abbas and Jensen 1975: 346). Similarly, in Brazil, marijuana users were lower-class males and a few extremely rich women and prostitutes until at least the 1970s (see Hutchison 1975). In other societies, ethnicity as well as gender separated drug users from non-users. For example, in Rwanda, the use of marijuana was almost entirely confined to male Twa. While some Twa women used marijuana, they did so differently than Twa men did and rarely used it in a manner that caused intoxication. Among the Rwandan Hutu and Tutsi, very few used marijuana (Codere 1975). The majority of working-class, adult males smoke marijuana in contemporary Jamaica. While some women smoke, they do so at much lower rates (Comitas 1975; Dreher 1982). In the more traditional rural communities of China, opium users are predominantly older males (OGD 1997). Qat chewing in Yemen and throughout the Middle East is highly gendered because custom prohibits females from using the drug (Weir 1985).

Historical evidence from core societies also suggests that drug use was more stratified prior to these societies becoming more modernized. For example, in Australia, aborigines were once prohibited from buying alcohol or entering pubs (Sargent 1990). Women were discouraged from drinking alcohol or smoking tobacco in pre-industrial Great Britain, America, and most European nations. Instead, they preferred opiate drugs and most opiate users were disproportionately middle-class women (see, Earle 1880; Mattison 1883; Parssinen 1983; Morgan 1974; Morgan 1981; Musto 1987; Courtwright 2001). As reported by Morgan (1981: 40), a doctor noted in 1891 that,

> a women is very degraded before she will consent to display drunkenness to mankind; whereas, she can obtain equally if not more pleasurable feelings with opiates, and not disgrace herself before the world.

Males, for their part, were the primary consumers of alcohol, choral hydrate and cocaine during the late 19th century (Morgan 1981; Musto 1987; Spillane 2000). As noted by several observers of the day, "as a rule, women take opiates and men alcohol" (quoted in Morgan 1981: 40). The drug-use history of Japan also supports our claim. During Japan's first amphetamine epidemic in the late 1940s and early 1950s, there were nine male methamphetamine users for every one female user. However, during the early 1980s, estimates held that male amphetamine users outnumbered females by only five to one. Recent estimates indicate a slightly higher proportion of women are now using methamphetamine (Greberman et al 1994).

Finally, social standing or class often limited use in pre-modern societies. At times, such limits were formalized as in the Inca Empire where coca use was limited to the ruling class (Phillips and Wynne 1980). At other times, simple economics prohibited the common masses from affording drugs. For example, alcohol was simply too expensive for the Rwandan Twa to afford (Codere 1975). Similarly, it was not until the Europeans transformed and modernized the production of coffee, tea and tobacco that prices were reduced enough for the common person to afford those drugs (see, especially, Courtwright 2001). As Courtwright (2001) argues, industrialization "democratizes" drug use by increasing the efficiency of drug production, thereby lowering prices.

Thus, as societies modernize, drug use becomes more affordable and therefore more attainable for the common masses. Moreover, with the greater rationalization of society, old customs that divide gender and ethnic groups begin to erode. We can see the transition in numerous societies as they modernize. For example, throughout southern Africa in the 1970s, there were two basic forms of marijuana use. In the first, found in most rural areas where use was common, use was limited to "mostly older men." However, in the urban areas that would be most modernized, "young people, males and females" used it (Du Toit 1975: 106). Similarly, while males are still more likely to use illegal drugs in most South East Asian nations, gender differences appear to be disappearing in many of these countries, at least in the urban areas (see Emdad-ul 2000).

In core societies, we see a greater "leveling" of drug use today. Although certain drugs are more common among one group than another, and, with a few exceptions, males tend to use more drugs than females, drug use is more equally spread across gender, ethnic and class lines than in less-modernized societies. For example, among British adolescents, males are more likely to use drugs; however, the differences between male and female use patterns are relatively minor (see Parker et al 1998; Measham 2002). There were virtually no differences in use patterns between various social classes or different ethnic groups (see Parker et al 1998: 87). The same is true in Japan and many Western European nations (see

Greberman et al 1994; Vaughn et al. 1995; OGD 1997; Mayock 2002). In the United States, in the intensive study of illegal drug users in Florida referenced before, gender differences for current and lifetime drug use were slight (Inciardi et al 1993: 81). Similarly, ethnic and age differences, while slightly stronger than the gender differences, were small (see Inciardi et al 1993). As the authors note (Inciardi et al 1993: 82), "the most striking aspect of the gender, age, and ethnicity analyses of current drug use is the small number and size of the apparent differences among the sociodemographic subsamples."

While the users in the Florida study were all "serious delinquents," similar patterns -- or the lack thereof -- hold in the general population. In a representative sample of youth from New York State, the prevalence of the use of alcohol and marijuana together in the past 6 months was similar in most gender, grade level, and racial/ethnic subgroups (Hoffman et al. 2000). Similarly, in a study of seventy-one adolescents in alcohol rehabilitation clinic, males and females did not differ in the percentage of subjects who used different drug classes or in the severity of their involvement with drug classes (Martin et al. 1993). More generally, based on 2001 data from the *National Household Survey on Drug Abuse* (SAMHSA 2003), there are few differences in monthly drug use based on gender or race. For example, while males are more likely to use cigarettes (27.1% versus 23.0% of females), marijuana (7.0% versus 3.8%), cocaine (1.0% versus 0.5%), hallucinogens (0.8% versus 0.4%) and "any illicit drug" (8.6% versus 5.5%), these differences are relatively small. Males and females have similar rates of monthly use -- within 0.1% of each other -- for crack cocaine, heroin, LSD, ecstasy, analgesics, tranquilizers, stimulants, methamphetamine, sedatives, PCP and any non-medical prescription drug (see SAMHSA 2003). Thus, while some gender differences exist, these are relatively small. Moreover, based on recent evidence, those existing gender differences in drug use are becoming even smaller. Gender differences in drug use among American youths ages 12 - 17 are much smaller than those of older users (SAMHSA 2002), possibly reflecting the coming of an even more egalitarian division of use in the near future.[79]

In addition to being few gender differences among American drug users, few differences exist between Anglo and African Americans with respect to their drug using patterns. These groups do not differ substantially (again, by less than 0.1%) in the monthly use of cocaine, heroin, LSD, analgesics, sedatives, PCP or "any illicit drug." Small differences exist in the monthly use of cigarettes (26.1% of whites versus 23.0% of blacks), crack (0.1% versus 0.4%), ecstasy (0.4% versus 0.2%), hallucinogens (0.6% versus 0.3%), inhalants (0.3% versus 0.1%), tranquilizers (0.7% versus 0.1%), stimulants (0.5% versus 0.1%), methamphetamine (0.3% versus 0.02%) and non-prescribed medicines (2.2% versus 1.6%). While there are differences, these are, once again, slight. Similarly, Hispanics and Anglo-Americans have very similar use patterns (see SAMHSA 2003). The biggest difference in drug use among ethnic groups in the United States is with the use of alcohol. While 52.7% of whites consumed an alcoholic beverage during the past month, only 35.1% of blacks and 39.5% of Hispanics did so. Thus, generally speaking, the differences, with the exception of alcohol use, are relatively small.[80] Moreover, these differences could very well be due to differences in class and educational levels. For example, when a multivariate logistic regression analysis is performed that predicts the use of any illicit drug in the past month, differences between Anglo, Hispanic and African-

---

[79] A notable exception to this general pattern occurs with the use of alcohol. Males consume significantly more alcohol than women do (54.8% of males versus 42.3% of females use alcohol monthly). In addition, Anglos are significantly more likely to have consumed an alcoholic beverage than ethnic minorities. However, this exception not withstanding, most differences are either very small or non-existent.

[80] Differences between these groups and other ethnic groups in the United States are more pronounced. For example, the rate of monthly use of any illicitsubstance is 10.5% for Native Americans, but only 2.9% of Asian Americans. Non-Hispanics of multiple races have the highest rate at 12.6% (see SAMHSA 2003).

Americans become insignificant once levels of education, income, work status, and age are controlled for.

Hence, as modernization occurs, we see fewer distinctions between groups in terms of who uses drugs and who does not. This is not to imply that no distinctions exist, because some remain. However, when compared to less-modernized societies, both cross-culturally and historically, modernized societies are more egalitarian with respect to intoxication. In general, the greater the level of modernization, the fewer distinctions in drug use between groups, and most of those differences that remain weaken in importance.

ASCRIBED VERSUS ACHIEVED STATUSES AND DRUG USE

As just noted, there are fewer distinctions between groups with respect to drug use in modernized societies than in pre-modern societies. That is, the correlations between drug use and broad demographic characteristics such as gender, age and class are not as strong in modern societies as in pre-modern societies. Yet, even in the most modernized societies, there are some differences between users and non-users. However, the nature of the distinction changes with modernization. As in other aspects of stratification, ascribed characteristics become less important than achieved characteristics in determining the allocation of resources (see Lenski 1966; Parsons 1977). The more fluid relations and greater economic diversification found in modern societies allows greater rates of social mobility among the various structural groups of a society. As social mobility increases, the strength of the structural parameters in determining who associates with whom, and who gets what, weakens (see Blau 1977). This statement does not imply that all differences between ascribed groups disappear in modernized societies. Any astute observer can attest to the difference in life chances and subsequent life choices between men and women and among various ethnic groups in most modernized societies. However, it is also undeniable that these differences have declined, and many that remain are due to the correlation between achieved statuses, such as education, income and occupation, and those ascribed characteristics. While discrimination undoubtedly still occurs, and is most likely the reason the ascribed characteristics correlate with the achieved characteristics, modernized societies legally prohibit discriminatory practices based on ascribed statuses. This was certainly not the case in pre-modern societies. Because of these changes, we can state proposition 6.0 as,

(6.0) The greater the level of modernization, the more drug use patterns are correlated with achieved statuses rather than ascribed statuses.

As the historical data noted above shows, drug use was highly structured by gender, age and ethnicity in many pre-modern societies. Today, using a cross-cultural comparison, the general pattern of the declining importance of ascribed characteristics in determining which groups use drugs as modernization occurs is evident in rates of tobacco use by gender in developed and lesser-developed nations. Tobacco use is highly gendered in the developing world.[81] According to WHO (2000), 700 million males smoke in developing nations while only 100 million females do. In India, for example, while nearly 65% of adult males smoke, only 3% of adult females do so (WHO 1999). Similarly, in South Korea, 68% of males and 7% of females smoke (Levinthal 2002: 251; WHO 1999). In Sri Lanka, the gender gap is even more pronounced. While 54% of Sri Lankan males smoke, less than 1% of females smoke (WHO 1999). These gender gaps in smoking are likely to continue in many developing countries.[82] For example, while smoking rates among male Korean teenagers was 30% in 1993, only 9% of

---

[81] Nepal may be an exception to the general pattern of large gender "smoking-gaps" in developing countries. In rural Nepal, it was estimated in the early 1980s that nearly 85% of males and 72% of females smoked tobacco. However, in urban Nepal, nearly 65% of males smoke while only 14% of females do so (WHO 1999).

[82] Bangladesh may be an exception to this trend. While over 60% of men and only 15% of women there smoke, females who smoke increased from 1% to 15% in the early 1990s (WHO 1999). Since male rates have remained stable, the gender smoking-gap in Bangladesh appears to be decreasing.

Korean female teenagers smoked. In Uzbekistan, 22.5% of teenage males smoke, but less than 1% of teenage females smoke (WHO/EURO 1997). In Malawi, nearly 20% of men but only 2% of women smoke (Davies 2003). Similarly, among Nigerian secondary school children in Lagos, 40% of the boys and 8.4% of the girls smoked tobacco (Gureje and Olley 1992). Since the majority of smokers begin while in their teenage years or earlier (DiFranza and Tye 1995; Warren et al 2000; WHO/EURO 1997), these numbers indicate that smoking in developing nations will likely continue to be a male-dominated pastime for the foreseeable future.

The story is much different in the developed world where the gender gap tends to be much smaller. In total, 200 million males and 100 million females are regular smokers in developed countries (WHO 2000). The gender gap in smoking is quite narrow in some nations. In Australia, for example, 29% of men and 21% of women smoke. In the U.S., 28% of men and 24% of women are regular smokers. A similar ratio is seen in the United Kingdom. In Sweden, females are actually more likely to smoke: 24% of females and 22% of male Swedes smoke regularly (Levinthal 2002: 251). These relatively small gender gaps will likely reduce more in the near future. In some nations, they may even reverse. According to WHO's European regional office's *Health Behavior in School-Aged Children* survey, young women are smoking more than young men in eight of the twenty surveyed countries. Fifteen-year-old women were more likely than their male counterparts to smoke in Austria, Denmark, France, Germany, Norway, Spain, Sweden, and the United Kingdom. The ratio was almost 1 to 1 in Greenland, Belgium, Finland, and Israel. Only in the lesser-developed nations of the former eastern bloc (Estonia, the Czech Republic, Hungary, Latvia, Lithuania, Poland, the Russian Federation, and Slovakia) did the gender ratios in smoking among fifteen-year-olds heavily favor males (WHO/EURO 1997).

Based on available data on smoking by gender for 35 nations and data concerning the status of women in those nations, there is a strong, inverse relationship between the status of women and the ratio of male smokers to female smokers.[83] For those nations for which data were available, the Spearman's rho correlation coefficient was -.717 (p < .001).[84] Thus, the greater the status of women in a country, the more equal is the number of male smokers relative to female smokers. As with all statistical findings based on international drug use data, this finding should be considered with caution. However, the strength of the correlation is telling, even keeping the customary cautions in mind. Since the status of women and modernization correlate positively (see, for example, Estes 1988), we can deduce from the above evidence that modernization reduces the gender gap in the use of tobacco.

We can also see the shift from drug use based on ascribed statuses to drug use based on achieved statuses in the histories of both the United States and Great Britain. As stated earlier, in the late 19th century, opiate use was most common among middle-class, middle-aged women (see Earle 1880; Morgan 1974; Morgan 1981; Musto 1987). For example, in a study based on information from 50 Chicago drug stores in the 1870s, 71.9% of habitual opium-eaters were female and 60.3% were between the ages of 30 and 50 (Earle 1880). Moreover, use was relatively widespread throughout the middle classes (Earle 1880; Mattison 1883). However, by the early decades of the 20th century, opiate use, as well as the use of cocaine, was "identified with wayward youth, irresponsibility, and crime"

[83.] The data for smoking were derived from the following sources: Levinthal 2002; WHO/EURO 1997; WHO 2000; Gureje and Olley 1992; Davies 2003. The groups in each of the nations are not necessarily equivalent. That is, some of the nations report the use of cigarettes among adults over the age of 12, while others report teenage smoking rates. While this is not an ideal situation, this inconsistency should not overly influence the basic point that the ratio of male to female smokers is higher in lesser-developed nations. Again, smoking is a habit that tends to be developed at young ages and, once developed, remains through the life course (however short that may be for smokers). Thus, the general pattern should hold regardless of age groups compared. I used Richard Estes' work concerning the social development of nations for the data concerning the status of women. The women's sub-index is a combined measure of percentage of age-eligible girls attending first-level schools minus the percentage of adult female illiteracy plus the age of the constitutional document or most recent amendment affecting the legal rights of women. I updated his measures to reflect the status of women in 1997 (see Estes 1988: 191).

[84.] The Pearson's correlation is similarly strong and inverse at -.696 (p < .001).

(Morgan 1981: 96). By mid-century, heroin users, and most users of illicit drugs, were typically poor and lived in urban areas (see Morgan 1981; Musto 1987; Parssinen 1983). Thus, between the late decades of the 19th century to the early decades of the 20th century, the characteristics of American and British opiate users changed rather dramatically. Use went from being rather widespread and correlated with gender to being relatively rare, weakly correlated with gender, but strongly correlated with social class. Similarly, in Great Britain, cocaine, when first popularized in the 1880s, was widely available through patent medicines and used by all social classes. By the 1920s, however, most cocaine users were either members of the "criminal classes," lower-class males or prostitutes. As Parssinen (1983: 173) notes, "cocaine offenders were much more likely to have a working-class occupation, or no occupation at all, than were other drug offenders." Finally, tobacco use provides another example, especially in the United States. Cigarettes were widely used by members of all social classes through the first half of the 20th century, and gender was the primary social characteristic that separated smokers from non-smokers. Even as late as 1982, nearly 40% of American males smoked regularly while only 30% of females did so (SAMHSA 2003). By the end of the century, however, males and females have nearly identical rates of tobacco use (28% of males and 24% of females), but income and education have become reliable predictors of use.

We can also see the process of achieved characteristics growing in importance with modernization in areas of other cultures that are rapidly developing. Namely, the process can be seen in the increasing division between rural and urban users in many developing societies. In urban areas of developing nations, where modernization is most evident, social class is critically important. In the rural areas, more traditional divisions based on gender and age tend to be of primary importance. In the Ukraine, for example, were opiate use had been limited to rural adult males, a recent study found that in the urban centers, social class was a strong predictor of use. According to the study, 65% of users were unemployed and an additional 21% were manual workers. Only 5% of users were college students or graduates (see, OGD 1997). Similarly, in the rapidly modernizing cities of China, drug users are found on either end of the income spectrum. Methamphetamine users tend to be "middle class youths who have more disposable income" while most Chinese heroin users are young, single, poorly educated and unemployed or underemployed (Kurlantzick 2002: 73). Social class plays a much smaller role in rural China were use is largely a habit of older males (OGD 1997; Kurlantzick 2002). In St Petersburg, Russia, the use of hallucinogenic mushrooms is popular among wealthy university students (OGD 1997). In Israel, heavy alcohol drinking is most commonly found among unemployed immigrants (Toch 1990), and illicit drug use is limited to either traditional users of hashish, who tend to be older males, and "Jewish criminal groups, Bohemia, and young people" (Javetz and Shuval. 1990: 428), all of whom are relatively poor. In Brazil, while cannabis use traditionally had been a habit of males, by the 1970s, use correlated with income as most users were from the lower social classes (Hutchinson 1975). In Pakistan, heroin use is strongly correlated with social class. Over half of Pakistan's heroin addicts come from among the poorest people in the country (Courtwright 2001).

Both history and contemporary patterns of use in developing nations highlight the change in importance of achieved relative to ascribed statuses in predicting who uses drugs. In core nations, however, we see two trends merge. In core nations, there are relatively few distinctions among various groups in terms of their drug use, as proposition 5.0 maintains. However, those distinctions that remain are based on achieved characteristics such as social class or education. We would therefore expect correlations between any demographic variable and use to be relatively weak in core nations. The strongest correlations, however, would be found between drug use and achieved statuses.

We can see this pattern using data from the *National Household Survey on Drug Abuse* from 1979 through 2002 and correlating gender and drug use and ethnicity and drug use over time.[85] For example, while the correlation between female and monthly marijuana use was (-.14) in 1979, it was only (-.07) in 2001. Thus, while males were more likely than females to use marijuana monthly in both years, the 1979 difference between male and female use was twice that of the 2001 difference.[86] Correlating these coefficients with "time" indicates that the differences in drug use between males and females and between whites and blacks have narrowed over time in the United States. Looking at the monthly use of cigarettes and marijuana and the yearly use of cocaine and "any illicit drug," there has been fairly consistent decreases in the strength of the correlations between the use of these drugs and gender. The correlation with these coefficients and time were (-.853) for cigarettes, (-.827) for marijuana, (-.644) for cocaine, and (-.661) for any illicit substance. Ethnicity follows a similar pattern. The correlations of the ethnicity–drug coefficients with time were (-.779) for cigarettes, (-.649) for marijuana, (-.941) for cocaine, and (-.939) for any illicit substance. Even the strength of the correlations between age and the use of these drugs decreased between 1979 and 2001. The correlations of the age–drug coefficients with time were (-.684) for cigarettes, (-.759) for marijuana, (-.884) for cocaine, and (-.941) for any illicit substance.[87] Thus, ascribed statuses play an increasingly minor role in determining who uses drugs in modernized societies.

[85.] Surveys were conducted in 1979, 1982, 1985, 1988, and 1990 - 2002 (see SAMHSA 2003). The analysis is based on data from the 1979, 1982, 1985, 1988, 1991, 1994, 1996, 1998, 2000 and 2001 surveys.

[86.] For gender, female was coded as "0" and male was coded as "1," thus the negative coefficient reflects that fewer women used marijuana monthly than did men. For ethnicity, the analysis was limited to whites (coded as 1) and blacks (coded as 0).

[87.] For gender, the correlations for cigarettes and time and marijuana and time were significant at the $p < .01$ level. The correlations for cocaine and any illicit substance with time were significant at the $p < .05$ level. For ethnicity, the marijuana–time correlation was significant at the .05 level, all of the other correlations were significant at the .01 level. For age, the cocaine and "any illicit drug" with time correlations were significant at the .01 level. The remaining two were significant at the .05 level.

Conversely, while the strength of the correlations between gender, ethnicity, age and drug use decreased between 1979 and 2001 in the *National Household Survey on Drug Abuse* data, they strengthened between income and drug use during this time. The correlations between the coefficients and time for income and the monthly use of cigarettes and marijuana and the annual use of cocaine and any illicit drug, were *positive* (.803), (.763), (.422) and (.110), respectively. While the last correlation was not statistically significant at conventional levels, the direction of the relationship still fits the general pattern. Thus, between 1979 and 2001, ascribed statuses played less of a role in determining the patterns of drug consumption in the United States and achieved statuses played an increasingly important role. Moreover, the strongest correlations between the use of any drug on a monthly bases and any demographic variable, including both achieved and ascribed characteristics, are found between the use of alcohol and income (r = .19) and education (r = .31). Other correlations that are moderately strong are between the use of any illicit drug and education (r = .17), any illicit drug and income (r = .12), education and marijuana use (r = .14) and income and marijuana use (r = .13) (see SAMHSA 2003). And, as just stated, many of these correlations have grown stronger over time.

Still, the correlations between use and most demographic characteristics, even achieved ones, are relatively weak in the United States and most other modernized nations. With the exception of the relationships noted above, all of the correlations between monthly, annual and even lifetime use and a number of demographic characteristics are all very weak. While, in general, the variables reflecting achieved statuses are more strongly correlated with use than those reflecting ascribed statuses, most of these correlations fall in the (-.05) to (.05) range (see SAMHSA 2003). Even the use of heroin, long thought to be one of the most class-based drugs, is uncorrelated with income (r = -.03) and education (r = .00). Very weak correlations, all less than (.10), are also found between income and the use of cocaine, crack, LSD, ecstasy, hallucinogens, analgesics, inhalants, tranquilizers, stimulants, methamphetamine, and sedatives. Consequently, in core nations, the two forces -- the leveling of use and use patterns being based on achieved statuses -- come together to weaken our ability to predict use based solely on demographic information.

Nevertheless, the general pattern appears to hold. As in other aspects of stratification, ascribed characteristics influence drug consumption patterns less as modernization develops. In fact, we can further specify proposition 6.0 by noting that in the most highly developed nations, use is primarily correlated with achieved statuses, though weakly so. In the most remote areas of the periphery, ascribed statuses still strongly pattern use. In those areas of peripheral and semi-peripheral nations that are most modernized, such as urban centers, use patterns are more similar to the core. In fact, the correlation between achieved statuses and use will be stronger in these areas than in most core nations. Therefore,

(6.1) Use will be moderately correlated with achieved statuses in core nations

(6.2) Use will be highly correlated with ascribed status in rural areas of peripheral or semi-peripheral nations.

(6.3) Use will be highly correlated with achieved statuses in the urban areas of peripheral or semi-peripheral nations.

Chapter Summary: The Contradictions of Modernity

This chapter has tried to specify the precise manners in which modernization and rationalization help pattern contemporary drug use. It has addressed the "what" of drug use. Modernized nations use a greater variety of drugs and use more drugs in combination than pre-modern societies. The "how much" of drug use was also discussed. Modernization increases use until the forces of modernization also bring about anti-drug legislation. After legislation, use is somewhat limited again. However, cross-culturally at any given time, modernization and drug use correlate positively. Finally, the "who" of drug use was discussed. Modernization levels the drug playing field as restrictions that once prohibited large segments of the population from using drugs become less relevant. Modernization also changes the dimension of stratification that separates users from non-users by decreasing the importance of ascribed characteristics and increasing the importance of achieved characteristics.

If we consider the above tendencies together, it becomes apparent that modernization creates contradictions concerning drug use. It simultaneously promotes use and a less-forgiving environment for use. It weakens our collective ability to control use while promoting our collective ability to worry about it. It promotes use while simultaneously promoting legislation that reduces use. It changes who uses drugs by weakening the structural constrains over use and "leveling" consumption across gender, ethnic groups and even social classes. It changes the dimension of stratification that patterns use, and in so doing, becomes a dimension of stratification itself. It is little wonder then why we as a society struggle with the issue of drug use. In the modern world, contradictions fill the world of drugs.

To this point, I have discussed how modernization and the related rationalization process help pattern contemporary drug consumption. These processes can account for much of the cross-cultural variation we see in drug consumption. They can also help account for long-term trends within nations over time. However, the explanation offered thus far cannot sufficiently account for variations in drug use over relatively short periods within a society. We will address this issue in the next chapter.

nations with drugs such as the opiates, cocaine and marijuana. Generally, as modernization leads to increased use and increased law, the use of some substances becomes criminalized, and the people who continue to use the criminalized drug become ostracized. As a consequence, we see in the modern world that, at least according to perceptions, "illicit drug use remains concentrated among street children, dropouts, petty criminals, prostitutes, the jobless, and other socially marginal groups" (Courtwright 2001: 199). While this may be true of the most dedicated illicit drug users, it overstates the case when it comes to occasional or even fairly regular use. As we saw in the preceding chapter, the correlation between social status and use is not overly strong, at least in core nations such as the United States. While official statistics such as arrest statistics often indicate an inverse relationship between illegal drug use and social status, this is as much a function of differential applications of law as it is about actual rates of use. Based on self-report data, there is little difference in terms of education or income between illegal drug users and non-users. Moreover, most of the differences that do exist indicate *a positive, not inverse, relationship*. Still, the perception that users are lower status remains. And, these social perceptions help determine the rates of drug use in the larger society.

Public perception influences the use of a drug because drugs that are perceived to be dangerous, either physically or socially, are less likely to be used by "normal" people. For example, based on the *Monitoring the Future* data (Johnston et al 2002), the extent to which American youth perceive a drug to be harmful, the extent to which they disapprove of those using a drug, and the extent to which they believe the use of a particular drug should be illegal are strong predictors of whether or not they use the drug (Goode 2003). When discussing the fluctuations usage rates for specific drugs over time, the *Monitoring the Future* authors state that,

> The usage rate for each individual drug . . . reflects many, more rapidly changing determinants specific to that drug: how widely its psychoactive potential is recognized, how favorable the reports of its supposed benefits

CHAPTER SEVEN
FLUCTUATIONS IN RATES OF USE OVER TIME

To this point, I have discussed how the modernization and rationalization processes influence contemporary drug consumption patterns. While these processes can help explain cross-cultural and long-term variations in drug consumption, they cannot sufficiently account for variations in drug use over relatively short periods within a society. Rates of drug use fluctuate over short periods in many societies. In the United States, for example, rates of use increased dramatically during the 1960s and 1970s, but then declined rather dramatically in the 1980s and early 1990s. In the U.S., and many other nations, drug trends appear to be cyclical, not linear. While the general trend over time has been for drug use to increase until anti-drug laws begin to curtail use to some extent, analyzing consumption over a shorter period reveals both increases and short-term decreases. These short-term fluctuations occur both before and after anti-drug legislation. Yet modernization and rationalization are, for the most part, linear trends. What, then, causes the cyclical nature of drug use?

THE MARGINALIZATION OF USERS AND THE RATE OF DRUG USE

Just as modernization changes the nature of stratification, it also influences the status of drug users themselves. While drug-using shamans in pre-modern societies were among the most revered, illicit drug users are among the most disreputable in modern societies. We have seen from a number of historic examples already presented that modernization alters who uses drugs. Often, use is initially an upper or middle-class activity. Over time, however, use becomes associated with people of lower social positions and relatively powerless groups. Such was the case in the United States, Britain, France and other European are, how risky the use of it is seen to be, how acceptable it is in the peer group, how accessible it is, and so on (Johnston, O'Malley and Bachman 2003: 6).

Social perceptions, therefore, influence the public's demand for the drug. Moreover, these social perceptions change over time and they can change quickly (see, for example, Johnston et al. 2003). Therefore, it is possible that the short-term fluctuations witnessed in rates of use are a function of changing perceptions of use. When use is perceived to be harmful, rates will decrease. As use is perceived to be relatively safe, use will increase. This appears to be straightforward enough. However, it leaves us with the troublesome question of what causes perceptions to change.

First, we know that it is not the behavior, in-and-of-itself, that causes social perceptions to vary. Smoking a cigarette is smoking a cigarette, regardless of the historic period in which the behavior occurs. Similarly, snorting cocaine is snorting cocaine, period. The behavior was the same in 1893 as it was in 1993. While the preferred method of using a particular drug may vary over time, that is not what is being argued here. Injecting a drug, for example, will usually be considered more dangerous than eating, drinking or even smoking that same drug. And, indeed, variations in the method of use could possibly alter the perceptions of the dangers of use (this is discussed later). However, the current argument holds the method of use constant. Doing so, a behavior like smoking marijuana is, objectively speaking, the same, regardless of when, by whom, why or where it is smoked. The smoke is inhaled and exhaled. Thus, holding the method of use constant, the objective behavior of using a particular drug is the same over time. It is a constant, and a constant cannot cause variation. Therefore, the objective behavior, in-and-of-itself, cannot account for variations in attitudes toward the behavior over time.

Second, it can certainly be argued that the perception of a drug is due to the objective harm associated with its use. Objectively, the more drug users suffer some adverse reaction or cause some objective problem for others, the less favorable drug use will be perceived. Therefore, all else being equal, the greater the toxicity of a drug, the less favorable it will be perceived. Moreover, regardless of the drugs actual toxicity, the more people using it, the more likely it will cause some objective harm. If enough people are drinking coffee, someone will eventually complain of "caffeine jitters." If enough people smoke tobacco, eventually someone will develop lung cancer. If enough people use cocaine, morphine or heroin, someone, somewhere, will fatally overdose. If enough people consume alcohol, someone somewhere will develop a liver problem, lose his job, beat his children or run someone over with his car. Simply put, the greater the use, the greater the chances of overuse or misuse. And, all else being equal, the greater the misuse of a drug, the less favorable the perceptions of its use. This logic appears to be straightforward enough. However, it simply does not fit the facts very well.

While it would be a mistake to discount totally the role a drug's objective harm plays in stirring social concern about its use, the objective harm associated with a drug is weakly correlated with the perceived harm of the drug. For example, the percentage of American high school seniors who said "drinking four or five alcoholic drinks nearly every day" posed a "great risk" was only 58.8%. Only 49.3% said using barbiturates regularly posed a "great risk," and only 64.8% said using amphetamine regularly posed a "great risk." Meanwhile, 53.0% said using marijuana regularly posed a "great risk" (see Johnston, O'Malley and Bachman 2003: Table 5). Thus, drinking alcohol or using amphetamine regularly is believed to be only slightly more harmful than smoking marijuana regularly. Using barbiturates regularly is allegedly less harmful than smoking marijuana. The empirical data on the relative dangers of these various substances do not match their perceived dangers. In fact, correlating the percentage of American high school seniors who said the "regular use" of a number of drugs posed a "great risk" with the objective harm of those drugs indicates a very weak, *inverse* relationship (r = -.062) between perceived and objective harm.[88] Similarly, the extent to which the public expresses concern over the "drug problem," is *inversely related* to the number of people using specific illegal drugs, including cocaine and marijuana, or the number of people using any illicit drug (see Hawdon 2001). Finally, a long tradition of work has demonstrated that the objective harm associated with a drug is weakly correlated with the drug's legality (see, for example, Brecher 1972; Goode 1997).

Moreover, the objective harm associated with a drug is relatively constant over time. Cigarettes are as dangerous today as they were when first introduced in the 19th century. While knowledge of the ill effects caused by smoking has certainly increased since the first cigarette was smoked, it has still been widely known that smoking is a dangerous habit for over 50 years. Injecting heroin was as dangerous in 1910 as it is today, more or less. While there is some variation with respect to the level of toxicity in drugs over time, especially in illicit drugs, these variations are typically relatively minor compared to the general impression of a drug's potential harm. For example, some "good" heroin may hit the street that is stronger than most users are accustomed to and therefore pose a greater threat of an overdose. However, it is not as if the majority of heroin users do not realize that heroin use can be dangerous. Similarly, it has been clear now for some time that smoking marijuana, while safer than using many drugs, should not be considered a healthy activity. With legal drugs, the variation in toxicity is almost non-existent. Fifty milligrams of Valium was the same in 1970 as it is today. So, the objective harm posed by a given drug is relatively stable across time. The perceived harm caused by the drug, however, is not.

---

[88] Data for the perceived seriousness of various drugs are taken from Johnston, O'Malley and Bachman (2003: Table 5). They are for high school seniors in 2002. The drugs used in the analysis include the regular use of marijuana, powered cocaine, heroin, LSD, amphetamine, barbiturates, tobacco (smoking one or more packs per day) and alcohol (drinking 4 or 5 drinks nearly everyday). The data for the objective harm of a drug was based on the extent to which the drug produced tolerance, emotional dependence, physical dependence, physical deterioration, anti-social behavior during use, and anti-social behavior during withdrawal. These data were derived from Seevers (1958).

Taken together, the weak correlation between objective harm and perceived harm and the relative constancy of a drug's objective harm over time renders the drug's objective harm an insufficient explanation for variations over time in the perceived harmfulness of a drug. While the objective harm associated with a drug may fashion our collective attitudes toward the drug to some extent, this cannot account for the wide fluctuations in perceived harm observed over time. Something else is needed. Thus, it is not the objective behavior nor the objective harm associated with the behavior that causes variations in perceptions. What, then, does?

What varies over time *and* is positively associated with public concern over a drug are the *subjective perceptions* of the drug's *potential harm*. We will call these subjective perceptions of the drug's potential harm the drug's *"subjective harm."* It is the *perceived danger,* both physically and socially, that is widely associated with a particular drug. While the legal status of a drug can influence the drug's subjective harm, history demonstrates that, more often, the drug's subjective harm determines its legal standing as much, if not more, than vice-versa. In addition, the subjective harm determines the drug's legal standing as much, if not more, than the drug's *objective danger.* Thus, it is possible that the greater the subjective harm associated with a drug, the less favorable the social opinions about the drug. And, the less favorable the opinions, the lower the rate of use. While this is probably true, it still begs the question. What, then, causes a drug's subjective harm to vary over time?

A drug's subjective harm, and therefore perceived danger, is largely a function of what group is associated with its use (see Helmer 1975; Szasz 1974). First, the subjective harm of a drug will be greater when people with fewer resources use it because they are less able to hide any problems it may cause. That is, drug use becomes more visible when society's less fortunate are the drug's primary users. The wealthy can afford more privacy and therefore hide their behaviors and indiscretions more successfully than the less wealthy. This can explain the stories of excessive drug use among the extremely rich. They can afford to hire handlers who protect them and take care of their daily needs while they float along in drug-induced oblivion. Such was reportedly the case with Elvis Presley and scores of other wealthy entertainers. Most people, however, cannot conceal their excessive behavior and indiscretions as effectively as the rich can and would therefore be subjected to forms of social control that would limit their use.

Next, the subjective harm and perceived danger of a drug will be greater when marginalized people are perceived to be using it simply because the marginalized are unconventional and the unconventional are subject to more social control. As Black (1976: 70) notes, "law varies inversely with the conventionality of the offender." Thus, the more marginal group using a drug, or, more accurately, the more marginal the group perceived to be using the drug, the greater the probability the drug will be legally prohibited. Moreover, the more marginal the group using the drug, the harsher the drug will be treated legally.[89] Therefore, for example, in a number of societies, including the United States, Japan and Australia, eating or drinking opium remained legal for decades after attempts to regulate smoking opium were made. This unequal treatment of the drug was in part due to the variation in the method of use. However, it was also due, in part, to the fact that middle class women ate or drank opium while poor Chinese smoked it. Similarly, Americans did not pass the 1937 Marihuana Tax Act, the first federal attempt to regulate marijuana use, until its use became associated with Mexican immigrants (see, especially, Helmer 1975). Globally, the popular drugs among the indigenous populations of lesser-developed nations have been criminalized while those preferred by Europeans, and later Americans, have remained legal (see, especially Szasz 1974). In general, as use becomes increasingly popular with people of lower social standing, the public's concern about use will typically heighten

---

[89] These general facts of law apply for other forms of social control as well (Black 1976). Thus, all else being equal, the amount of social control varies inversely with the conventionality of the offender.

It is important to emphasize here that the *actual* rates of use by a relatively powerless group are not necessarily important. Instead, the *perceived rates* of use determine the social reaction to the drug. This was the case when white Americans feared the "increasing use" of cocaine by African-Americans during the early years of the 20th century. Based on the best available evidence, however, the "number of cocaine users at the time was very small . . . and no particular concentration of blacks were observed" (Helmer 1975: 48; also see Ashley 1975; Musto 1987). The fact that few African-Americans used cocaine was unimportant. The fact that many people *believed* that African-Americans used cocaine was important. As W.I. Thomas noted, beliefs are real in their consequences. The belief that, as Dr.Christopher Koch wrote in a 1914 *Literary Digest* article, "most of the attacks upon white women of the South are a direct result of a cocaine-crazed Negro brain" (cited in Goode 1999: 276) was enough to convince some southern police officers that cocaine-using African-Americans could withstand .32 caliber bullets and they should switch to .38 caliber bullets (Musto 1987).

Looking at the legislative histories of drugs, a general pattern emerges that supports the argument that the drug's subjective harm is directly related to the extent to which marginalized people are perceived to be the drug's users. The often-repeated life-course of a drug is that it is introduced to a society, or "rediscovered" by a society, and quickly becomes fashionable among the relatively elite.[90] As is often the case with the behaviors, lifestyles, and fashions of the elite, the "trend" spreads to those of inferior social positions. This was the case with both coffee and tea in Europe and elsewhere (see Courtwright 2001). It was the case with chloral hydrate, morphine, marijuana and cocaine in the United States and Europe of the late 19th century (see Parssinen 1983; Morgan 1981; Spillane 2000). For example, as late as 1898, Thomas Crothers described the

---

[90.] It should be noted that the use of folk drugs and folk medicines does not follow this pattern. However, in most modernized societies, these folk drugs have been eliminated from the daily experiences of most citizens (see, for example, Szasz 1974).

typical cocaine user in the United States as a professional male over the age of 30. However, he also noted, "persons of the tramp and low criminal classes who use this drug are increasing in many of the cities" (cited in Spillane 2000: 90). By the early decades of the 20th century, cocaine use had spread to the American population at large. The pattern happened again in the United States when drugs like marijuana and cocaine were "rediscovered" in the 1960s. These drugs became popular with middle class college students before their use diffused to the less-privileged classes.[91]

As use spreads throughout the population, the concern over use intensifies. As use becomes more prevalent among lower status persons, many high status persons quit using the drug. A drug may fall from favor among the elite for several reasons. Since they typically begin using a drug before lower class people do, elites may tire of it and turn to a "new" drug. Alternatively, as the drug becomes more commonly used, elites may maintain the social distance between themselves and commoners by not using the drug that has now become popular with commoners. On the other hand, as use becomes associated with negative behaviors and inferior social positions, elites may quit due to concern for their social reputations. Regardless of their motives, however, the general tendency is for those of the upper social classes to stop, curtail, or more effectively conceal their use of a drug as lower status groups begin to use it. At this point, the anti-drug sentiments become reinforced as users are now disproportionately of lower status. Use now becomes synonymous with the "disreputable" group perceived to be using it. The co-defining of a drug with a "disreputable" group further elevates the social costs associated with using the drug and therefore reinforces the process. Eventually, the negative definition of the drug becomes so strongly held that most people, regardless of their social status, will avoid the drug for fear of being defined as a social outcast or outsider.

---

[91] See Goode (1972: 36 - 37) for a review of studies from the late 1960s and early 1970s that clearly established the significant, positive correlation between social class and marijuana use.

While this process can and has occurred with legal drugs (currently, cigarettes provide a good example), the process is strengthened further when drugs are legally prohibited. The legal prohibition adds yet another social cost to the use of the drug. Once a drug is criminalized, the drug not only reflects low status, it *can cause* one's status to be lowered. As a drug is criminalized, fewer people will use the drug since use can earn them a criminal status. Who, then, is left using the drug? The devoted, hard-core users are left. The "addicts" are left. The marginalized are left. Once this process occurs, the continued use of the drug by the marginalized further marginalizes them from the dominant society.

Thus, *the more the dominant society perceives a drug is being used by a marginalized group, the greater the perceived harm associated with that drug.* In addition, *the greater the perceived harm caused by a drug, the less likely that drug will be widely used in a society.* These propositions seem straightforward enough, and, as mentioned earlier, there is empirical evidence supporting them. Yet there is still something missing from the explanation. It is true that as drugs spread from the upper to lower classes, social concern and the negative status of using a drug increase. It is also true that, use will likely decrease as concern increases. Nevertheless, we are left wondering why relatively elite people often begin the drug cycle. While the model of a drug flowing from the rich to the poor and the status of the drug decreasing as it does fits the scenarios of the 19th century, it does not fit very well in more recent times. Drug use was strongly associated with lower status persons in 1950s America. Social attitudes were extremely anti-drug, especially in the early part of the decade. The passage of the Boggs Act in 1951 and the Narcotic Control Act in 1956 illustrate the strong anti-drug sentiments held by America's power elite. These acts provided lengthy mandatory sentences or the death penalty for drug traffickers. As Morgan (1981: 147) notes when discussing the 1950s, "on the surface, the consensus against drug use and for enforcement seemed stronger than ever." Why then, did middle class youth begin using drugs?

The answer lies in the fact that what is acceptable changes over time. While the objective behavior itself is a temporal constant, the social definitions used to evaluate that behavior as either acceptable or unacceptable change. Similarly, while the reaction to unacceptable behavior may generally be to marginalize the person or persons engaging in it, what is considered unacceptable is not constant. We therefore must ask the question, "What causes the definitions of acceptable and unacceptable behavior to change over time?" The most likely culprit is the aspect of the modernization process that fluctuates most frequently, social mobility.

Social Mobility and Rates of Drug Use

In an earlier paper, I proposed a theory to account for the cyclical nature of deviant behavior (see Hawdon 1996b). That argument can be summarized as follows.[92] Rates of drug use are determined, at least in part, by the scope of the *deviance structure*. The deviance structure sets boundaries on what is consider acceptable and unacceptable behavior. That is, it is the society's moral boundary. The more narrowly defined the deviance structure, the greater the number of behaviors that are defined as unacceptable. The more augmented the structure, the fewer behaviors that are defined as unacceptable and the more behaviors that are tolerated. Like all social structures, the deviance structure shapes rates of interaction among people and groups (see Blau 1977). Those who are labeled as "deviants" are "outsiders," "criminals," "weirdoes" or some other type of person that "the rest of us" should avoid. These definitions, like those of "lower class" or "rich," will determine, in part, with whom one interacts and the resources that are available for their use. The deviance structure divides the population into two groups: "deviants" and "non-deviants" or "us" and "them." Since a deviant label will adversely affect one's access to resources, most people will try to avoid

---

[92.] The following argument is from my 1996 *Sociological Spectrum* article "Cycles of Deviance: Structural Change, Moral Boundaries, and Drug Use, 1880 - 1990." Readers who would like more detail about the processes involved are directed to that article.

falling into the "deviant" or "them" category, all else being equal. Thus, rates of drug use will increase when the deviance structure expands to the point that drug use is no longer considered deviant or considered to be "less deviant" than before. Use will decrease when the deviance structure constricts to the point that drug use is considered deviant or considered "more deviant" than before.

Social mobility, defined as all movements of people between and among structurally defined positions (see Blau 1977),[93] can be measured by changes in the occupational structure, unemployment rates, immigration rates, residential mobility, changes in attendance rates in educational institutions, changes in the age structure of the society, changing divorce rates and changes in religious or political affiliations. All of these changes bring people of different upbringing, education, cultural assumptions, values, religious orientations and political attitudes together thereby weakening the structural parameters that pattern rates of inter-group contact.[94] As Noble (2000: 37) points out, rates of social mobility "measure the enduring patterns of institutionalized segregation of social strata." Social mobility reflects the relative inflexibility or permeability of social divisions and indicates the salience of stratification or heterogeneity as determinants of interaction.

Social structures are therefore affected by rates of mobility (see Blau 1977). In this respect, the deviance structure is similar to any other social structure. However, unlike other dimensions of stratification -- such as income or wealth -- or heterogeneity -- such as ethnicity or gender -- that are based on somewhat concrete, objective factors, the deviance structure is determined by normative definitions. Whereas the objective resources of income and wealth

---

[93]. I use a more liberal definition of social mobility than found in economics. For example, social mobility often refers "to movements by specific entities between periods in socioeconomic status indicators" (Behrman 2000: 72). Such movements are types of social mobility. However, movement in terms of other graduated parameters, such as educational attainment, also reflects mobility. Similarly, movement between or among horizontally differentiated positions also constitutes mobility. My definition is similar to the one developed by Blau (1977).

[94]. Actually, if changes "bring people together" or "drive them apart" depends on the nature of the change.

determine one's class standing, one's position on the deviance structure is determined by one's objective behavior *and* the *social definitions regarding that behavior*. Given the more subjective nature of the criteria used to define the deviance structure, it is likely to be more flexible than most social structures. Thus, social mobility can alter the deviance structure more quickly than it alters other structures.

The process through which social mobility changes the deviance structure begins when greater mobility weakens the social or physical barriers that had once hindered interaction among differing social positions. As inter-group contact increases and divergent groups are integrated into the social whole, the heterogeneity of the group increases. This increased heterogeneity "reduces discrimination against outgroups" (Blau 1977: 81) and produces pluralism. Thus, "heterogeneity, by obscuring the criteria of discrimination, results in behaviors formerly defined as deviant gradually becoming accepted" (Hawdon 1996b: 187).[95] This process permits group members to engage in a variety of behaviors that, during earlier times, would have earned them a deviant label. Conversely, when mobility is low, inter-group contact decreases, prejudice against the "out-group" increases, and behaviors that were once tolerated are increasingly defined as deviant. In this situation, fewer people would likely engage in behaviors such as drug use because these can now result in them being labeled "deviant." Yet, those few who still engage in the behavior will be defined as "deviants;" therefore, "although the number of people engaging in the objective behavior may have actually decreased, the rate of deviance, as a socially constructed definition, will have increased" (Hawdon 1996b: 188).

Next, as discussed in earlier chapters, social mobility also increases the number of group associations available to a person. As more groups become available for membership, the exit costs associated with leaving or being ousted from any specific group decreases. The lowered exit costs reduce the group's

---

[95] It should be noted that the acceptance of a behavior that is associated with a group does not necessarily mean the group will be accepted (see Hawdon 1996b: 187 - 188).

ability to demand conformity. Moreover, exposure to more groups results in exposure to differing and competing value systems. The co-existence of competing value systems often results in normative ambiguity, which ultimately undermines mechanisms of social control (see, for example, Shaw and McKay [1942] 1972: 170). Then, as the group's ability to control its members decreases, individuation increases. The greater freedom enjoyed by the individual coupled with an ambiguity concerning "proper behavior" results in increased variation in behaviors, including those that may have once been considered deviant. Conversely, when rates of social mobility slow, the group can more easily define its moral boundaries and demand more conformity from its members. As a result, the behavior of individuals is limited and behaviors such as drug use will likely decrease (see Hawdon 1996b: 188 - 190). The argument can be summarized as follows.

> Social mobility alters the deviance structure of a group by increasing the rate of inter-group associations and the number of structural parameters that intersect. As structural change occurs, so do cultural definitions of proper and improper behavior. Consequently, when rates of social mobility are accelerated, the deviance structure changes in ways that permit individuals to engage in behaviors that were formerly defined as deviant. Conversely, when rates of social mobility slow, the deviance structure changes to prohibit an increasing number of behaviors. Consequently, more individuals will earn the label of deviant despite a reduction in the number of people actually practicing those behaviors selected for deviant definitions (Hawdon 1996b: 190).

Social mobility is therefore the motor behind the cyclical patterns of drug use. Its influence is most pronounced on illegal recreational drug use; however, the relationship holds for the use of drugs in other settings and for those drugs that are legal but socially criticized such as tobacco and alcohol. While modernization produces a general trend of increasing drug use, the general process of modernization cannot account for short-term changes in drug-using behaviors. Instead, the related process of social mobility, which fluctuates within the longer-term process of modernization, produces the short-term increases and decreases in drug use witnessed within societies.

Empirical Evidence

Social mobility's influence on drug use is evident in the histories of numerous cultures. In the remainder of this chapter, I will present five examples that demonstrate the relationship. I will review the history of drug use and social mobility in the United States, Great Britain, Japan and modern Israel. I also review the very recent history of several Eastern Europe nations. While the histories of any number of nations could be reviewed, these cases were selected for theoretical and practical reasons. Theoretically, these cases represent extremes with respect to the consumption of drugs in modernized nations. In the first two cases, drugs have been widely used for centuries. In fact, the United States and Great Britain are among the world's leading drug-consuming nations. In Japan, Israel and Eastern Europe drug use, at least illegal use, has been more effectively controlled until relatively recently. In addition, these four examples should provide sufficient variation in cultures to demonstrate the generality of the theory. While the United States and Great Britain have obvious cultural similarities, they differ significantly from Japan, Israel and Eastern Europe. Practically, these nations were selected because data were available. Although the data on drug use and social mobility are flawed for these nations, data for other nations would be more seriously flawed. Let us now turn to the evidence.

Social Mobility and Drug Use in the United States

In general, the United States has experienced three periods of relatively high rates of drug use, 1860 - 1920, 1960 - 1980, and 1992 to the present.[96] Between these periods, the United States had relatively low rates of use, at least

---

[96.] I provide supporting evidence for this argument elsewhere by analyzing rates of use and social mobility in the United States between 1880 and 1980 (Hawdon 1996b). Some of the evidence presented here was first presented in that article. Additional evidence is presented here.

by American standards. The "high" periods correspond to periods with high rates of social mobility. Conversely, the "low" periods correspond to periods with low rates of social mobility.

We have already discussed how rapidly rates of drug use increased in the United States between 1881 and 1920. Indeed, rates of drug use at that time were among the highest ever observed in the United States (see, for example Musto 1987; Morgan 1981; Hawdon 1996b). During this same period, America was also experiencing unprecedented rates of social mobility. For example, America witnessed its highest rate of immigration in history during this time, as 23.4 million immigrants came to America (U.S. Bureau of the Census 1975). Largely due to the influx of immigrants, the U.S. population grew at an annual rate of 2.1% between 1870 and 1913 (Maddison 1995: 102). Urbanization also increased at unprecedented rates (see Hawdon 1996b). In addition to the influx of newcomers, the American economy was being transformed. For example, between 1880 and 1920, there was a 42.3% decrease in the percentage of Americans who worked in the agricultural sector (U.S. Bureau of the Census 1975). Between 1870 and 1913, the manufacturing sector grew by nearly 22% and the percentage of the population working in the service sector nearly doubled (see Maddison 1995). With these changes, more Americans were going to, and staying in, school. The average years of education per person aged 15 to 64 doubled from 3.92 to 7.86 between 1870 and 1913 (Maddison 1995: 37). Thus, during the first American "drug epidemic" that occurred in the late 19th century and early 20th, the United States went from being a relatively homogeneous, rural, agricultural society to a heterogeneous, urban, industrial one" (Hawdon 1996b: 196).

Social mobility began to wane after the 1920s. The two World Wars and the Great Depression limited social mobility and social change. When the period of 1920 through 1960 is compared with the forty years prior to 1920, immigration rates decreased by over 188%. Moreover, between the mid-1920s and the late 1950s, most immigrants to the United States were Europeans (U.S. Bureau of the

Census 1975; also see Model and Ladipo 1996). America had therefore already been exposed to most of these newcomers' cultures. Rates of urbanization also slowed, from an 81.6% increase between 1880 and 1920 to a 15.2% increase from 1920 to 1960. Similarly, during the "down cycle" of drug use, the rate of occupational change was approximately one-half what it had been in the previous era (see Hawdon 1996b). Drug use, like social mobility, also decreased during this forty-year period. As Musto (1987: 251) says, "from being easily available and commonly used, heroin, morphine, and cocaine almost faded out of nonmedical situations."

Beginning in the1960s, however, drug use and social mobility increased again. While use increased slowly in the late 1950s and early 1960s, it seemed to explode in the mid-1960s. In 1966, for example, only 1.8% of youth aged 12 to 17 had ever used marijuana. Beginning in 1967, however, use increased steadily. By 1970, nearly 10% of the youth aged 12 to17 had used marijuana. Similarly, while only 5.1% of young adults aged 18 to 25 had ever used marijuana in 1965, by 1970, nearly 20% of these young Americans had at least experimented with marijuana (see SAMHSA 2002). In addition to marijuana use, LSD use increased. Between 1966 and 1970, the annual number of new hallucinogen users rose almost sixfold, from 168,000 to 956,000. LSD use primarily accounted for this increase (SAMHSA 2002). Cocaine, PCP, methamphetamine, heroin and tranquilizer use also increased rather dramatically in the late 1960s (see Morgan 1981: Musto 1987). As Inciardi (1992: 33) states, "the use of drugs seemed to have leapt from the more marginal zones of society to the very mainstream of community life."

Drug use continued to increase during the 1970s. The percentage of young adults aged 18 to 25 who had ever used marijuana increased steadily until 1981. Similarly, the incidence of cocaine use, for example, generally rose throughout the 1970s and peaked in 1980 at 1.7 million new users (SAMHSA 2002). The use of stimulants, tranquilizers, sedatives, pain relievers, PCP, LSD and a host of other illicit drugs rose sharply in the early 1970s and modestly in the late 1970s (see SAMHSA 2002). Generally speaking, illicit drug use peaked in the United States in 1979 (see Johnston et al. 2003). By 1979, over two-thirds of the youth population had tried marijuana, nearly thirty percent had used cocaine, and one-fourth had used a hallucinogenic substance at least once in their life.

As in the previous era of high rates of drug use, social mobility increased during the second "drug epidemic." Beginning in the mid-1960s, the McCarran-Walter Act relaxed restrictions on residents from Latin American and other independent Western hemisphere countries (see Model and Ladipo 1996). As a result, immigration rates began to increase. These rates, which never exceeded 1.5 per 100,000 population between 1930 and 1960, increased to 1.7 per 100,000 population during the 1960s and reached 2.1 per 100,000 population during the 1970s. This represented the highest annual rate of immigration since the Great Depression (U.S. Bureau of the Census 1991). There were also significant changes to the American economic structure during this era. The percentage of the population employed in agriculture decreased by 78% while employment in the manufacturing sector decreased by nearly one-third (see Maddison 1995: 39). This was the first time in U.S. history that percentage of workers employed in the manufacturing sector shrunk. Now, Americans were working in the service sector, where employment had increased by 38.3%. Increasing numbers of Americans also moved from the urban areas to the suburbs. More Americans also attended colleges, more women joined the labor force, and the Civil Rights Movement helped open educational and occupational opportunities to America's ethnic minorities. As a result, the U.S. poverty rate, which is inversely related to occupational mobility (see, for example, Passarides and Wadsworth 1989), declined steadily throughout the 1960s, reaching 12.1 percent, or 24.1 million individuals, by 1969. For the next decade, the poverty rate fluctuated between 11.1 and 12.6 percent (U.S. Bureau of the Census 2003). These were among the lowest rates of poverty witnessed in the United States since detailed data tracked it. Intergenerational mobility also increased between the early 1960s and the mid-1980s as the association between class of origin and class of destination weakened (see Featherman and Houser 1978; Hout 1984; Hout 1988). Indeed, between 1962 and 1973, the effect of one's origin on their destination decreased by over 28% among American men (Hout 1984: table 2). Thus, in short, "social mobility increased notably during the 1960s and 1970s" (Hawdon 1996b: 199).

As in the 1920s, however, rates of social mobility began to decrease in the 1980s. The average annual percentage of change in the U.S. population due to death, births and migration decreased from 12.4% during the 1960s and 11.4% during the 1970s to 7.4% during the 1980s. The overall annual rate of change in the American population was lower in the 1980s than during any other decade in the twentieth century except the depression-plagued 1930s (see, U.S. Bureau of the Census 2000). Immigration rates also began to decline. Moreover, not only were fewer newcomers arriving in America, those already living here moved less than in the previous decade; America's flight to the suburbs had slowed. In addition, the economy was now fully a service-oriented economy. As a result, the rate of economic change subsided. The rate of occupational change in the 1980s was approximately 22% lower than it had been in the 1970s (see US Bureau of the Census 1991). While the average annual level of occupational reallocation of workers was 9.5% in the 1970s, it was only 8.0% in the 1980s (Moscarini and Vella 2002). Even economic mobility in terms of wages and earnings, which has generally been stable over time in the United States, decreased during the 1980s. Buchinsky and Hunt (1996), for example, found a significant decrease in year-to-year mobility for both wages and earnings over the period 1979 to 1991, at least among young adults. Overall, occupational mobility due to structural changes decreased in the 1980s (see Hout 1988). And, for the first time since the 1950s, university and college attendance decreased and rates of poverty significantly increased (US Bureau of the Census 1991). By 1983, the number of poor individuals had risen to 35.3 million individuals, or 15.2 percent of the population. And, the poverty rate remained above 12.8 percent for the next ten years, eventually climbing to 15.1 percent, or 39.3 million individuals, by 1993 (US Bureau of the Census 2003). While more Americans fell into poverty during the 1980s, the wealthy became richer; thus, inequality also increased sharply during the 1980s (see Juhn, Murphy, and Pierce 1993; Buchinsky 1994). Consequently, the various structural parameters that divide Americans -- measured by immigration, occupational mobility, poverty, inequality, educational attainment and migration -- strengthened and became increasingly salient during the 1980s.

Once again, as social mobility began to decrease, so did rates of drug use. Contrary to media and political claims made at the time, drug use declined rather substantially in the United States during the 1980s. For example, between 1979 and 1985 the number of young adults who had ever used marijuana and cocaine decreased by 12% and 11% respectively. This trend continued throughout the decade and into the early 1990s. While nearly 20% of youth ages 12 to 17 had used marijuana at least once in their life in 1979, the rate of use among these teens decreased to 11.5% by 1991 (see SAMHSA 2002). Between 1979 and 1990, the number of marijuana, cocaine, and hallucinogen users decreased by 23%, 32%, and 52% respectively (Hawdon 1996b; Hawdon 2001).[97]

In the early 1990s, however, rates of social mobility again began to increase. First, the average annual rate of change in the population increased, nearly matching the rate of the 1970s (see, U.S. Bureau of the Census 2000). Much of the change was due to resurgent rates of immigration. In 1994, for example, net international immigration accounted for 30% of the total increase in the population (Deardorff and Montgomery 2003). The annual average net international migration figure of 759,000 during the 1990's was well above the annual average of 634,000 for the previous decade. Moreover, as in the 1960s, the newly arriving immigrants were largely from nations that had not previously sent large numbers of immigrants to the United States (see U.S. Bureau of the Census 2000). In addition, for the first time in American history, the growth in the Hispanic population was numerically larger than the growth in the White,

---

[97.] Trends are based on National Institute on Drug Abuse data (1986; 1988; 1990; 1994).

non-Hispanic population (Deardorff and Montgomery 2003). Once again, Americans were being exposed to, and interacting with, relatively unfamiliar cultures. Next, as the economy expanded after the recession of the early 1990s, occupational mobility, as measured through occupational reallocation, increased slightly and then stabilized throughout the remainder of the decade. While not rising sharply, the 1990s saw a reversal of the decade-long trend of decreasing mobility (see Moscarini and Vella 2002). Partially because of this reversal, rates of poverty declined during the 1990s after increasing substantially during the 1980s. While over 15% of the U.S. population was poor at the beginning of the decade, only 11.3% were by 2000 (US Bureau of the Census 2003).

As rates of mobility increased in the 1990s, so did rates of drug use. Rates of marijuana use began increasing again during the early 1990s, especially among early adolescents. The annual number of first-time marijuana users nearly doubled from 1990 to 1995, reaching levels similar to those in the mid-1970s when the number of new marijuana initiates peaked. This number has remained relatively stable since that time (SAMHSA 2002). The number of Americans over the age of 12 who use marijuana regularly has increased by approximately 22% since 1992 (see SAMHSA 2002). The use of other drugs also increased during the 1990s. The number of cocaine initiates steadily increased between 1992 and 2001 after declining consistently over the previous decade. The use of ecstasy increased steadily after 1992, as did the use of methamphetamine, pain relievers and sedatives. Likewise, during the late 1990s, heroin use rose to a level not reached since the late 1970s (SAMHSA 2002). Thus, in the United States, as the rate of social mobility increases, so does drug use. The model fits the data well in the United States. Does it fit in other cultures?

Social Mobility and Drug Use in Britain

The history of British drug use since the 19th century follows a pattern similar to that in the United States. It was noted earlier that Britain's first "drug epidemic" occurred between the 1830s and early 1900s. As in the United States, opium and morphine were the popular drugs. For example, from 1827 to 1859 the average consumption of opium rose from 600 mg. to 1,410 mg. per person per year. Opium was available in solid and liquid forms, and mainly taken by mouth (Marks 1990). While this increase in use is impressive, rates of use rose at even faster rates between 1855 and 1905 (Parssinen 1983: 31). In addition to the narcotics, cocaine was widely used by the turn of the century, although it never achieved the popularity in Britain that it did in the United States. Recreational use was especially prevalent in "certain industrial, dockside and farming regions, and had some popularity in limited middle-class and bohemian circles" (Ruggiero and South 1995: 18). Use remained relatively high until approximately 1910, when rates of use slowly declined. After 1920, however, use decreased rather sharply. As Ruggiero and South (1995: 18) state,

> Between the end of the war and the mid to late 1920s, various forms of drug use in Britain had a degree of visibility and sensation attached to them that had far less to do with their actual prevalence than their treatment in the press and cinema of the day. In reality, opiate and cocaine users were to be found in diminishing numbers under treatment by a doctor, in the opium dens of Limehouse or among the cocaine 'fast set' of the West End.

While drug users were cast as folk devils during this time, there was, as Downes (1976: 89) states, "an almost nonexistent addiction problem." Drug use "of any magnitude or significance was fading away" (Ruggiero and South 1995: 18).

Use remained relatively rare in Britain from the late 1920s until the late 1950s. As in the United States, use was largely contained to the working poor of urban areas. The "British System," which allowed physicians to prescribe drugs while treating patients for addiction, was considered a success at controlling drug use and preventing the emergence of new drug problems (Downes 1976; Ruggiero and South 1995). Although the "British System" received credit for controlling use, researchers agree there was little use to control (see, especially, Downes 1976; also see Marks 1990). Rates of use remained low through the

1940s and into the 1950s. Late in that decade, however, the drug scene began to change. Still, despite evidence that marijuana and heroin use was increasing slightly during the late 1950s, the Brian Committee's 1961 report concluded that the supply and use of these drugs were quite negligible (Ruggiero and South 1995). Consequently, Britain's first "cycle" saw high rates of use between 1830 - 1920, with especially high rates of use between 1830 and 1870. Then, low rates of use between 1920 and 1960 completed the first cycle. Did rates of social mobility parallel rates of drug use during these times?

Britain: Social Mobility, Wave One

When discussing the increased popularity of opium in English medicine during the 1830s and 1840s, Parssinen (1983) notes two reasons that reflect the influence of social mobility on the use of drugs. First, during the cholera outbreaks of 1831-32, 1848-1849 and 1853-54, the East India Company promoted the use of opium, a practice learned in India where cholera was endemic and had been treated with opium for centuries. Thus, contact with other cultures introduced a new use for a well-known drug. The second reason for opium's popularity was the continued English practice of self-medication. However, social mobility changed the nature of this practice from using traditional folk medicines and herbs to opium. Parssinen (1983: 26) states,

> Herbalism was the predominant form of self-medication so long as most Britons lived in villages and small towns, close to the countryside and attuned to rural folk traditions. Rapid urbanization changed that. By mid-century, the majority of the population lived in cities, cut off both from physical proximity to wild herbs, and from the culture which encouraged their use. At the same time, the population density of urban areas brought many more people into contact with chemists' shops than previously. While self-medication continued to thrive in Victorian Britain, the forms that it assumed changed significantly.

The rapid urbanization, itself a form of social mobility, was appreciable throughout this period. According to Jason Long (2001), Victorian Britain experienced the most rapid and thorough urbanization the world has ever seen. The rural exodus, while occurring to some extent in the 18th century, began rapidly increasing in the 1830s and 1840s. Then, between 1851 and 1871, the urban population grew by an average of 23.6% per year, with the highest growth rates occurring between 1861 and 1871 (see Alford 1996: 15; also see Mingay 1986; Long 2001). There was also a shift in the regional distribution of the English population. The percentage of people living in London and the urban centers of the Southeast grew by approximately 25% during the mid-19th century (Alford 1996). New industrial centers, such as Middlesbrough, Cardiff and Barrow-in-Furness, "that hardly existed at all in 1831," emerged and grew at impressive rates (Mingay 1986: 4). This rapid urbanization and greater regional concentration of the population were due to changes in the English economy. During the mid-1800s, the English economy was transformed from agriculture and light manufacturing to heavy industry and financial services (see Hobsbawm 1968; Chase-Dunn 1998; Maddison 1995; Alford 1996: Mingay 1986). For example, the 1871 census was the first to register a decline in the agricultural workforce, from 37.6% in 1820 to 22.7% in 1870 (Maddison 1995: 39). In contrast, between 1820 and 1870, there was a 28.6% increase in the number of persons employed in the manufacturing sector. While agricultural work still accounted for a large segment of the workforce, "industrialization and industrialism were the new forces with which the old order had to come to terms" (Alford 1996: 6; also see Mingay 1986).

Due to the expansion of industry during the mid-nineteenth century, England was clearly the world's economic leader. England absorbed about a quarter of world imports of food and raw materials and accounted for nearly one-third of the world's exports (Alford 1996). The majority of these exports were manufacture goods, and their volume of growth grew at an average of 4.9% per year between 1820 and 1870 (Maddison 1995: 74). Britain was also the largest provider of trade-related services, and "its joint factor productivity growth was better than that of the US economy" (Maddison 1995: 61). Based on the total changes in the English labor force, Alford (1996) notes that the structural change in the economy averaged 1.2% per year between 1851 and 1871. Real per capita GDP growth was at an impressive 1.2% per year, and England's population grew at relatively high rates between 1820 and 1870 (Maddison 1995: 62; Alford 1996: 15).

In addition to the mobility directly related to economic changes, the wealthier and additional Britons were also experiencing social mobility in other ways. They were also moving about the country more easily as over 21,000 miles of rail were laid before 1870 (Maddison 1995: 64; also see Mingay 1986). The expanding rail system not only spurred economic expansion and residential mobility, it provided greater access to news sources and other means of communication, greater movement between towns and cities, and increased contact between rural and urban peoples. As Mingay (1986: 56) notes, "rural isolation was vastly reduced and country people felt that for the first time they were really part of the national life."

In general then, rates of social mobility were very high in Britain between the 1830s and 1870s. This time also corresponds to rapidly increasing rates of drug use. However, social mobility began to slow in Britain after 1870, and, not coincidently, the rate of drug use began to slow also. With respect to mobility, urbanization, for example, continued to increase but at a slower pace. While urbanization increased by an average of 25.6% per year between 1861 and 1871, it dropped to a rate of 22.8% per year in the 1870s and 18.8% per year during the 1880s (Alford 1996: 15; also see Long 2001). As Alford (1996: 15) notes, "the highest rates of growth occurred up to the 1880s but thereafter the rate fell off quite sharply." Similarly, the percentage of the population employed in the manufacturing sector increased by 28.6% between 1820 and 1870, but it increased by only 4.1% between 1870 and 1913 (see Maddison 1995). The rate of structural change, which averaged 1.31% between 1861 and 1871, averaged only 0.97%

between 1871 and 1911 (see Alford 1996).[98] Thus, social mobility began to slow at the same time drug use began to stabilize and decrease.

After 1920, when rates of drug use were extremely low in Britain, social mobility virtually halted. Between 1913 and 1950, for example, the rate of population growth was half what it had been during the previous 100 years. Rates of urbanization also slowed, and changes in the economic structure of the nation were minimal. Between 1913 and 1950, for example, the percentage of the population employed in the manufacturing sector increased from 44.1% to 44.9%. As a result, the volume of growth in the rate of exports was *zero* during that period (Maddison 1995: 74), as Britain's percentage of world exports of manufactured goods and capital goods decreased (Alford 1996). While the percentage of workers employed in the service sector increased, the increase was a modest 13% (see Maddison 1995: 39). Similarly, the number of people going to colleges and universities, which had increased steadily through the early decades of the 20th century, began to subside. The average years of education per person aged 15 to 64 doubled between 1870 and 1913; however, it increased by only 20% between 1913 and 1950 (Maddison 1995: 37). Geographic mobility even became more difficult as the length of railway lines across Britain decreased between 1913 and 1950 (see Maddison 1995: 64).

Thus, the model fits the available data well for Britain's first wave. When social mobility increased during the first half of the nineteenth century, drug use also increased. As social mobility slowed its pace towards the end of that century, the increasing rates of drug use also slowed. Social mobility then began to slow even more as World War I approached and then virtually halted after the Great War. Drug use became less common too and by the time World War II approached, illicit drug use was rare. Yet, as in other European countries, Britain would experience another "epidemic" in the second half of the twentieth century.

[98.] The rate of structural change was high between 1891 and 1901, at 1.29% per year. However, it never exceeded 0.90% in any of three other decades between 1871 and 1911 (see Alford 1996: 14).

Drug Use: Wave Two

Rates of drug use began to increase again in Britain during the early 1960s. Heroin use increased modestly at first and more quickly in the latter part of the decade. Between 1961 and 1968, for example, the number of heroin addicts reported to the Home Office increased from approximately 50 to over 1,000 (Ruggiero and South 1995; Marks 1990). Cannabis use also increased during the 1960s and quickly became Britain's favorite illicit drug as it spread among middle-class youth (Hartnoll et al 1989). Similarly, LSD became popular in the newly emerging youth counterculture. Amphetamine use was common among the Mods during weekend parties (Cohen 1980) and eventually spread to other working-class youth (Hartnoll et al 1989). Rates of use of most illicit drugs continued to increase through the 1970s. For example, between 1973 and 1977, officially registered narcotic users had increased to 4,607. Similarly, barbiturate use also increased rather rapidly, especially among the opiate-using population (Ruggiero and South 1995).

Use continued to increase until the end of the 1970s when approximately 20% of young adults had used marijuana at least once in their life (see Mott 1989). Rates of use then began to decrease. By the mid-1980s, for example, lifetime cannabis use among young adults had decreased to approximately 12% (see Mott 1989; ISDD 1993). Based on available data, drug use remained stable until the late 1980s (see Hartnoll et al 1989; ISDD 1993; Parker et al 2002; Measham, Newcombe and Parker 1994). For example, past-year and past-month cannabis use remained virtually unchanged between 1982 and 1988 (Home Office1982; Home Office 1988), and, in 1985, ecstasy "was a scarce item" (Ter Bogt et al 2002: 160).

Beginning in the late 1980s, however, rates of cannabis, amphetamine, ecstasy and other drugs began to increase, especially among young adolescents. As the Institute for the Study of Drug Dependence (ISDD 1993: 27) states after surveying the drug prevalence research of the 1980s and early 1990s, putting these fragments of evidence together suggests the relatively stable drug use patterns of the mid-80s were disturbed in the late '80s and that by the '90s there was increased use of established drugs like cannabis, solvents, amphetamine and magic mushrooms, and an upsurge in the use of ecstasy and LSD.

Similarly, Parker and his associates (Parker et al 1998: 13) state about the mid-1990s, "when compared with similar surveys (of the general population) undertaken in the mid 1980s we have a substantial increase, almost a tripling, in the number of people reporting experience of drugs." For example, data from the 2001/2002 sweep of the British Crime Survey shows that in 1994, 28% of adults had used some illicit drug at least once in their life. This percentage increased to 29% in 1996, 32% in 1998 and 34% in 2000 (Aust, Sharp and Goulden 2002; also see Boreham and Shaw 2001; also see Office of National Statistics 1999; Graham and Bowling 1995; Jeffery, Klein and King 2002; EMCDDA 1998). Likewise, the first *Youth Lifestyles Survey* in 1992-1993 found 22% of 14 to 25-year-old Britons had used an illegal drug in the past year. The corresponding percentage in the second survey (1998-1999) was 32% (Flood-Page et al., 2000). While the majority of drug users were cannabis smokers, the use of other drugs also increased during this time. Cocaine use, while still not overly widespread, became more noticeable (Aust et al 2002), and ecstasy use became popular among adolescents involved in the "House" scene (Ter Bogt et al 2002; Measham et al 2001). Increases in illicit drug use were also reported in Scotland (Fraser 2002; Office of National Statistics 1999b). Rates of use, therefore, increased until the late 1990s. By that time, it became clear that adolescent drug use in the United Kingdom had been rising throughout the decade, and, by the mid-1990s, young Britons were the most drug-involved youth among twenty-six European countries (ESPAD 1997).

Rates of use appear to have stabilized somewhat since the mid-1990s. For example, cannabis use, by far the most widely-consumed drug, has more or less stabilized in all age groups since 1994 (Jeffery et al 2002; Drug Scope 2002;

UNODC 2004). Moreover, the use of amphetamine, the second most favored illicit drug (see Parker, Williams, Aldridge 2002; UNODC 2004; Measham et al 2001), declined from 7.9% of 16-29 year olds in 1998 to 5.2% in 2001 (Jeffery et al 2002; also see Aust et al 2002). Similarly, the regular use of ecstasy remained stable from 1994 to 2001 (Jeffery et al 2002), but then increased slightly between 2001 and 2003 (UNODC 2004). Next, while there has been a slight increase in overall rates of use among the adult population, it appears fewer British youth are using drugs. A survey of illicit drug use among 15 - 16 year olds attending schools showed a decline in most forms of drug use in the United Kingdom (Plant and Miller 2000; also see Jeffery et al 2002; Parker et al 2002; ESPAD 1997; ESPAD, 2001). Plant and Miller (2000) report 41% of 15 to 16 year old youth used marijuana in 1995, while only 37% did so in 1999. Likewise, amphetamines were used by 13% of youth in 1995, but by less than 9% in 1999. Ecstasy use also declined by nearly 40% between 1995 and 1999 among the youth population. Similar trends of decreasing rates of use among young people were reported in the *British Crime Survey* (Aust et al 2002). Thus, while there have been slight fluctuations, these changes have been relatively minor. Overall, prevalence rates fell between 1995 and 1999 in the United Kingdom (UNODC 2002b), and have apparently stabilized since.

British patterns of drug use are therefore similar to those seen in the United States. Use began to increase slowly in the 1960s and then rapidly in the 1970s. Then, rates of use decreased. The increase in the 1970s was not as substantial as it had been in the United States; similarly, the decrease in the 1980s was not as dramatic either. Plus, the decrease did not last as long. Use again increased from the late 1980s until the mid-1990s. It then decreased slightly until the late 1990s and has stabilized since. Do these general trends fit the theory?

Social Mobility: Wave Two

Like rates of drug use, social mobility in Great Britain was relatively limited from 1920 until the early 1950s. However, social mobility, and the social changes it causes, increased again in the mid-1950s. As Bédarird (1979: 249) notes,

> Around 1955-56 English society was suddenly hit by a wave of change and from then onwards driven on a new course that was to transform the atmosphere of the country within a few years. This happened without warning. . . . The ingrained conformism of decades was driven by shocks both from without and within. While the young rebelled, literature and the arts were seized with a new creative spirit. In the mist of the calm, the country suddenly burst out in all directions.

These cultural changes were propelled by Britain's economic recovery and changes in the social structure.

While the two World Wars devastated the British economy (see, for example, Kennedy 1987), the recovery, both in terms of the economic growth and the speed with which it occurred, was "a miracle." The British economy began to grow substantially after the war. Beginning in the late 1940s and continuing through the 1950s, the average annual rate of growth of output per capita in Britain reached 2.4%, outpacing America's 1.6% growth rate. Similarly, exports increased by 28% between 1958 and 1963 (see Kennedy 1987; Hawdon 1996c). As a result, the British population was again enjoying prosperity. GDP per capita, which had increased by only 26.5% between 1913 and 1950, increased by 75.1% between 1950 and 1973. The growth rate in export volumes reached an annual average of 3.9% between 1950 and 1973, far outpacing the annual rate of growth between 1870 and 1950 (see Maddison 1995: 23, 74, respectively).

The economic recovery was accompanied by another dramatic change in the occupational structure. While the percentage of workers employed in manual occupations had decreased by only 9.1% between 1931 and 1951, it decreased by nearly 16% between 1951 and 1971 (Halsey 1988: 24). Instead of manual occupations in the manufacturing and agricultural sectors, an increasing number of British laborers were finding non-manual employment in the service sector (see Maddison 1995: 39). Moreover, an increasing number of women entered the paid labor force, especially during the 1960s. While only 37% of British women over the age of 16 were economically active in 1961, by 1971, 43% worked outside the home. By 1981, nearly half of all women over age 16 were economically active (Halsey 1988).

These economic changes resulted in greater mobility across social classes. Although circulation mobility remained limited in Britain, structural mobility increased during the second half of the 20th century (see Macdonald and Ridge 1988). For example, nearly 88% of professional and managerial workers in 1953 remained in those categories by 1963. Of those who were in the professions or management in 1971, 79% of the men and 76% of the women apparently remained in such jobs through the following ten years. However, because of the growing number of professional and managerial jobs, by 1981 nearly 38% of professional and managerial workers had been recruited from other occupational classes. The 1963 figures were 24% of men and only 18% of women. In fact, in 1981, 21% of men in management and the professions were manual workers ten years earlier. The comparable figure in 1963 was only 16% (Noble 2000).

Thus, increased prosperity, coupled with Labour policies, helped "ease the rigidities of the hierarchy" as many of the very poor Englanders "moved up into the ranks of those with some income and some social status" (Bédarird 1979: 225). Halsey (1988: 33) argues,

> The post-war period to the mid-1970s was one of economic growth, full employment, and prosperity against an international background of rapid dismantling of imperial power and relative decline of economic productivity. The managerial, professional and technical classes waxed and the industrial working class waned. Women, particularly married women, moved increasingly into paid employment.

In general, the economic situation in Britain after the Second World War induced social mobility, which, in turn, helped promote other structural changes. For example, levels of education expanded as more children extended schooling and attended universities (see Maddison 1995; Halsey 1988; Halsey 1988b).

While the total number of students enrolled in full-time higher education increased by a mere 11% between 1924 and 1940, it increased by 76.8% between 1940 and 1954. Yet the wave of increased education was just beginning. The number of students enrolled in full-time higher education increased an additional 213% between 1950 and 1963 (see Halsey 1988b: 270). The rate of increase continued during the 1960s as "the total student population in full-time higher education doubled" between 1963 and 1971 (Halsey 1988b: 271; Office of National Statistics 2003). The number of pupils at school in England rose from 4.7 million in 1946/47 to 8.4 million in 1975-1976 (Office of National Statistics 2003). There was also an exodus from the cities, beginning in the 1950s and quickening during the 1960s and 1970s. For example, by 1961, Greater London was experiencing depopulation and, between 1971 and 1981, the industrial centers of Birmingham, Manchester and Derby lost between 19% and 25% of their population (see Wood 1988). Meanwhile, the suburbs of the West Midlands and Mersey increased in population by 16.6% and 17.1%, respectively (Halsey 1988; Wood 1988).

In addition to these domestic changes, Britain was also undergoing radical changes in the international arena. Once clearly the hegemonic power, Britain's relative position in the world system had obviously slipped. Britain withdrew its colonial presence in India and Palestine in 1947 to concentrate their limited resources on protecting vital imperial interests such as the Suez Canal and Arabian oil. Even strategic moves such as this could not halt the dismantling of the English Empire. By 1956, when the English could not protect their interests in the Suez, the Empire had largely disintegrated. Further attempts to maintain the colonies, such as the repression of the Mau Mau rebellion in Kenya, only delayed the inevitable (see Hawdon 1996c; Chirot 1986). While the loss of their colonial possessions represented the end of British hegemony, it stimulated social mobility at home by attracting an influx of new immigrants. Although there was a net migration loss of 2.3 million people between 1900 and the mid-1980s, refugee Jews and Irish immigrants continued to come to England as they had in the early part of the century. In addition to these familiar groups, an influx of ethnic minorities from Africa and the West Indies arrived from the "New Commonwealth" after former colonies were granted, or won, their independence. Although the influx of minorities led to the passage of the Immigration Acts of 1962 - 1967, these had the "paradoxical consequence of increasing the flow of Indian and Pakistani families to settle permanently" (Halsey 1988: 15). A pattern of chain migration emerged where more and more inhabitants of the former colonies immigrated to Britain to fill the relative labor shortage. Soon their relatives and friends followed. As a result, the number of ethnic minorities in Britain increased during the 1960s (Office of National Statistics. 2003b). The influx of new ethnic groups challenged the long-time English assumption that minority populations were homogeneous. Instead, they found that these new groups were culturally varied. Not only did they differ from the English majority, they differed from each other (see Halsey 1988).

Consequently, social mobility in many forms increased dramatically after World War II in Britain. The economy underwent structural change thereby opening, at least to some extent, the rigid British class structure. More women had entered the paid workforce, thereby increasing their social contacts. Increasing numbers of youth were attending institutions of higher education, thereby encountering others from different parts of the country and different social classes. People fled the urban centers and moved to the suburbs. New groups of people from far-off lands, complete with their unique culture and customs, had entered the country. The social structure of Britain had changed, and, with this structural change, normative changes occurred too. The moral boundaries of Britain, which had constricted tightly after World War I and were widely shared by the bourgeois and the working classes (see Bédarird 1979), had expanded. When discussing the moral tradition in England, Lloyd (2002: 354) states,

At the beginning of the century the moral attitudes of the unenfranchised people at the bottom of the social scale were less restrictive than those of the classes concerned about respectability who made up the great bulk of the electorate. By the 1930s, the ideal of respectability was accepted throughout society, but by the 1960s it was much less universally accepted and some sections of the middle class, especially around London, were consciously uninhibited and regarded freedom from restraint as a good thing for its own sake. The greater freedom, or laxer sense of social discipline, showed itself in fashions and styles.

Greater freedom was also evident as "the traditional meanings and experiences attached to masculinity, femininity, adulthood, childhood, work, leisure and learning all changed as the division of labour altered" (Halsey 1988: 33). Greater freedom also meant increased rates of drug use.

Once again, however, rates of mobility began to slow as the 1980s approached, just before rates of drug use began to subside. The severe recession of 1979 - 1981 significantly altered the fortune of British workers and reintroduced an era of large-scale unemployment. By 1980, an excess of 3 million people were out of work (Doogan 1996), and, by 1986, unemployment had doubled its 1979 level (Lloyd 2002). It was not until 1986 that job gains began to reduce the levels of unemployment. Of the 3.2 million people out of work in October 1986, nearly 60 percent were "long-term unemployed" (Doogan 1996). In the southeast, where unemployment was noticeably lower than in the northern regions, employers received an average of 11 applicants for every notified vacancy (Doogan 1996). With unemployment increasing, the rate of job changing decreased. During the 1970s, ten million job changes occurred each year. By the mid-1980s, however, only six million annual job changes were occurring. By 1983, occupational mobility fell by nearly a quarter (see Doogan 1996). Because of the recession and slowing rates of structural mobility, wage inequality rose sharply in the 1980s (see, for example, Machin 1996; Prasad 2000). According to figures provided by Prasad (2000), for example, measures of inequality were a mere 0.04 between 1975 and 1980 but increased to 0.20 during the 1980s. In addition, migration decreased. Even into the early 1990s, the prolonged recession "produced 'gridlock of immobile households' in which, according to some estimates, a quarter of owner-occupiers were unable to sell their homes" (Doogan 1996: 215).

Moreover, the relative immobility reflected in the occupational structure was mirrored in terms of educational enrollments. While the number of pupils at school doubled between 1946 and 1976, the declining birth rates during the late 1970s led to a decline in pupils in the early-to-mid 1980s (Office of National Statistics 2003). Rates of university attendance also decreased. Fewer students were therefore enrolled in school at all levels of education. As fewer young people attended school, fewer students met new groups and associates. Similarly, fewer immigrants were coming to and settling in the United Kingdom. The total number of Grants of Settlement declined sharply in 1981 and 1982 (Jackson and Bennett 1998). They then decreased again by approximately fifteen percent (Dudley, Turner and Woollacut 2003). The early-to-late-1980s, therefore, was a period of relatively low rates of social mobility in Great Britain. Correspondingly, rates of drug use also decreased.

Beginning in the late 1980s, however, the British economy began to recover. With this recovery, rates of mobility began to increase. First, unemployment began to decrease noticeably in 1987. However, the improvement was not evenly spread across the various regions of the United Kingdom. Instead, the job market improved considerably in the southeast, while the improvement was much less noticeable in the north (see, for example, Lloyd 2002; Doogan 1996). These regional economic discrepancies led to increased migration. While net migration was one-third the rate of the 1971–1981 rate during the 1980s, it increased by a factor of four during the 1990s, actually surpassing the net migration rate of the 1970s (Office of National Statistics 2003b). Rates of immigration also began to increase again (Office of National Statistics 2003b). Next, inequality continued to increase after 1987; however, it did so at a much more modest rate than in the early 1980s. Measures of inequality indicate that inequality increased in the 1990s at half the rate of the early-to-mid-1980s (see

Prasad 2000). The greater economic opportunities and lower rates of inequality re-opened educational opportunities. After the number of students enrolled in higher education decreased in the mid-1980s, by the early 1990s, enrollment rates began to increase again. More importantly, the type of student going to university changed rather dramatically in the 1990s as a greater number of students from the lower social classes began to continue their education. Among students from families working in unskilled occupations, the percentage of youth attending institutions of higher education more than doubled between 1991 and 1997. In 1991-1992, only 6% of students from the unskilled classes went for higher education. By 1997-1998, 14% of such students did so (Office of National Statistics 2003). Thus, the "Thatcher Recovery" stimulated rates of social mobility. Apparently, it also stimulated increasing rates of drug use.

However, the recovery, like the increasing rates of drug use, was rather short-lived as economic problems re-emerged in the mid-1990s. Although the British economy flourished in the late 1980s, it did so because of an increased willingness for the government to incur new debt. While unemployment decreased after 1987, the growth was "not accompanied by development that might have provided more secure jobs for the future" (Lloyd 2002: 448). As a result, by the mid-1990s, unemployment returned to the levels of the early 1980s "without visible signs of economic recovery" (Lloyd 2002: 448). Economic problems also slowed rates of social mobility by decreasing internal migration. After interest rates doubled from 7.5 percent to 15 percent, the property market, as well as consumer confidence and the willingness and ability to migrate, was severely weakened (Doogan 1996; see Office of National Statistics 2003b). Thus, the relative boom years of the late 1980s quickly gave way to another slump in the mid-1990s. Finally, since the early 1990s, pupil numbers have increased relative to the 1980s, but remain well below their peak level of the 1970s (Office of National Statistics 2003). Consequently, since the mid-1990s, rates of mobility -- as measured by inequality, occupational changes, rates of migration, and educational attendance -- decreased and then stabilized. As the theory would predict, rates of drug use in the United Kingdom decreased and then stabilized.

Consequently, in the United Kingdom, as in the United States, rates of drug use and rates of social mobility co-vary. From the early 19th century until the present, when rates of social mobility increased, so did rates of drug use. Conversely, when mobility slowed, rates of drug use soon decreased. This pattern holds on both sides of the Atlantic and over a period of nearly 200 years. Does it also hold in the Pacific?

Social Mobility and Drug Use in Japan

Given their cultural similarities and obvious historic ties, it is not overly surprising that the United States and Great Britain have similar drug histories. Nor is it surprising that the influence of social mobility on rates of drug consumption is the same in these two westernized, advanced-capitalistic societies. Can a similar pattern be seen in the non-Anglo-Saxon nation of Japan? The Japanese do not consume drugs at the rate with which Americans and Brits do. In fact, relatively speaking, Japan has low rates of use. In addition, Japan, despite the strong influence the United States has had on its culture and social structure since 1945, is culturally and structurally different from both the United States and Great Britain. Does the theory also fit the patterns of Japanese drug use?

It should be noted that the data on drug use in Japan is not of high quality. Official government sources or rehabilitation workers publish most available data. There is no available household survey such as exists in the United States or Great Britain. The research that exists, at least that which exists and is published in English, relies heavily on arrest records. While I recognize the limitations and problems associated with using official data, and especially arrest records, there does seem to be agreement concerning the general patterns of use among the few scholars who have bothered to look at drug use in Japan. Therefore, while the usual cautions are noted, the best available data are used to provide an additional, albeit tentative, test of the theory.

A Brief History of Drug Use in Japan

With the exception of alcohol, Japan was not a major drug-consuming nation prior to World War II (see Vaughn et al, 1995; Tamura 1989; Greberman and Wada 1994). For example, in the 1930s, "3,600 of Japan's 70 million inhabitants were known drug addicts, whereas 50,000 of the 130 million inhabitants of the United States were addicts" (Vaughn et al 1995 495). These numbers reflect, in part, the long Japanese history of anti-drug laws and harsh treatment for drug offenders. For example, Japan prohibited opium smoking and the non-medicinal use of opium as early as the Edo or Tokugawa Period (1600 - 1867) (see Merrill 1942). The Japanese anti-opium sentiments stemmed from their awareness of the economic, political and social problems associated with opium in China (Yokoyama 1992). The Japanese held such strong anti-drug sentiments that they refused to sign the 1857 Japanese-Holland Trade Agreement until opium trading was explicitly prohibited. Similarly, the trade agreement between the United States and Japan, signed in 1858, banned the importation of opium into Japan (see Vaughn et al, 1995). Until the late 19th century, violators of Japanese drug laws were executed. Similarly, during the Meiji Restoration (1868 - 1912), the use, possession and distribution of opium was punished very harshly (see Yokoyama 1992). The strong anti-drug position held by the Japanese government, the equally strong anti-drug sentiments of the Japanese population, and the traditional culture based on "conformity, deference to authority, and filial piety" (Vaughn et al 1995: 494-495) effectively limited the use of illicit drugs. Japan's isolation from, and lack of commercial interests in, the opium trade also shielded Japan from the temptations of drugs.

However, this situation began to change during the 1930s. Once Japan occupied Manchuria and established the state of Manchoukuo, the government encouraged the cultivation, distribution and use of opium in Japanese-occupied lands. While drug use remained tightly controlled in Japan, it was promoted for profit in their newly conquered territories (see Vaughn et al 1995; Greberman and Wada 1994). Then, once World War II started, Japan's domestic position on intoxicants changed. As the war effort drained resources and labor from the small Island nation, the government began to contract with pharmaceutical companies to produce stimulants for their soldiers and factory workers (Kato 1969; Tamura 1989; Vaughn et al. 1995; Yokoyama 1992). Throughout the war, the Japanese government aggressively encouraged the use of stimulants to inspire their citizens' "fighting spirits."

> Pilots were expected to fly planes for many hours beyond their physical capacity; soldiers were expected to fight as long as days at a time with no rest; submarine commanders and midshipmen were required to endure months of maritime service on meager rations; factory workers laboured in subhuman conditions with deteriorating and broken equipment. Taking stimulants to enhance performance was a mark of patriotism (Vaughn et al 1995: 497).

The Japanese government, therefore, provided an air of legitimacy to amphetamine use and drug use in general. As a result, "there was an explosive stimulant epidemic" (Tamura 1989: 83) and the country experienced its first significant drug problem (Greberman and Wada 1994; Kato 1969; Vaughn et al 1995). Between 1946 and 1956, for example, it was estimated that up to 2 million Japanese regularly injected amphetamine (Reid 1998: 75). While stimulants were by far the most widely used illicit drug in Japan in the 1950s, the use of narcotics also increased during this time (Tamura 1989; Vaughn et al 1995; Yokoyama 1992). While military veterans swelled the number of users, use was not limited to them. Students, entertainers, service workers and juvenile delinquents also began experimenting with drugs (see Kato 1969; Vaughn et al 1995).

The Japanese Government responded to the increased rates of amphetamine use by passing the Stimulant Drug Control Law of 1951. This law was revised, adding harsher punishments, in 1954 and 1955 (see, Tamura 1989; Vaughn et al 1995; Yokoyama 1992; also see Greberman and Wada 1994). In addition, the government launched targeted educational campaigns in communities, drug prevention classes in schools and police crackdowns on the drug-trafficking "Boryokudan" gangs (JICA 1992; Greberman and Wada 1994; Vaughn et al 1995). These efforts helped shift public sentiments toward stimulants, and, by the late 1950s, stimulants had "acquired such a bad reputation in the Japanese press that major drug dealers started to traffic narcotics instead" (Vaughn et al 1995: 499; also see Kato 1969; Yokoyama 1992). As a result, stimulant use decreased after the mid-1950s.

The 1960s marked what researchers have called the "transition period" in Japanese drug history. Beginning in the early 1960s, the use of tranquilizers and narcotics, especially methaqualone and morphine, increased. Cannabis use also began to increase slightly, and by 1963, the use of solvents had become popular (Greberman and Wada 1994; Kato 1969; Tamura 1989; Vaughn et al 1995). Japanese pharmaceutical companies increased their production of barbiturates, sedatives and tranquilizers as these drugs became more widely used, both legally and illegally, throughout Japan (Kato 1969; Tamura 1989). By the end of the 1960s, the preferred illicit drug had changed from stimulants to narcotics and cannabis (Kato 1969; Vaughn et al 1995).

Despite the increasing popularity of non-stimulant drugs and change in the drug of choice, overall rates of drug use remained relatively stable in the 1960s. For example, heroin-related offenses peaked in 1962 at slightly less than 2,400 (Tamura 1989) and then decreased steadily throughout the 1960s (see Vaughn et al 1995). However, arrests for opium offset the decline in heroin arrests. Cannabis offenses also increased steadily over the course of the decade and solvents became popular recreational drugs (Tamura 1989; Vaughn et al 1995); however, stimulant use decreased substantially during this time. Overall then, the choice of drug changed during the 1960s, but the rate of use remained stable. While this rate of use was much higher than Japan had traditionally witnessed, it was lower than during the "first stimulant period" of the 1940s and early 1950s.

This situation changed in 1969 when the "second stimulant drug period" began. Beginning in 1969, arrests for stimulant drug offenses increased again, going from fewer than 1,000 in 1969 to nearly 15,000 by 1973. Cannabis offenses also more than doubled between 1969 and 1972 (see Vaughn et al, 1995: 502). Rates of use -- at least if estimated with arrest data -- continued to increase until the mid 1980s. In 1981, approximately 40,000 drug offenses were cleared by arrest (see Vaughn et al 1995). Once again, stimulants became the most widely used illicit drug in Japan (see Tamura 1989; Reid 1998; Suwanwela and Poshyachinda 1986). By the mid-1980s, as many as 600,000 Japanese were regular stimulant users (Vaughn et al 1995).

Beginning in the mid-1980s, however, rates of use began to decrease. Between 1984 and 1993, for example, arrests for drug offenses decreased by approximately 40% (Reid 1998). Similarly, in 1996, it was estimated that between 400,000 and 600,000 Japanese used stimulants regularly (Reid 1998), a similar number of stimulant users as in 1985. Thus, despite nearly a 4% increase in the population between the mid-1980s and mid-1990s, the number of stimulant users remained unchanged during that time. This decline lasted until the mid-1990s after which rates of use stabilized. Then, use rates increased slightly in the late 1990s (UNODC 2000; UNODC 2002; Drug Policy Alliance 2002).

Consequently, Japan experienced two waves of relatively high rates of drug use, at least by Japanese standards. Unlike the United States and Great Britain, however, Japan's periods of relatively high rates of use have both occurred since World War II. The first period of high use was immediately following World War II. Rates then stabilized, but at relatively high levels. Then, beginning in 1969, rates of use increased rapidly again. Rates then decreased in the mid-1980s and, despite a slight increase recently, have remained generally stable since then. Do these periods of high rates of use correspond to periods of high social mobility?

Social Mobility in Japan since World War II

By the end of World War II, Japan was in chaos. The war had devastated Japan economically, socially and emotionally. In addition to the estimated 1.2 million soldiers who had been killed, Japanese civilian casualties approached

500,000 killed and 625,000 seriously injured. Thousands of others were missing after the frequent fire raids and two atomic bombings. In addition, about 360,000 Japanese captured by the Russians in Manchuria, Korea, and the Kuril Islands were still missing. This massive loss of labor, coupled with the relentless pounding of Allied bombs, left Japan's economy nearly ruined. In 1946, for example, real national income was only 57% and real manufacturing wages were only 30% those of 1934-1936 (Kennedy 1987). Coal production had been halved, and the cotton industry had lost nearly 80% of its productive capacity. Japanese exports were at 8% their 1934-1936 level (see Allen 1981; Kennedy 1987). In addition to their economic problems, Japan was now an occupied nation. In January 1946, it appeared that Japan would never be a major world power again. However, the Japanese did posses a well-educated, cohesive, highly skilled and motivated workforce. As in Europe, America provided the needed financial capital and, as in Europe, Japan provided the human capital. As a result, the Japanese rebounded, albeit slowly at first (see Allen 1981; Allen 1958). Japan's crash and rebound as an economic powerhouse is well documented. However, the rebound also created social upheaval and social mobility that led to changes in the deviance structure.

First, the war displaced many of the survivors. In the 66 major cities that endured air attacks during the war, approximately one-half of the dwellings had burned to the ground. This loss represented one-quarter of the nation's total housing units (Allen 1958). Consequently, residential mobility was extremely high immediately following the war. The movement of people may have come from necessity, but it came nevertheless.

Next, the Japanese economic system was fundamentally restructured after the war. First, as part of the economic recovery plan and a means of averting a communist revolution by radicalized rural groups (see Moen 1999), Japan instituted land reforms in 1946 - 1947. Over one-third of the farmland changed hands during the first four years following the war (Kosai 1997; Moen 1999). Between the end of the war and 1950, the cultivated area farmed by tenants was reduced from 46% to approximately 10% of the total cultivated area (Moen 1999). Thus, independent landholders dominated the agricultural sector (Kosai 1997). The land reforms not only changed the structure of the agricultural sector, they also created new social relations in Japanese rural society (Moen 1999). Second, part of Japan's economic recovery resulted from the weakening of the giant *Zaibatsu*.[99] These powerful families, working with the Japanese state, had monopolized many of Japan's major industries. The American occupiers were distrusting of the Zaibatsu since they represented a link to Japan's Imperial past. The weakening of the Zaibatsu's control over the economy lessened concentration in the industrial sector. Thus, both the agricultural and industrial sectors became increasingly fluid and competitive (Allen 1958; Kosai 1997).

Third, the increased competition that resulted from the economic transformations stimulated growth. This heightened rate of growth then received an additional catalyst when the Americans entered the Korean conflict (Allen 1981; Allen 1958; Kosai 1997). For example, America's increased orders for trucks saved the Toyota Corporation, which was on the verge of bankruptcy (Kennedy 1987). The high rates of economic growth ushered in additional changes in Japan's economic structure. For example, immediately following the war, more than 10 million workers gave up their jobs in agriculture and forestry and moved into the manufacturing and service industries. As a result, the percentage of full-time farmers fell from 50% of the workforce to 34.8% during the 1950s (Moen 1999). As manufacturing expanded and heavy industry began to catch the more traditional light industry in terms of exports (Kosai 1997), occupational mobility increased. While Japan is famous for "permanent employment" practices, this perception may have been overstated. Permanent employment, which did increase during the prosperous years of the 1950s, was mostly enjoyed by employees in large companies. It was very unlikely among blue-collar occupations, those working in small firms or younger workers (Cheng and Kalleberg 1997; also see Cheng and Kalleberg 1996). In fact, the rate at

---

[99]. The Zaibatsu were large family enterprises that rose to prominence during the Meiji era.

which a typical 30 year-old Japanese male changed jobs was extremely high by Japanese standards during the period 1946 to 1955 (see Cheng and Kalleberg 1997).

Fourth, these economic changes increased demand for skilled labor. Industry's call for labor and promise of higher wages increased residential mobility again. Between 1950 and 1960, Japanese farm families provided the urban industrial sector with a stream of low-wage labor. Each year during this time, nearly thirty thousand young men left the rural farming areas to work in the urban factories.

Fifth, as in Europe and the United States, "educational mobility" increased. The relatively high rates of occupational mobility, growth in manufacturing employment and the labor laws enacted in the first two years of occupation improved workers' wages and their quality of life (Kosai 1997). As a result, income became more equally distributed and educational opportunities opened to greater numbers of Japanese youth. Japanese youth took advantage of these improvements as more young people attended university. For example, in 1950, 390,687 students were enrolled in university-level programs. By 1955, 609,685 students were attending university classes. Therefore, in a five-year period, the number of students leaving home to attend university increased by 56%.[100] Moreover, by 1955, nearly 11,000 Japanese women were enrolled in universities. Compared to just five years earlier, the number of Japanese women attending universities jumped nearly fourfold.

In addition to occupational, residential and educational mobility, changes in the family structure also led to a more fluid society. As is often the case following major wars, the Japanese divorce rate increased dramatically after 1945. While divorce rates ranged from 0.63 to 0.80 per 100,000 population from the late 1920s until World War II, they increased to over 1.0 per 100,000 population in 1947 and remained above 0.92 per 100,000 population until 1953 (Japanese

---

[100] Throughout the following sections, I use several volumes of the United Nations' *Statistical Yearbook* (1948 - 2000) for the educational enrollment figures.

Ministry of Health 2004). Thus, divorce rates were 20 to 40% higher immediately following the war than they had been prior to the war. [101]

Finally, because of the military occupation of Japan, Americans and American influences were everywhere. While Japan has traditionally limited the number of foreigners in their country (see, for example, Fuess 2003), their military defeat ensured outsiders would be among the general population. In addition to the exposure to American culture, the American occupation, which lasted from 1945 until 1952, helped usher in land reforms, economic decentralization and a democratic political structure. The chaos of war and the reconstruction of the nation afterward had done more to transform Imperial Japan into a westernized nation than the Meiji reforms.

These transformations all resulted in heightened rates of social mobility. The destruction of residential housing and the land reforms stimulated residential mobility. The economic decentralization stimulated occupational mobility, which, in turn, stimulated educational mobility. The adoption of Japan's constitution created a more open political structure. As these changes elevated rates of social mobility, cultural norms and definitions of acceptable behavior changed. As the "deviance structure" expanded, drug use increased.

By the late 1950s, however, the pace of change began to slow somewhat. Although the Japanese economy continued to modernize and averaged an annual GDP growth rate of over 10% per year during the 1960s (see Kennedy 1987; Bartram 2000), overall occupational mobility was relatively limited. The occupational mobility that occurred during this time was mostly limited to *within-firm movements* (see Cheng and Kalleberg 1996). Moreover, "permanent employment" was extended to more workers in blue-collar occupations and small firms during the 1960s. Thus, on average, a 30 year-old Japanese male working

---

[101.] I consider divorce a source of social mobility because it represents a change in the status position of the persons involved. That is, they become "single" instead of "married." One's status of married or single then determines rates of interaction among groups of people. Moreover, divorce creates new social relations and unfamiliar roles (see, for example, Cherlin 1981). It is therefore a source of social mobility in the sense the term is used here.

in the non-farm sector would have worked for approximately three-and-one-half employers between 1946 and 1955. This average dropped to just above two employers between 1955 and 1970 (see Cheng and Kalleberg 1997). Thus, for young workers, the average number of job changes *decreased* by approximately 28% between 1955 and 1970. A similar trend, although not as pronounced, occurred for older workers.

Next, although Japan continued to develop its manufacturing sector -- and, in fact, came to dominate the world's production of several commodities ranging from cameras to musical instruments and other electronic products -- the pace of the transformation from an agricultural to a manufacturing society remained steady during the 1960s. Approximately 48% of Japanese workers were employed in the agricultural sector in 1950 (Maddison 1995), 30% were agricultural workers in 1960 and 18.6% were farmers in 1969 (see Bartram 2000). These figures represent a 37.5 and 38% reduction in the number of agricultural workers each decade. While these decreases represent substantial change, the pace of change was relatively constant. Moreover, they do not compare to the rate of change that occurred after 1969.

A similar phenomenon occurred with respect to education and changes in the family structure. While enrollments into college boomed immediately after the war and through the mid-1950s, they increased at a much slower rate in the late 1950s. While enrollments increased by 56% between 1950 and 1955, they increased by only 6.9% between 1955 and 1959. Similarly, after a substantial increase in the early 1950s, the number of Japanese women attending university classes increased by only 8.7% between 1950 and 1955. With respect to the family, divorce rates, which rose noticeably during the late 1940s and early 1950s, began to decrease. In 1952, for example, there were 0.92 divorces per 100,000 population. By 1957, divorce rates had fallen to 0.79, and by 1963, divorce rates were at a 30-year low of 0.73 per 100,000 population (Japanese Ministry of Health 2004). The Japanese family structure, although still undergoing the changes associated with modernization, was nevertheless stabilizing to some degree in the 1960s.

Consequently, while social mobility remained high in the late 1950s and throughout the 1960s, it was lower than during the "stimulant epidemic" of the late 1940s and early 1950s. Similarly, while the drug of choice changed during the "period of transition" (see Vaughn et al 1995), rates of drug use remained relatively stable. Although use was relatively high by Japanese standards during the 1960s, it remained below the levels observed in the late 1940s and early 1950s. Thus, the 1960s saw both rates of use and social mobility decline and then stabilize. Both of these trends changed rather abruptly at the end of the decade.

As stated earlier, the second Japanese drug "epidemic" began in the late 1960s. Once again, it was also in the late 1960s that significant structural change began occurring again in Japan. Economically, the pronounced labor shortages that developed in the early 1970s provided one source of change. In 1971, for example, there were 1.19 job openings for every applicant. By 1973, there were 2.26 openings per applicant (Bartram 2000). Because of this increased demand for workers, rural workers moved into the urban areas to pursue new jobs and better economic opportunities (Bartram 2000; also see Fuess 2003). During the 1960s, there was a 38% reduction in agricultural workers. There was a 54.3% reduction in the number of Japanese agricultural workers during the 1970s (see Bartram 2000). Thus, the amount of change that occurred in the 1970s was nearly one-and-one-half times that which occurred in the previous decade. The job-boom also attracted workers from other countries. Despite the government's attempts to limit the number of foreign workers entering the nation, there was an influx of illegal immigrants to Japan during the 1970s (Fuess 2003; Iyotani 1998). In addition, as jobs became increasingly plentiful, occupational mobility again increased. This mobility was especially high among men aged 16 to 20 (Cheng and Kalleberg 1997).

Enrollments in institutions of higher education, which began to increase substantially again during the mid-1960s, grew at unprecedented rates after 1969.

Between 1964 and 1972, for example, university enrollments went from 985,077 to 2,007,870, an increase of 103.8%. The number of female students increased between 1964 and 1970 by an astonishing 132.6%. Enrollments continued to increase throughout the "Izanagi Boom" of the early 1970s (see, for example, Fuess 2003; also see UN *Statistical Yearbook* 1972) and throughout the entire decade of the 1970s. Between 1970 and 1979, overall enrollments increased by 34% and female enrollments increased by over 55%. By the end of the 1970s, nearly 2.5 million people were enrolled in Japanese universities or university-level educational institutions (UN *Statistical Yearbook*, 1981).

Another source of mobility during this time occurred in the family structure. As in Europe and the United States, Japan experienced a baby boom after the war, especially between 1947 and 1949 (Kumagai 1995). By 1969, these "boomers" created a "new student youth movement" (Tamura 1989: 86). This movement of youth, now between 20 and 22 years old, altered social relations and, in so doing, altered the deviance structure and social definitions of morality. They provided a mass of people of prime drug-using age who helped usher "society's hedonism" (Vaughn et al 1995: 502). Another change in the family structure also altered social relations. As in other modernized nations, the presence of "stem families" consisting of three or more generations became less common and nuclear families became increasingly common. Divorce also became more common as divorce rates nearly doubled between the mid-1960s and the early 1980s (Kumagai 1995; Japanese Ministry of Health 2004). These new familial arrangements created new and un-institutionalized social relations that were contrary to the long-standing Japanese norms of conformity and filial piety.

When discussing the "second stimulant period," Vaughn and his associates offer similar reasons for the sudden increase in drug use that occurred in 1969 Japan. While these authors do not use the current theory per se, their argument is certainly consistent with it. The primary causes these authors give for the increase in Japanese drug use represent forces of social mobility. They (Vaughn et al 1995: 503) write,

> As the 1970s began, Japan's economy grew at a feverish rate, bringing prosperity, a new student youth movement, and an increased demand for stimulants. . . . Many Japanese were entering the middle class; hence, the economic prosperity of the Japanese people created more disposable income for cannabis purchases. Improving economic conditions also allowed more Japanese to travel abroad, where they had more opportunities to purchase and import cannabis.

Economic mobility, increasing college enrollments, a youth movement and even foreign travels reflected changes in the structural conditions of Japanese society. As the structural dimensions that determine rates of interaction changed, so did the nation's moral boundaries. As in the United States and Great Britain, rates of drug use increased in Japan during times when social mobility altered Japanese social structures.

Yet, as in the United States and Great Britain, rates of drug use began to decrease when rates of social mobility decreased. As noted above, rates of use began to decrease around 1984 and continued to do so until the late 1990s. Social mobility also began to decrease during the 1980s and continued to do so until the late 1990s. First, economically, mobility began to slow. Although Japan was still experiencing an economic boom in the 1980s, rates of mobility failed to keep pace. For example, while rates of mobility increased by 12% between 1955 and 1975, they increased by only 1% between 1975 and 1995 (Kanomata 1997). Similarly, while the percentage of Japanese workers employed in the agricultural sector decreased by 54.3% during the 1970s, there was only a 16% reduction in the percentage of workers in the agricultural sector during the 1980s. Moreover, unemployment rates never exceeded 2% between 1969 and 1980, however, in 1985 they reached 2.6% and remained above 2% through the rest of the 1980s and 1990s (see Bartram 2000; UN *Statistical Yearbook* 1991). Even college graduates were having a difficult time finding employment once the recession of the 1990s hit (see Fuess 2003).

Next, as in other modernized nations, the baby boom of the late 1940s and early 1950s was followed by a dramatic decrease in fertility during the 1960s and 1970s. In fact, fertility rates fell below levels of reproduction during the 1960s and were barely above them during the 1970s (Kumagai 1995). Thus, while Japan's population increased by 55.4% between 1945 and 1975, it increased by only 12% between 1975 and 1995 (see UN *Statistical Yearbook* 2000), and the rate of population growth in Japan has steadily decreased since 1973 (Fuess 2003). In addition, the family structure further "stabilized" when Japanese divorce rates declined in 1984. After doubling between 1963 and 1984, divorce rates began a downward trend for the fist time in two decades (Japanese Ministry of Health 2004). Divorce continued to decline appreciably -- by nearly 20% -- until 1988. Rates remained stabile until the early 1990s then began increasing again modestly (Japanese Ministry of Health 2004).

Because of the aging population, college enrollments began to decrease. For example, college and university enrollments *decreased* by 2.3% between 1984 and 1985 (UN *Statistical Yearbook* 1991). Although enrollments increased again, largely due to the difficulties in finding employment, the rate of increase was significantly lower during the late 1980s than it had been in the 1950s, 1960s, and 1970s. During the 1990s, the college-aged population actually shrank (Fuess 2003) and, as a result, enrollment rates suffered (UN *Statistical Yearbook* 2000). Thus, by 1984, the "boomers" who created the "new student youth movement" of the early 1970s were nearly forty years old and well past their prime drug-using age. And, as the "boomers" aged, fewer youth were replacing them in the deviance-prone age group. Moreover, fewer youth were leaving home to experience the liberalizing effects of college.

By the dawn of the 21st century, Japan had completed another "cycle of deviance." Once again, drug use, particularly "deviant drug use," became common when rates of social mobility increased. Their use was less prevalent when rates of mobility decreased. When Japan experienced a tremendous amount of structural change immediately following World War II and between 1969 and 1984, rates of use were high. When the pace of change slowed in the late 1950s, rates of use stabilized. When rates of mobility decreased after 1984, rates of drug use decreased. Once again, the theory fits the available evidence.

Israel: Social Mobility, Drug Use and the Six-Day War

The changes in Israel caused by the 1967 Six-Day War further illustrate the relationship between social mobility and drug use. Like the Japanese, Israeli Jews never approached the American or British levels of drug use. However, like in Japan, Israel experienced a "drug epidemic," at least by their standards, and, as in Japan, this epidemic occurred when the nation underwent rapid structural change due to increased rates of social mobility.

Prior to the Six-Day War, hash smoking was nearly non-existent in Israel, at least among the Jewish population. In fact, based on available evidence, drug use in general was very limited in Israel during the modern nation's early history. While a few urban Arab workers smoked hash, some European immigrants used morphine, and opium smoking was somewhat common among the small pockets of Turks, Iranians and Salonikans, drug use was not widespread nor was it considered overly problematic. As Palgi (1975: 214) says, "until the Six-Day War in 1967, hashish smoking was on a very limited scale, almost completely identified with unskilled or socially marginal groups among Eastern Jews and the urban Muslim population." Even Jews from areas where hashish smoking was common and accepted rarely used illicit drugs. According to Moroccan Jewish informants living in Israel, hash smoking or eating was not an accepted or practiced custom in their community, regardless of where they lived. When asked why, they responded, "it was not Jewish behavior" (cited in Palgi 1975: 211). Similar to the Eastern European Jew's rejection of drunkenness, Jews kept their social lives separate from Muslims, and using hashish was viewed as an Arab tradition.

After the Six-Day War, however, drug use, especially the recreational use of hashish, began to spread to the Israeli-born population of European origin (Palgi 1975; also see Javitz and Shuval 1982). Use became popular among the urban, middle-class youth. Hash smoking became especially prevalent in the kibbutzim (Ben-Yehuda 1986). By the early 1970s, the percentage of young who had tried an illicit psychoactive drug had reached nearly five percent. After this "dramatic increase" in the early 1970s, use rates, at least among Israeli youth, remained stable until the early 1980s (Ben-Yehuda 1986; Javitz and Shuval 1982).[102]

Did rates of social mobility increase in Israel after the Six-Day War? The answer is an unconditional "yes." First, Israel's stunning victory expanded their territory to include all of the lands of Biblical Israel. The Israelis captured the Sinai and Gaza Strip from Egypt, East Jerusalem and the West Bank from Jordan and the Golan Heights from Syria. While much of this land had relatively few inhabitants, the territorial expansion added significant numbers of non-citizens to the Israeli population. Because of the expansion, approximately 1 million non-citizen Arabs were now living in Israel (E. Cohen 1989). As Kimerling (1989: 254) says, "after 1967, a mass of Arab residents, lacking civil rights, were unexpectedly included and integrated in the political and economic system." Given their religious, cultural and obvious political differences, the incorporation of these non-citizens radically altered the Israeli economic, political and social systems (see Parker 1996).

Second, the war helped Israel emerge from the recession of 1966 - 1967. The boom of the early 1960s ended in 1966 and signs of recovery were slow in coming in 1967. For example, between 1966 and 1967 per capita GNP decreased by 1.8% and unemployment reached an all-time high at 12.4% during the first

---

[102] Virtually no systematic evidence exists regarding the use of psychoactive drugs among the general population of Israel. A number of studies have attempted to estimate the rate of use among adolescents. Nevertheless, according to Ben-Yehuda (1986: 500), "the epidemiological picture is clear and consistent." Use among adolescents rose after the Six-Day War and remained stable until the early 1980s at between 3 - 5 percent of the youth population.

quarter of 1967 (Kanovsky 1970). While the war caused damage and temporary economic loss, it stimulated growth shortly after the hostilities ceased. First, the capturing of the Sinai opened additional oil reserves to the Israelis, and oil production and revenues increased (Kanovsky 1970). Next, the war stimulated the development of the Gulf of Agaba city of Eilat. This small port city, always an important link to the Red Sea, expanded in size and importance once the Israelis built a "land bridge" from Eilat to Haifa and an oil-pipeline from Eliat to Ashkelon. These investments in Israel's infrastructure quickly paid dividends by decreasing the price of imports and exports and stimulating industrial production. While industrial production was 20% lower in June of 1967 than it had been in May, it returned to pre-war levels by the end of July. After July, "the advance (in industrial production) was both unusually rapid and almost uninterrupted" (Kanovsky 1970: 100). Similarly, the seizure of the Golan Heights provided a buffer from Syrian military outposts and an opportunity to better utilize the Jordan River. Agricultural production immediately benefited from the dredging and irrigation projects that quickly reclaimed 12,500 acres in the fertile Hula Valley (see Kanovsky 1970; *UN Statistical Yearbook* 1970). In part because of the increases in industrial and agricultural production, GNP grew by 14.2% in 1968, and per capita GNP increased by 11.0% (Kanovsky 1970; E. Cohen 1989). In addition, unemployment dropped to 5% by the end of 1968 and continued to decrease until it reached 2.6% in 1973 (see UN *Statistical Yearbook* 1974). Private consumption, which had increased at a modest 2.1% in 1967, jumped by 11.1% in 1968 (cited in Kanovsky 1970: 87). Israel was enjoying a "psychological stimulus of victory and (an) air of optimism" that were "powerful forces accelerating the economic revival and rapid rate of expansion characterizing the postwar economy" (Kanovsky 1970: 86).

Third, the economic boom that followed the Six-Day War also altered the Israeli economic structure by creating labor shortages. The shortages generated pressures to permit Arab laborers and other immigrants to join the workforce. In 1968, for example, Palestinians were officially allowed to work in Israel, provided they received permission from the Military Government and Ministry of Labor. By the spring of 1969, 15,000 territory workers were enrolled to work legally in Israel. By 1973, the number of territory workers had increased to 65,000 (Cohen 1989). Increasing numbers of women also joined the labor force (Kanovsky 1970). In addition to the new Arab and female workers, "Israel was flooded with volunteers from Europe and North and South America" (Ben-Yehuda 1986: 499). Then, large numbers of Soviet Jews began to immigrate to Israel in the early 1970s (Kimerling 1989). Israel was experiencing unprecedented rates of economic mobility, both for its citizens and its non-citizens.

Fourth, as growth rates expanded, economic mobility increased and unemployment rates reached historic lows, the typical growth in higher education followed. In 1966, slightly over 20,000 students attended Israeli universities. University enrollments more than doubled after the war. By 1969, 49,076 students enrolled. By 1972, 70,431 students were attending university classes (U.N. *Statistical Yearbook* 1973). Thus, in a seven-year period, university enrollments increased by 246%. Part of this impressive increase in enrollments was due to more young Israeli females attending universities (U.N. *Statistical Yearbook* 1973).

Finally, the capturing of historically significant land, coupled with the publicity Israel received during the war, caused a wave of tourists to visit the country. Between January and April of 1968, for example, 30% more tourists visited Israel than had during the same period in 1967. Approximately 400,000 tourists came to Israel that year. This number is 24% higher than in 1966, the previous peak year for tourists (Kanovsky 1970). Rates of tourism continued to increase until the economic boom ended with the 1973 Yom Kippur War and the ensuing oil embargo.

Thus, immediately following the Six-Day War, Israel was transformed. Economic prosperity returned and brought greater mobility. A sense of optimism replaced "the feeling of pessimism and even gloom which had been prevalent during the recession, and especially during the weeks of tension preceding hostilities" (Kanovsky 1970: 86). The prosperity and optimism stimulated consumer spending and a torrid economic expansion. More young Israelis received a college education. Large numbers of Arabs were assimilated into the workforce, even if they were still excluded from the political realm. Several other groups with less-than-familiar cultures also entered the country as volunteers, paid foreign workers, political refugees and tourists. As they came, they brought new attitudes and behaviors. Most importantly for our purposes, these "outsiders" brought new ideas about drug use. Indeed, many believe that young Israelis were first introduced to recreational hashish use by the westerners who flocked to Israel in the late 1960s (see Ben-Yehuda 1986; also see Palgi 1975; OGD 1997).

This point I do not contest. In fact, it seems not only plausible, but also most likely that "outsiders" introduced marijuana to Jewish youth living in Israel in the late 1960s. However, I would add to this explanation the causal role that social mobility played. The increase in mobility that followed the Six-Day War was precisely what enticed these westerners to come to Israel and compelled Israel to assimilate millions of Arabs, Americans and Europeans. Social mobility brought the young Israelis together in universities and cities where they met the westerners and learned the recreational practice of drug use. Finally, social mobility created an air of affluence, which, in turn, afforded the luxury of engaging in recreational drug use. Rates of use then stabilized after the Yom Kippur War ended the rapid expansion and stabilized rates of mobility.

Drugs and Social Mobility in the Soviet Republics and Eastern Bloc Nations

The drug histories of the United States, Great Britain, Japan and Israel demonstrate the relationship between social mobility and rates of drug use. We can see the relationship between rates of social mobility and rates of drug use in other cultures also. The recent histories of the former Soviet Republics and eastern bloc nations also support the theory. Social mobility increased in the former republics and eastern bloc nations once the Soviet's rule was lifted. Reform and independence brought radical changes in the economic, political and social systems in all of these nations. The transformation from a socialist to capitalist economy and a totalitarian state to a democracy radically altered social relations and the social structure. Although much of this mobility was "downward," it was, nevertheless, mobility. Despite widespread unemployment that plagues these countries, the changes in the economic and political structure caused millions of persons to leave their old status positions and find new ones. As the theory states, the direction of mobility is irrelevant, it is the rate of mobility that matters (see Hawdon 1996b).

The social structure of the Soviet Union -- and, indeed, all former "communist" countries in Europe -- was characterized by self-perpetuation and limited mobility (Zickel 1989; Glenn 1996). While claiming to be "classless," these states were rigidly divided. Party membership determined both mobility and life-chances. Access to higher education, the major path to political and social advancement, had always been restricted and declined dramatically after World War II (OECD 1995). Moreover, the "period of stagnation" that coincided with Brezhnev's tenure (1964 - 1982) further curtailed movement socially and politically (Glenn 1996). Moreover, the sluggish Russian economy reduced opportunities for social mobility, thereby accentuating differences among social groups and widening the gap between the powerful *nomenklatura* elites and the rest of society (Glenn 1996; Zickel 1989; Thompson 2004). During the Soviet era, therefore, mobility was limited and any advancement was slow (Zickel 1989).

The Gorbachev reforms and then the formal demise of the Soviet Union in 1991 brought a measure of freedom to Russia's people. However, it also severely weakened the welfare system that had provided a strong safety net. Under the Soviet system, Russians were guaranteed employment, basic medical care and a relatively handsome pension. Moreover, they received government subsidies for food, clothing, shelter, and transportation (Thompson 2004; Glenn 1996; Zickel 1989). These state-provided securities were largely removed once communism ended. For the average citizen, social and economic conditions worsened considerably in the early post-communist era (Glenn 1996; also see Lipsmeyer

2003). Although some components of state support remained close to their Soviet-era levels, there was an overall decline in living standards and widespread downward mobility (see OECD 1995: 123). While part of the decreasing standard of living was due to the rapid marketization of the economy and the adjustments associated with it, it was also due to the greater rate at which Russians changed jobs in the early 1990s. Numerous government and heavy industry jobs were eliminated. Unemployment hit the agricultural, energy and light manufacturing sectors as well. However, the business class and service sectors expanded (Glenn 1996; OECD 1995). In 1995 alone, over 18 million Russians either left old jobs or began new ones (see IOL 1996).

In addition to the far-reaching economic and political transformations, the dismantling of the Soviet Union induced other changes as well. First, existing housing was increasingly privatized in the mid-1990s, and more types of dwellings became eligible for privatization (OECD 1995). In addition, officials abolished the internal passport system in 1993 that had effectively restrained internal migration during the Soviet period. As a result of housing privatization and eased internal migration, large numbers of Russians changed residences in the early and mid-1990s (Glenn 1996).

Next, the Russian Federation also absorbed millions of immigrants beginning in the late 1980s and continuing through the 1990s. In 1993, Russia became a UN "country of first resort" for foreign refugees fleeing their home countries. As a result, thousands of Azerbaijanis and Armenians came to the Russian federation countries when conflicts broke out in their homelands in the late 1980s. Similarly, thousands of Turks fled Uzbekistan after the 1989 massacre occurred there. Large numbers of non-forced immigrants also arrived from fellow *Commonwealth of Independent State* nations such as the former Soviet republics of Kazakhstan, Ukraine, Tajikistan and Turkmenistan. The decline in border security since the dissolution of the Soviet Union also made illegal immigration easier in many areas. In 1994, for example, approximately 28,000 recent immigrants were living illegally in Moscow. In total, as many as 2 million people are believed to have immigrated to Russia in 1996 alone (Glenn 1996).

Third, the liberal educational policy of 1992 decentralized education. The new policy permitted tremendous local autonomy and, as a result, the rigidly standardized Soviet educational system lost its centralizing and unifying force. Now, students learned formerly prohibited ideologies such as individualism. Local and non-Russian histories influenced the new curriculums (Glenn 1996). Higher education underwent a radical transformation in the post-Soviet era as well. In particular, a growing demand for knowledge about business and the workings of capitalism attracted new schools and new students. In the first two years after the Soviet Union dissolved, for example, more than 1,000 business schools and training centers were established (Glenn 1996).

These radical changes ushered in an unprecedented cultural rebellion. A new youth culture emerged that adopted the nonconformist dress, music and anti-establishment values of young people in the West (Glenn 1996). This youth culture also began using drugs like their western counterparts. Especially in the larger cities, LSD and natural hallucinogens hit the street in the early 1990s and their use quickly spread among the urban youth. Cocaine also entered the nation's drug scene in the early 1990s. By 1993, clubs and bars, all protected by Russia's growing organized crime groups, featured a variety of drugs including opiates, hallucinogens, cocaine and cannabis (see Poznyaka et al 2002; Glenn 1996). By the mid-1990s, use had increased dramatically. According to Glenn (1996), the number of users increased by at least 50% a year every year until the mid-1990s. Based on figures from the Russian *Center for the Study of Drug Addiction*, an estimated 2 million Russians used some illicit drug in 1995. This represents more than twenty times the total number of drug users recorded ten years earlier in the entire Soviet Union. Drug trafficking and internal production boomed as Russia became the biggest drug market among all of the former Soviet republics. Between 1993 and 1995, for example, the annual seizure of illegal drugs increased from thirty-five to ninety tons (Glenn 1996).

Rates of use continued to increase throughout the decade. By 1999, nearly 4% of the adult population, or nearly 5 million people, were annual users of cannabis. Moreover, use rates were reportedly still increasing (UNODC 2002). Use was highest among adolescents and the new business-classes. According to estimates, 15% to 30% of pupils in the Moscow and St. Petersburg areas were regularly using drugs (OGD 1997; also see Glenn 1996). While these groups are still the primary drug users in Russia, the drug scene now includes a sizable number of housewives and manual workers (Glenn 1996). Thus, in Russia, as in the United States, Great Britain, Japan and Israel, high rates of social mobility led to high rates of drug use.

Of course, the dismantling of the Soviet empire had major repercussions in all of Eastern and Central Europe. Indeed, all of the former Soviet-ruled or Soviet-dominated nations followed the pattern of greater political freedom, downward economic mobility for a large segment of the society and radically altered social relations after the formal dissolution of the Soviet Union. It is therefore not surprising that these states have also reported dramatically increased rates of drug use (see UNODC 2002; UNODC 2002b; Cortese 1999). In particular, rates of marijuana and opiate use increased in Hungary (WHO Regional Office of Europe 2001a ; Elekesa and Kovács 2002), Latvia (WHO Regional Office of Europe 2001b), Bulgaria (WHO Regional Office of Europe 2001c;), the Czech Republic (WHO Regional Office of Europe 2001d), Poland (WHO Regional Office of Europe 2001e), Slovenia (WHO Regional Office of Europe 2001f; Flaker 2002), Romania, Moldova (Elekesa and Kovács 2002), Belarus, the Ukraine (Poznyaka et al 2002), Tajikistan, Kyrgyzstan (Drug Policy Alliance 2002) and Turkmenistan (Kerimi 2000) once Soviet domination ended (also see UNODC 2004).

The dramatic mobility caused by the collapse of the Soviet system has apparently led to increases in drug use in all of the affected nations. However, as the theory would predict, the effect of the rapid socio-political and economic changes on rates of drug use are most apparent in the relatively affluent and more modernized areas of the former Soviet Union and the eastern bloc. For example, when the Czech Republic and Slovak Republic became politically independent and implemented radical changes in their social systems, a decade of increased drug use followed. According to OGD (1997), while only a handful of Czech residents used opiates when the Soviet Union dominated Czechoslovakia, heroin appeared on the Czech and Slovak markets shortly after the Soviets departed. Once heroin hit the market, use spread quickly among the urban youth (OGD 1997). In addition to heroin, marijuana, methamphetamine and hallucinogens also became popular. By the late 1990s, researchers estimated that 17% of adults in the Czech Republic and 12% of the Slovakia population were using illicit drugs (ESPAD 2001; Ladislav et al 2002; also see UNODC 2000; UNODC 2004). The respective figures were even higher for the population of adolescents. According to the data from the ESPAD survey carried out in 1999, 35% of young Czechs and 19% of young Slovaks used marijuana (see ESPAD 2001; also see Ladislav et al 2002). Likewise, all available evidence suggests that the use of illicit drugs grew rapidly in Estonia, Latvia, Lithuania and Poland when these nations regained their independence and began post-socialist market and political reforms. In fact, rates of use in these nations are now close to the Western European level (Lagerspetza and Moskalewicz 2002). Again, these are among the more modernized nations of the once Soviet-dominated region.

While the formerly Soviet-dominated nations still have lower rates of use than most Western European nations (see UNODC 2004), use increased once the relatively immobile Soviet system weakened. As these nations entered the capitalistic modern world system, their patterns of mobility and, consequently, drug consumption began to mirror those of their western neighbors. This trend is especially pronounced among youth in the "more modernized" states.

CONCLUSION

This chapter explains why modernization and rationalization's influence on drug use is not linear. The root cause of the curvilinear relationship is the complex interplay between modernization, informal social control, law and collective definitions of use and users. Modernization liberates individuals from the strict restrains of religion. It also frees individuals from the restrictive forces of informal controls. Yet, it also leads to greater social concern and, eventually, more law. Law, while offering less control over individuals than other control mechanisms, limits the ability to use drugs. The passage of drug laws, however, is only part of the story. When these laws are passed and what drugs they regulate are a function of social definitions. The groups perceived to be using drugs influence the social definitions that determine the law's content. Apparently, the subjective harm associated with use decreases when society is changing rapidly. As this occurs, moral standards tend to be more relaxed, users are less marginalized and use carries fewer social costs. Consequently, rates of use tend to increase during times when social mobility increases. Conversely, rates of use decline as rates of mobility decline. The moral boundaries are more clearly and stringently defined and moral standards tend to be more tightly enforced. The subjective harm associated with drug use increases. As social perceptions turn increasingly negative, the cost associated with use increases. People are therefore more likely to avoid using illicit drugs when rates of social mobility are low. These forces interact and create a cyclical trend in drug use within the longer-term general increase in use caused by modernization and rationalization.

While the co-variation between rates of mobility and use is not perfect, it does appear to be strong. Moreover, the relationship is consistent across varied cultures and histories. Social mobility, regardless if its source is economic, political or social, alters relations and, in so doing, alters the social structure. As the "structure of deviance" changes, so do definitions of drug use and social appraisals of users. Social mobility, therefore, is the engine of change, and social mobility increases with modernization.

CHAPTER EIGHT
APPLYING THE THEORY

To this point, drug consumption has been discussed at a macro level. Cultural variations in levels of development -- the result of the modernization and rationalization processes -- have been used to explain the cross-cultural and historical variations in drug use. The process of rationalization has altered social definitions of drug use. Drugs have been moved from the realm of the sacred to that of the profane. As this occurs, controls over drug use begin to weaken. Drugs are viewed as one more natural object that people can manipulate or use to manipulate something else. The functions of drug use become more personal and less symbolic. Moreover, drug use becomes more democratic. That is, the social distinctions that separate users from non-users become blurred. Drug use is no longer the privilege of the political, religious or economic elite only. Moreover, those distinctions that remain between who uses drugs and who does not are based primarily on achieved rather than ascribed statuses. The result is that people living in modernized nations use more drugs than do people living in lesser-developed nations. Moreover, more people use drugs when rates of social mobility are high since social mobility weakens the structural parameters that separate "conforming" and "deviant" members of society.

While this explanation accounts for variations across time and space at a macro level, some readers will undoubtedly be left asking "so what." What, they may wonder, can this explanation tell us about *who uses drugs*? How can this explanation predict which specific individuals will use drugs? How can this explanation guide policy? What are the practical implications of all this? This chapter is for these readers. Let us consider some prominent individualistic theories of drug use. Then, I will demonstrate how the macro-level perspective developed here can account for some of the same facts explained by these theories.

Theories Explaining Drug Use by Individuals

Broadly speaking, sociologists explain variations in illicit drug use with two general classes of theories. Although most theories are eclectic, we can classify them based on their central explanatory variable or variables. The first theoretical set uses social-psychological variables and personality traits to account for drug-using patterns. These theories rely on attitudinal variables to account for variations in drug use. The second theoretical set emphasizes social relations and how the setting of drug use attracts or deters potential users. In general, these theories rely on behavioral variables to explain drug-using patterns.

Attitudinal Theories of Drug Use

There are several variants of attitudinal theories. Prominent attitudinal theories include Kaplan's (1975) *self-degradation / self-esteem theory,* Gottfredson and Hirschi's (1990) "general theory of crime" or "*self-control theory*" and Jessor's (1979; 1987) *problem-behavior proneness theory.* These theories locate the primary causal factors for drug use within the individual's psyche and emphasize individualistic factors. It is the individual's perception of him or herself that generates the motives and rationalizations for using illicit drugs. These theories are well known and have received considerable empirical support. I will therefore review them only briefly.

Self-degradation / Self-esteem Theory

A subset of the social-psychological theories is comprised of several *inadequate personality theories.* Inadequate personality theories argue that individuals suffering from some personality flaw or inadequacy are most likely to use illicit drugs. Use is a defense mechanism and a means of compensating for their flaws (see Wurmser, 1980). Kaplan's (1975) *self-degradation / self-esteem theory* is a prime example of this line of thought. Kaplan maintains that all adolescents seek acceptance and approval for their behavior. However, when their behavior is unacceptable to their parents, teachers or conforming friends, adolescents experience psychological distress. This distress produces feelings of self-rejection and, if left unresolved, eventually will produce low self-esteem. Some adolescents with low self-esteem adjust their behavior. Others, however, withdraw from the source of stress. These latter adolescents are likely to develop a disposition toward deviance. Once such a disposition develops, the adolescent drifts into a deviant peer group. Unlike the adolescent's old friends, parents or other conforming adults, their newfound peers reward their behavior. Using drugs becomes a source of status among their new peers and alleviates their sense of rejection, at least temporarily. Thus, low self-esteem leads to a disposition toward deviance, participation in a drug-using peer group, and, eventually, drug use. Several empirical tests have found support for Kaplan's model, especially when tested in a longitudinal setting (see Kaplan and Fukurai 1992; Kaplan and Johnson 1991; Kaplan, Johnson and Baily 1986; 1987; Vega, Apospori, Gil, Zimmerman and Warheit 1996; Vega and Gil 1998; Miller et al 2000; contrast Jang and Thornberry 1998).

Self-control Theory

Another version of the inadequate personality theory is Gottfredson and Hirschi's (1990) "general theory of crime" or "*self-control theory*." According to the authors, drug users, and criminals in general, lack self-control. Lacking self-control, drug users are unable to control their behavior adequately. They "tend to be impulsive, insensitive, physical (as opposed to mental), risk-taking, short-sighted, and nonverbal" (Gottfredson and Hirschi 1990: 90). In essence, self-control theory is a variant of rational choice theory. Those with low self-control are more likely to value the rewards of deviance over the punishments associated with it because they fail to calculate the negative outcomes of their behavior properly. As Gottfredson and Hirschi (1990: 95) say,

> So, the dimensions of self-control are, in our view, factors affecting calculation of the consequences of one's acts. The impulsive or short-sighted person fails to consider the negative or painful consequences of his acts; the insensitive person has few negative consequences to consider; the less intelligent person also has fewer negative consequences to consider.

Thus, those with low self-control emphasize the immediate rewards associated with drug use or other deviant behaviors and fail to recognize the potential dangers or pains associated with the behavior. The "general theory of crime" has produced numerous attempts to test. Although several authors note that the theory overstates its more general aspects, there is empirical support for the claim that low self-control is related to drug use, delinquency and crime (e.g. Arneklev, Grasmick and Bursik 1999; Gottfredson and Hirschi 1990; Hope and Damphousse 2002; Leeman and Wapner 2001; Mason and Windle 2002, Turner and Piquero 2002; Vazsini et al 2001; contrast Wang et al 2002).

Problem-behavior Proneness

Jessor's (1979; 1987) *problem-behavior proneness theory* asserts that certain personality types are attracted to deviant behavior, including drug use. According to the theory, there are three systems of psycho-social influence: the personality system, the perceived environment system, and the behavior system. Within each system, variables reflect either "instigations" to problem behavior or "controls" against it. Together the systems generate "proneness." The more prone to problem behavior, the greater the probability the adolescent will find drug use attractive and engage in the behavior. Two distal and one proximate "structures" comprise the personality system. The "instigation structure" includes the individual's value toward academic achievement, independence and peer affection. The "personal beliefs structure," which centers on "beliefs about the self, society, and self in relation to society (Jessor and Jessor 1977: 20), includes the variables alienation, social criticism (the degree of acceptance or rejection of the values, norms, and practices of the larger society), and self-esteem. The "personal control structure" includes the adolescents' attitudinal intolerance of deviance, religiosity, and perceptions of positive or negative functions of the problem behavior. Next, the "distal environment structure" and the "proximal environment structure" constitute the "perceived environment system." The distal environment structure includes the adolescents' perceived strictness and perceived sanctions for transgressions from parents and friends. The proximal environmental structure includes the social support for problem behavior available in the social environment. Low parental controls, low compatibility between parent and peer expectations, high peer versus parent influence, low parental disapprovals of different problem behaviors, and exposure to friends' approval for engaging in problem behavior all increase the likelihood of deviant behavior. Finally, the behavior system includes the "problem-behavior structure" and the "conventional behavior structure." Problem behaviors include the use of alcohol, marijuana, and other illicit drugs as well as engaging in other deviant behaviors. The conventional behaviors that control deviance include involvement in religious organizations and school.

If the three psycho-social systems generate "proneness," the adolescent is more likely to have "a concern with autonomy, a lack of interest in the goals of conventional institutions, like church and school, a jaundiced view of the larger society, and a more tolerant view of transgression" (Jessor and Jessor 1980: 109). As with the self-esteem theory, the problem-behavior proneness theory has received considerable empirical support (see Jessor and Jessor 1977; 1980; Donovan, Jessor, and Costa 1991; 1993; 1999).

Behavioral Theories of Drug Use

While social-psychological theories emphasize individualistic factors and attitudinal variables, sociological theories of drug use typically stress structural factors and behavioral variables. As Goode (1999: 100) notes when discussing sociological theories of drugs, "the most crucial factor to be examined is not the characteristics of the individual, but the situations, social relations, or social structure which the individual is, or has been located." Although there are others, the major behavioral theories are *differential association theory, social learning theory*, and *subcultural theories*. Although there are differences among them (see Goode 1999), they are highly interrelated. I will therefore discuss them as a whole.

Differential Association / Social Learning Theory and Subcultural Theories

Differential association / social learning, subcultural and *selective interaction theories* of drug use all emphasize the socialization process. In addition, they all maintain that crime is learned through intimate interactions and underscore how associating with deviant peers increases the likelihood of drug use. Sutherland's (1939) *differential association theory* is the fundamental theory upon which all these theories are based. The central tenet of Sutherland's theory is that individuals learn the specific motives, directions, techniques, and rationalizations for crime and deviance from intimate interactions. Through interactions with significant others, individuals are taught definitions of behavior that either favor the violation of law or favor conformity. The principle of differential association states that a person becomes delinquent because of an excess of definitions favorable to the violation of law over definitions unfavorable to the violation of the law (Sutherland 1939).

Akers (1977; see also Akers 1992) and his associates (Akers et al 1979) extend Sutherland's basic theory by using principles of behavioral psychology to specify the process by which individuals learn to deviate. According to Akers, behavior is modeled through the application of rewards and punishments. Certain groups, however, reward deviant behavior thereby teaching the individual that such behavior is "good" or "acceptable." Thus, drug use is a result of exposure to and participation in groups that use illicit drugs and define this behavior as "good." For example, Becker (1953) defines specific factors that must be learned before one becomes a regular marijuana user. According to Becker, initiates must learn the techniques of use, how to recognize the effects of the drug, and the how to define those effects as pleasurable. Thus, peers socialize users into use. Moreover, this socialization occurs in a stable peer group or a subculture. As Akers notes, the group provides "the environments in which exposure to definitions . . . and social reinforcements for use of or abstinence from any particular substance takes place" (Akers et al 1979: 638). Eventually, this process, like all socialization processes, results in a transformation of the user's identity and normative belief system.

Others, most notably Johnson (1973) and Kandel (1980; also see Kandel and Yamaguchi 1993; 1999; 2002) and Thornberry (1987; Thornberry 1996; Thornberry et al 1991; Thornberry et al 1994; Krohn et al 1996), have elaborated the processes through which individuals become involved in deviant peer groups. Johnson (1973) argues that alienation and isolation from the parental subculture leads to involvement in a deviant subculture. Kandel (1980) claims youth drift into a delinquent peer group. Thornberry (1987; 1996) claims weakened social bonds (see Hirschi (1969) for a discussion of "bond theory") lead to involvement. In any case, all of these related perspectives note the importance of being socialized into drug use. Moreover, this socialization process most frequently occurs in a peer group, and, once involved in a delinquent subculture, adolescents will likely engage in a range of deviant and criminal activities that reinforce their newly formed delinquent identity and belief system. Consequently, individuals who associate with drug-using peers have a high probability of using drugs. This assertion is one of the most frequently supported claims in research on illicit drug use. In fact, associating with drug-using peers is by far the most important and consistent predictor of drug use.

Although not technically a subcultural theory, I (Hawdon 1996; 1999; 2004) have specified the types of activities that delinquent youth are likely to engage in with their deviant peers. Youth, like adults, do not engage in single activities; instead, they engage in *routine activity patterns* (RAPs) which vary in the amount of social control inherent to them. Individuals who are engaged in RAPs with high levels of social control are constrained from drug use, even if they want to engage in the behavior. Conversely, RAPs with low levels of social control free the individual to partake in use. The amount of social control inherent to a RAP is determined by the visibility (the extent to which the activities occur under the supervision of conforming adults or in public places) and instrumentality (the extent to which the activities are goal-oriented in a traditional sense) of the activities that comprise it. This re-conceptualization of Hirschi's (1969) notion of "involvement" has received empirical support when tested on a nationally representative sample of high school seniors (Hawdon 1996), a sample of college freshmen (Hawdon 1999) and a sample of middle school children in South Carolina (Hawdon 2004).

Summarizing What We Know

Reviewing these theories and other work attempting to correlate personality characteristics and behaviors with drug use (see, for example, Hawdon 2004, Goode 1999; McBroom 1994), we find that drug users, as compared to non-users, tend to be independent, receptive of new experiences, risk takers with low levels of self-control, accepting of deviant behavior and normative violations, receptive to uncertainty, hedonistic and rebellious. They tend to be nonconforming. They tend to have little interests in conventional institutions like church and school. Yet, they tend to have broad intellectual interests and tend to be critical of the larger society. In general, with respect to their personalities, a "single summarizing dimension underlying the differences between users and nonusers might be termed conventionality-unconventionality" (Jessor and Jessor 1980: 109). Behaviorally, drug users frequently interact with other drug users in deviant subcultures. They associate with those engaged in similar behaviors so their drug use is defined positively instead of negatively. Moreover, they select leisure activities that remove them from the public eye and permit their drug use to go undetected.

A model combining the various psycho-social factors outlined above with the behaviors known to be associated with drug use can do a reasonably accurate job of predicting who does and who does not use drugs. For example, in a model that included only some of the factors listed above, I recently was able to classify 291 of 317 (91.8%) middle-school children correctly with respect to their drug use. Overall, the partial model explained 46.3% of the variance in illicit drug use among these adolescents (see Hawdon 2004). Numerous other researchers have achieved similar successes in predicting youthful illicit drug use. While the insights such models offer us are important and intriguing, these models cannot explain cross-cultural or historical patterns in drug use. Again, as I argued earlier, why is it that American, British, Canadian and Australian youth have such lower levels of self-control than do Japanese or Israeli or Saudi Arabian youth? Why are children in developed nations more likely to associate with drug-using peers than those in lesser-developed nations?

The answer to these questions lies in the relationship between rationalization, modernization and the various psycho-social factors that are related to drug use. That is, the processes that I have discussed throughout this text not only lead to the macro-level social changes and the macro-level patterns of drug consumption described here, they also foster the micro-level "causes" of drug use. While undoubtedly a "distal cause," the processes of modernization and rationalization radically alter the worldviews, normative systems and institutions of the cultures that experience them. They therefore indirectly alter the socialization processes children experience and, by doing so, pattern their beliefs, attitudes and behaviors. While data limitations obviously prohibit any critical test of such an argument, a supporting argument can be sustained. Are the attitudes and behaviors associated with drug use consistent with the cultural and structural aspects of the rationalization and modernization processes? Specifically, can these micro-level correlates of use be connected with the central aspects of the modernization process? Can they be connected to the scientific worldview and the ideology of individualism?

Science and Attitudinal Theories of Drug Use

Let us first consider the personality traits of drug users. First, we should notice that many of these traits are "good middle class values" that "good parents" try to instill in their children. What "good" parent does not want his children to be independent? What "good" parent wants her child to shy away from new experiences? Moreover, some of these traits are official goals of many modern universities. Any academic, especially those in modern American state universities, knows to promote "tolerance of diversity." We strive to have our students demonstrate "critical thinking." We like when our students have "broad intellectual interests." So, it appears, middle class values and the modern university promote drug-using personalities. More to the point, some of these characteristics are consistent with the worldview that evolved with the progression of rationalism and modernization. That is, they are consistent with science and individualism.

In what manner do these traits share an affinity with the scientific worldview? First, does science not teach us to be receptive and open to new experiences? Science is, by its nature, experiential. "Truth," according to science, can only be verified through observation and observation is based on experiences. Science seeks new experiences to test its theories and to establish the accuracy of those theories. New experiences lead to conditioning statements that specify our scientific understanding of various processes. A scientist *should be open to new experiences* and the scientific worldview encourages this openness.

Second, science promotes broad intellectual interests. Many scientific discoveries come from applying the basic insights of one discipline to a problem in another. For example, for a functionalist to say that "society is like an organism" requires her to know the rudimentary characteristics of biological organisms. One cannot understand molecular biology without a basic understanding of organic chemistry. Comprehending the science of economics requires a basic understanding of psychology and sociology. Sociologists incorporate history, anthropology, economics, political science and psychology in their explanations. Renaissance thinking, forever linked with the rise of science as a dominant worldview in the west, encouraged flexibility and broad interests. Indeed, the "Renaissance man" is one who has a mastery of several disciplines and skills.

Third, science is not, indeed cannot be, dogmatic. While science does not like uncertainty and strives to reduce it, it is, nevertheless, more receptive to it than other worldviews. Religion, for example, eliminates uncertainty. God is God, period! Why does the sun rise? God wants it to, period! End of question. While there is debate over the fundamental doctrines of all religions, the "higher religions" develop universal cosmologies that explain and justifying traditions and the world in a unified, orderly fashion based on "universal truths" (Habermas 1979; Weber [1922] 1964; Weber [1922] 1978; Parsons 1977). These "universal truths" leave little room for uncertainty. If something is "universally true," there cannot be exceptions and, therefore, all uncertainty is gone. When exceptions exist, however, there is uncertainty. Therefore, religion, as a worldview, does not deal well with uncertainty. Uncertainty is explained as "part of God's plan" or by saying that "God works in mysterious ways." The basic beliefs that "God exists," "God has a plan" and "God works" are certain and cannot be seriously questioned without questioning the foundation of the philosophy. Conversely, uncertainty is at the heart of science and scientific exploration. It is uncertainty that leads science to investigate phenomena in the first place. It is uncertainty that leads science to re-investigate its initial findings. Uncertainty is especially common in the more observational sciences such as the social sciences, meteorology, astronomy and evolutionary biology. Even more than the experimental sciences like physics and chemistry, observational sciences rely on probabilistic explanations and probabilistic explanations are exactly that, probabilistic. By definition, they are filled with uncertainty. A "deviant prone" individual *probably will* use drugs. There are no guarantees that he or she will. There is uncertainty in our scientific explanations and it is tolerated. Any scientist familiar with the basic tool of the observational sciences knows that her statistical equations will include an error term. Thus, science promotes being receptive to uncertainty.

Fourth, science teaches us to be critical of tradition and our society. As explained by Weber and Habermas and discussed in Chapter Four, the greater objective understanding of the natural and social worlds associated with the rise of rationalism replaces the once unquestionable religious understandings of the world. As this occurs, we are encouraged to seek a more objective understanding to fill the void. To obtain that understanding, however, we must question tradition. As Habermas (1984b) argues, the domination of science as a means of explaining the world results in cultural traditions being constantly criticized and renewed. According to science, *tradition should not simply be accepted as being "true."* Tradition, and everything else, must be examined. With science, nothing is sacred. Everything is open to question, investigation and debate. We are, according to science, suppose to question. Indeed, what science teacher does not want to promote "critical thinking" among his students?

Fifth, science promotes tolerance. As discussed earlier, science results in competing validity claims with no universally accepted normative guidelines to judge their "truth." With science, moral traditions become relative. The unified worldviews that once secured the cohesion of social life are weakened and, "the authority of the holy is gradually replaced by the authority of an achieved consensus" (Habermas 1984b: 77). There is, according to science, no way to establish which moral code is "right" or "correct." Thus, science teaches us to be tolerant of others. The frequently cited goal of social science is "cultural relativism." It is inappropriate for me as a scientist to express disgust at the Siberian reindeer herdsmen's practice of drinking intoxicating urine. I would be guilty of ethnocentrism if I showed displeasure with the infanticide practiced by numerous cultures throughout history. According to the scientific worldview, these moral reactions to practices and behaviors unfamiliar to my culture need to be explained, not judged. Therefore, science promotes tolerance of others, their "deviance" and their "normative transgressions." After all, as a scientist, who am

I to say their cultures are "wrong." Since I have adopted the scientific perspective, I cannot pass moral judgment because science is devoid of morals.

Finally, science challenges religious belief. As Weber ([1919] 1946: 154 notes when discussing the rationalization process, eventually "the tension between the value spheres of 'science' and the sphere of 'the holy' (was) unbridgeable." While science and religion can be integrated (see Barbour 1997), attempts to merge the worldviews are often strained. Many try to avoid the subject all together. For example, The National Academy of Sciences (1999: ix) avoids the issue by stating,

> Scientists, like many others, are touched with awe at the order and complexity of nature. Indeed, many scientists are deeply religious. But science and religion occupy two separate realms of human experience. Demanding that they be combined detracts from the glory of each.

In general, however, there is an inverse relationship between the strong belief in science and strong religious beliefs. In fact, the existing literature consistently concludes that very few scientists are religious, and most do not hold to any conventional theistic religious beliefs (Bergman 1996). Some researchers have gone as far as saying that "the religious scientist is an oxymoron" (Goss 1994: 105).

Hence, scientists tend to be open to new experiences, have broad intellectual interests, be receptive of uncertainty, critical of tradition, accepting of normative violations and not overly religious. That is what their dominant worldview teaches them to be. Of course, these are some psychological traits of drug users too. Thus, drug users tend to have personalities consistent with the scientific worldview. Maybe their alleged cocaine use made Albert Einstein and Sigmund Freud the outstanding scientists they were.

Yet, astute readers will likely object that I have selected only certain drug-user traits. They may argue that I have ignored the more "unsavory" traits of drug users. They would be correct because so far, I have. However, remember that the rationalization and modernization processes lead to more than the rise of science. These processes also promote individualism. In turn, individualism promotes a valuing of independence, rebellion, nonconformity, hedonism and low self-control.

Individualism and Attitudinal Theories of Drug Use

As argued in Chapter Four, the rationalization process results in "the substitution for unthinking acceptance of ancient customs of deliberate adaptation to situations in terms of self-interests" (Weber [1922] 1978:30). As rationalization and modernization advance, personalities become more autonomous and interests become increasingly defined in private, as opposed to collective, terms (Habermas 1984b; Parsons 1951). Moreover, as explained earlier, modernization leads to individuals having more associations from which to choose. The greater options frees the individual from group controls (see Simmel 1955; Simmel [1908] 1971; Simmel [1908] 1971b). As the group's ability to control the individual decreases, the individual's freedom increases (see Hawdon 1996b). Thus, the rationalization and modernization processes promote individualism. In the most modernized societies, individualism, or the ability to "be one's own person," has become a dominant goal and cultural norm. Many have argued, for example, that individualism is the central mores of American society (see, for example, Bellah et al 1985).

Assuming Bellah is correct and individualism is a major American mores (and presumably most western societies if Weber, Habermas, Parsons and other theorists are correct), some additional psychological correlates of drug use can be understood by the macro-level theory of drug consumption. These correlates are a result of the individualism promoted by the processes of rationalization and modernization. Individualism as an ideology contends that the individual is fundamentally "good," and the corrupt and dysfunctional group is the source of "evil" (e.g., Nietzsche [1887] 1964; Rousseau [1762] 1979; also see Rawls 1971;

Avineri and De-Shalit 1992). Individuals must therefore be protected from corruptive groups, and the rights of individuals take precedent over the group's wishes, desires or rights. In such an ideological system, there is primacy placed on being unique, different and on "standing out" from the crowd. Thus, those persons subscribing to the doctrines of individualism are more likely to be nonconforming. Conformity, by definition, unites individuals through similarities. Individualism, by contrast, separates them through differences. It is difficult to maintain that you are a unique individual if your behavior, appearance and belief system are conforming to all those around you. Similarly, the greater one values individualism, the greater they value independence. This statement is nearly, if not fully, tautological and should need little supporting argument. By definition, individualism places an extreme value on being independent.

Next, individualism promotes rebelliousness. Rebelliousness, after all, is a relative term. Whether rebellion is defined as "good" or "bad" depends on one's perspective and relative position in the status hierarchy. For example, from an American perspective, those who rebelled against British rule in the 18th century to establish their own nation were "patriots" and "founding fathers." The British, of course, had different terms for these folk. In the 1980s, the El Salvadorian *Contras* rebelling against their communist government were "freedom fighters," not "rebels." Conversely, those who rebel from American world domination in the 21st century in hopes of establishing a worldwide network of Islamic fundamentalist states are "terrorists." So, whether one is a rebel or pioneer depends on the perspective of the labeler. We must therefore ask, "from what are drug users rebelling?" Well, this depends on the relative position of the drug being used in the current status hierarchy of drugs. Legal drug users are not rebelling, of course. They are conforming to accepted normative standards. Illegal drug users, however, are rebelling. They are rebelling against the authority structure as currently comprised. Is such behavior not encouraged by individualism? Individualism, at least in an extreme, contends that no one has the right to tell the individual what to think, feel or do. Any group demand for conformity is illegitimate and repressive. From an individualistic perspective, one *should rebel* from such inauthentic authority. To quote the great promoter of individualism,

> I hold it that a little rebellion, now and then, is a good thing, and as necessary in the political world as storms are in the physical. Unsuccessful rebellions, indeed, generally establish the encroachments on the rights of the people, which have produced them (Thomas Jefferson to James Madison, 1787. ME 6:65).

Illegal drug users are expressing their individualism, their "inalienable right" so to speak, when they rebel against what they obviously perceive to be an illegitimate authority that condemns their practice as being "wrong" or "illegal."[103]

It can also be argued that individualism, if extreme, promotes hedonism and low levels of self-control. While individualism has served western societies well, its nature has changed over time. In the United States, for example, individualism has assumed at least four primary forms: religious, political, instrumental and expressive (see Bellah et al 1985). Each of these types of individualism granted people greater freedoms over successive realms of life. However, each successive type, while granting freedom over some dimension of thought or behavior, also required the individual to connect to a group. That is, while granting freedoms, it also demanded responsibilities. For example, as Bellah and his associates (1985) explain, the political freedom expressed by its champions such as Thomas Jefferson freed individuals from governmental domination, but required them to participate fully in self-government. He called not only for political freedoms but also for a bond between the individual and his country. As Jefferson said when discussing the duties of citizens, "Love your neighbor as yourself, and your country more than yourself" (Thomas Jefferson to

---

[103]. Davis (1967) makes the argument that drug use is a natural extension of middle-class values such as individualism. Use symbolizes an attack on normal forms of consciousness and a disregard for, or rebellion from, normal society.

Thomas Jefferson Smith, 1825. ME 16:110). Similarly, Edison's instrumental individualism granted freedom for individuals to pursue economic riches; however, the path to such riches lay in self-discipline and righteous living. These types of individualism, and the pursuit of individual interests they engendered, helped the west modernize and led to incredible inventions and social institutions that serve the collective well. However, I would argue that individualism has continued to evolve. It is not a giant leap of logic to take individualism to the extreme. If I have the right to worship as I chose, participate in self-government, pursue economic interests and express myself as I please, why do I not have the right to act as I please? In essence, if taken to an extreme, individualism can become rampant. Jerry Rubin, the self-professed founder of the "Yippie" movement, expressed such thinking when he proudly wrote, "Yippies do whatever we want to do whenever we want to do it" (Rubin 1970: 84). It was expressed by recording star Sheryl Crow when she sang, "if it makes you happy, it can't be that bad." Thus, by this logic, becoming intoxicated is perfectly acceptable, regardless of when or where or how it is induced, provided it is what you wanted to do. Similarly, intoxication "can't be that bad" provided it made you happy. While extreme, these are expressions of individualism devoid of any moral connection to the collective. These are expressions of individual rights devoid of responsibilities. These are expressions of a doctrine that holds pleasure or happiness is the sole good in life. That is, these are expressions of hedonism. While I do not mean to argue that western individualism as practiced by the majority of the population is synonymous with hedonism, it is not a stretch to argue that individualism, if taken to the extreme, is related to hedonism. Moreover, I do not believe it is a major stretch to argue that some members of western society have taken individualism to an extreme.

Accepting that extreme individualism leads to hedonism, it is a short extension from hedonism to low levels of self-control. As stated earlier, self-control theory is a variant of rational choice theory in that those with low self-control are more likely to value the rewards of deviance over the punishments associated with it because they fail to calculate properly the negative outcomes of their behavior. If we assume that most drug users find their use pleasurable and therefore rewarding, hedonism would elevate the relative weight of the drug's rewards and decrease the relative weight of any future negative consequence of using the drug. Even if the drug's effects are only mildly pleasurable for a brief moment and devastatingly harmful in the long term, hedonism, which calls for the pursuit of immediate pleasure, would alter one's calculation of the rewards and punishments associated with use. Thus, individualism leads to low self-control by promoting hedonism, which, in turn, alters the perceived rewards and punishments of behaviors such as drug use. It is important to remember that the rationalization process increases instrumental rationality, not substantive rationality (see Weber [1922] 1978; Weber 1947). It is difficult to ignore the power of intoxicating drugs to alter one's perceptions, reduce inhibitions or stimulate the senses. Assuming one finds these pleasurable (as millions of people apparently do) and that one's goal is to achieve pleasure, drug use is instrumentally rational. Drugs are quick and efficient means to pleasure. Whether use is the "correct means" or a "morally good means" is not at issue. These questions are matters of substantive rationality, not instrumental rationality. From an instrumentally rational perspective, drugs meet the bill, and rationalization promotes instrumental rationality, not substantive rationality. It can also be argued that individualism directly promotes low self-control. Again, whether one is demonstrating "low self-control" or expressing her "individualism" by not obeying authority is a matter of perspective. I am sure numerous Civil Rights activists who risked bodily harm and their lives by practicing passive resistance would not equate their disobedience with low levels of self-control.

Consequently, many of the psychological correlates of illegal drug use are consistent with the ideologies promoted by the rationalization and modernization processes. Science promotes an eagerness for new experiences and an acceptance of uncertainty. It encourages broad intellectual interests. Science teaches its followers to be critical of traditions and to examine the larger society around them. Science promotes tolerance by making moral perspectives relative. Similarly, individualism demands its adherents to be nonconforming, unconventional and to value independence. It justifies rebellion against any authority perceived to be illegitimate. In its extreme, individualism promotes hedonism and low levels of self-control.

Therefore, the rationalization and modernization processes can be linked to the attitudinal correlates of drug use. So, why is it that American, British, Canadian and Australian youth have such lower levels of self-esteem and self-control than do Japanese or Israeli or Saudi Arabian youth? Why are children in developed nations more likely to have "deviant prone personalities" than those in lesser-developed nations? The answer is because their cultures are more modernized and more rationalized. Thus, their cultures encourage personality traits that are associated with drug use. Therefore, youth in the most rationalized and modernized nations on earth are more likely to develop the personalities and attitudes of drug users than those living in less-developed nations.

But, what about the behavioral factors that are correlated with drug use? Specifically, is there reason to suspect that the modernization and rationalization process would likely lead to the creation of a drug-using subculture? Indeed there is. As Weber argued, the rationalization process de-mystifies the world. The domination of scientific thought over religious thought has secularized the world and, by doing so, removed sacred mystery. In its place, we have substituted profane understanding. This process has left us disenchanted (see Weber [1919] 1946). And, it is this disenchantment that led to, and continues to lead to, the development of a drug subculture. It is this disenchantment that lures millions of youth to the subculture's world of illicit drug use.

Rationalization, Disenchantment and the Creation of a Drug Subculture

A considerable amount of journalistic and academic work has been dedicated to the emergence of the drug subculture in the United States. While the subculture can be traced to the opium dens, snow parties and hash clubs of the late 19th and early 20th centuries, the subculture became national and recognizably more unified with the emergence of the Beats in the 1940s and 1950s. The Beats were, in part, an outgrowth of the bohemian literary and artistic tradition. The "movement" was also a drug subculture since it is undeniable that drugs were a central part of their art and philosophy. Moreover, the Beat movement was much broader in scope than simply a discrete artistic form. It was a lifestyle that was "manifested in expressive forms and rituals that not only referred to artistic work but also to specific attitudes, behavior patterns, dress codes and the like" (van Elteren 1999: 72). Although I do not want to overstate the cohesiveness of the group or its philosophy, there was a core ideology that can be identified from the writings of many of its key members. The Beat's central leaders, Jack Kerouak, Allen Ginsberg, William Burroughs and Neil Cassady for example, were typically of middle-class origin (although Kerouak and Cassady were of working-class descent). These men, including their large supporting cast of novelists, poets, jazz musicians and youthful followers, were united by what they frequently referred to as a "new consciousness" and a "new vision" (see Ginsberg 1980; also see Prothero 1991).

Their ideology was influenced by Eastern mysticism, nihilism, Oswald Spengler's *Decline of the West*, jazz music and drugs. Their lifestyles were expressions of traditional bohemian values such as spontaneity, expressive values, creativity and individualism (see Brake 1980; Prothero 1991; Watson 1995; Spates and Levin 1972; van Elteren 1999; Matza 1961; Holmes 1958; Bent 1988). The subculture's focal concerns or "spirit" were withdrawal from the dominant society, disaffiliation from traditional family life, society and career structures, existentialism, (Brake 1980), romantic primitivism, expressive authenticity (Matza 1961) and narcissistic male bonding (van Elteren 1999). Their bohemian roots and nihilistic tendencies led to a glorification of individualism and a rejection of socially-defined obligation and arbitrary social norms. In short, they were alienated and rejected what they saw as the "stifling conservativism" of

American corporate society. Avoidance and, ultimately, withdrawal from mainstream society was a central goal of the stereotypical Beat (see Watson 1995; Brake 1980). As Brake (1980: 91) says when describing the Beats,

> Like many Bohemians they developed a radical critique of what they saw as wrong with society, but not why . . . Instead they voluntarily espoused poverty, disaffiliated from family, career and prospects in any conventional sense, and withdrew from a society they detested.

Their withdrawal from society protected them from what they believed were the corrupting influences of consumer capitalism (van Elteren 1999). Moreover, it permitted them the freedom to pursue artistic endeavors. For the Beats, the production of "authentic" art was critically important and highly valued. The Beats sought to produce art that flowed from the "self" and believed that to create "pure art" one had to be spontaneous, unfettered and individualistic. As a result, one could not be overly concerned with material goods. Being involved in the capitalist system, formal education, the state, family or organized religion were trappings of conventional society that stymied spontaneity and therefore obstructed creativity (see Watson 1995). "Pure art" could not be produced under such circumstances. Thus, to be "true to oneself" one had to, as Ginsberg said, "get Beat down to a certain nakedness, where you are actually able to see the world in a visionary way" (quoted in Watson 1995: 4). Given this perspective, Beats were anti-materialistic, anti-capitalistic, anti-technology and anti-mechanization. In short, they were anti-modern. They rejected the rationalization and modernization processes that had transformed American society and, from their perspective, limited people's abilities to pursue "authentic emotions" and new experiences.

The Beats had "an enormous hunger for experience" (van Elteren 1999: 90), and their rejection of and withdrawal from mainstream society permitted them the freedom to participate in life's endless variety of experiences. Since norms were relative and arbitrary, Beats rejected the social obligations norms imply and instead pursued a life filled with new experiences. They were preoccupied with sex, drugs, religious experimentation, travel and criminality. Their desire for new experiences and rejection of American materialism created empathy for society's less materially fortunate. As Matza (1961) van Elteren (1999), Prothero (1991) and others note the Beats cultivated a "romantic primitivism" that saw a simple way of life close to nature such as that found in pre-industrial societies as being superior to modernized societies. This ideology resulted in a sense of solidarity with the oppressed people and classes of society. This solidarity, in turn, led to the veneration of the poor, street people and criminals in Beat ideology. The "folk" or "fellanheen" as Kerouac referred to them, were romanticized as being instinctual, cunning and "in tune with the cosmos" (van Elteren 1999: 80; also see Bent 1988). The "fallanheen" were the "real people" and "authentic." They embodied the rejection of the "massification" of modern society and therefore were to be emulated. Thus, according to Beat philosophy, one could begin to realize the relativity of existence and the "new vision" only by shunning the wealth of modern American society and striving to experience all facets of life. By becoming like those who society had "beat up" and "beat down," one could clear their consciousness and seek clarity. As John Clellon Homes stated, "Beat implies the feeling of having been used, of being raw. It involves a sort of nakedness of mind, and, ultimately, of soul: a feeling of being reduced to the bedrock of consciousness" (Holmes 1952). This romantic primitivism led the beats to seek out and associate with those traditionally shunned by the dominant culture: jazz musicians, junkies, drifters, prostitutes, criminals, hobos, street people and migrant workers.

Associating with the socially disenfranchised and hanging out in the ghettos with people deemed socially inferior by most of society allowed the Beat's access to "alternative worlds" and provided the Beats with ample new experiences (Prothero 1991: 213). Yet their desire for experience was "not simply in order to satisfy themselves, but in the hope of discovering a New Vision." (van Elteren 1999: 90). This "new vision" could be realized by seeking unconventional experiences. They therefore experimented with the limits of human perception through meditation, seeking and accepting psychosis and drugs (Matza 1961). For the Beats, their drug use and other experimentations were extensions of their desire to withdraw from the conventional and transcend everyday reality. Their drug use was also an expression of individuality, a means of stimulating spontaneity, a catalyst for discovering truth and a way of enhancing creativity. Their drugs of choice were marijuana and hallucinogens such as peyote and synthetic mescaline. However, heroin, morphine and other opiates, Benzedrine, barbiturates, and, of course, alcohol were also popular (Polsky 1967; Matza 1961; van Elteren 1999; Watson 1995).

This brief history of the Beats glosses over many of the group's subtleties.[104] However, it is not meant to be a narrative documentary of the Beats. As I said, the history of the Beats, both as a subculture and a literary movement, has been discussed thoroughly elsewhere (see, for example, Harney 1991; Lee 1996; Lee 1996b; McNeil 1996). For our concerns here, the question is "why was Beat ideology so attractive to young people in the 1950s?" Why did an ideology that rejected "everything sacred about Eisenhower's America," as a 1950s *Life* article argued (quoted in Prothero 1991: 206), become popular?

The Beats did reject "everything sacred about Eisenhower's America" and that, in part, was the attraction for the more rebellious youth of the time. And, contrary to many stereotypes of 1950s America, the number of rebellious youth was increasing. The 1950s have been described as "a time of fatuous complacency, mindless materialism and stultifying conformism" (see Bawer 2004). Americans, eager to find stability and security after the Great Depression and Second World War, flocked to the suburbs and enjoyed unprecedented economic prosperity. However, many white, middle-class youth were growing

---

[104] I recognize that the Beat movement was not monolithic. Several types of Beats can be identified including the "hipster," the "beatnick" the "hot" and the "cool" (see, for example, Brake 1980; Matza 1961; van Elteren 1999). These distinctions, while important for understanding the full scope of the movement, are largely unimportant for our purposes and will not be considered.

discontent with the suburbs their parents seemed to love (see, for example, Echols 2002; Gitlin 1987; Lipsitz 1990). The angst-ridden ideology of the Beats, popularized through the prolific writings of their "leaders," became attractive to a wave of young people who had heard about the "revolution" and wanted to join the movement (see Maynard 1991). John Clellon Holmes noticed and discussed the appeal of the Beat counterculture for the growing mass of disillusioned youth in 1950s America (Holmes 1958). Similarly, Mel van Elteren (1999: 76) writes,

> The Beats' stance of social disengagement and their underground culture of disillusionment, expressed in café scenes where pot smoke intermingled with blues and jazz music, created a 'bliss of indifference.' This scene attracted many young and even middle-aged people to the Beat scene who in the past would not have entered Bohemian culture.

While not as ideological, artistic or prolific as Kerouak, Ginsberg or Burroughs, these younger "Beatnicks" provided the critical mass of followers that made the "movement" newsworthy. More importantly, they were a relatively large group of middle and upper-class whites outside the jazz world who, through their exposure to Beat writings and philosophies, were introduced to recreational marijuana use. These youth eventually helped the practice diffuse through American youth culture in general (Polsky 1967). In fact, the Beat's legacy can be traced through the recent history of the drug subculture. The Beats are the ancestral predecessors of the Hippies, Deadheads and Phishheads of today (see Cooper 2000).

The Subculture as Religion

While deviant behavior in general can undoubtedly be alluring to "deviant-prone" youth, I would argue that the Beat's attraction for discontented middle-class youth was more than the pursuit of deviant behavior. In addition to the "fun and exiting" elements of the culture, Beat ideology offered a source of enchantment. The Beat's alienation, as well as that of many middle-class youth, was a direct result of the disenchantment Weber argues rationalization creates.

The Beats saw American culture as disenchanting and devoid of authentic connections with the realm of the sacred. They rejected established religious institutions as well as popular postwar alternatives such as evangelicalism or "positive thinking." As Prothero (1991: 209) argues, "for this beat trio (Kerouak, Ginsberg and Burroughs), neither positive thinking nor evangelical Christianity could make sense of God's apparent exodus from the world." As they saw it, America's god was materialism and mechanization, despite the adherence of most Americans to traditional religions. These sentiments were reflected in much of the early writings of Kerouac, Ginsberg and Burroughs (see Prothero 1991). For example, Ginsberg proclaimed, "There is a God / dying in America" (Ginsberg 1984: 105).

While the Beat's rejection of dominant American values and institutionalized religion was widely noted and criticized by their contemporary commentators, there has been growing recognition that the Beats were not only a literary movement but also a religious movement. As Prothero (1991: 208) argues,

> If, as Miller argues, transcendentalism represented a religious revolt against 'corpse-cold' Unitarian orthodoxy, the beat movement represented a spiritual protest against what the beats perceived as the moribund orthodoxies of 1950s America.

I concur with Prothero. The Beat's attempt to find a "new vision" was an attempt at re-enchantment. Holmes argued that, despite their excesses, the Beats and their philosophy exhibited "a perfect craving to believe," "a quest," "a desperate craving for affirmative beliefs" (Holmes 1952: SM 10 - SM 14). Similarly, Ginsberg spoke of a "new consciousness" (see Ginsberg 1980). As Prothero (1991: 21, emphasis in original) contends,

> the beat's *flight from* the churches and synagogues of the suburbs to city streets inhabited by whores and junkies, hobos and jazzmen never ceased to be a *search for* something to believe in, something to go by.

Indeed, Ginsberg expressed the Beat's sense of disenchantment and search for re-enchantment in his signature poem *Howl*. The poem begins:

> I saw the best minds of my generation destroyed by
> madness,
> starving hysterical naked,
> dragging themselves thru Negro streets at dawn
> looking for an angry fix,
> Angelheaded hipsters
> burning for the ancient heavenly connection to the starry
> dynamo in the machinery of night,
> who poverty and tatters and hollow-eyed and high,
> sat up smoking
> in the supernatural darkness of cold-water flats
> floating across the tops of cities
> contemplating jazz,
> who bared their brains to Heaven under the El and saw
> Mohammedan angels staggering on tenement roofs
> illuminated . . .

Ginsberg often argued his poetry, and particularly *Howl*, was religious in nature and professed that everyone was "one self" and shared one consciousness. Kerouac also expressed religious sentiments -- albeit a highly pluralistic religion -- that is evident in several works, especially *Mexico City Blues* and *Dharma Bums*. They also argued their lives were dedicated to finding a "new vision" and an "ancient heavenly connection."

Through their writings, the Beats offered an alternative to orthodoxy. Their brand of "religion" attempted to contact the sacred through intuition and feeling, insisted on the sanctity of everyday life, sanctified the nonconformist and "aimed to create a spiritual brotherhood based on shared experiences, shared property, shared literature, and an ethic of 'continual conscious compassion'" (Prothero 1991: 220). Their spirituality was based on "the sacralization of everyday life and the sacramentalization of human relationships" (Prothero 1991: 214). Thus, every moment was sacred, especially if shared with friends. The Beats, therefore, provided a path to "enlightenment" and "enchantment:" live for the moment and share the moment with a community of like-minded souls.

The Hippies who eventually replaced the Beats as the baby-boomers began coming to age in large numbers during the mid-1960s[105] took the mysticism of the Beats even further. Ideological offspring of the Beats, Hippies also glorified and professed individualism, expressive values, subjectivity, dissociation from formal education and career structures, anti-materialism and anti-modernization. They glorified drugs, sex, rock music and their own community (see Spates 1976; Miller 1991; also see Brake 1980; Cavan 1972; Davis and Munoz 1968; Partridge 1973; Brake 1980; Kallen 2001). Like the Beats, their ideology called for self-exploration through drug use, mysticism or religion (Brake 1980). At least among Hippie "heads," drugs were "a means of self-realization or self-fulfillment, and not as an end in itself" (Davis and Munoz 1968: 160; Miller 1991; Partridge 1973). While praising the virtues of individualism, the Hippies, like the Beats before them, were communally oriented. Hippies preached love, togetherness and passive resistance as a means of rebelling against the dominant culture they rejected. While simple hedonism was part of the reason drug use spread so quickly among the Hippies, drugs, or "dope" as Hippies preferred, were also an expression of rebellion and a means of gaining spiritual insight. Their emphasis on subjectivity "opened the self to experience, assisted by drugs, and by religious and mystical even magical explanation" (Brake 1980: 102).

The religious nature of the subculture becomes very explicit when we move to a discussion of the Hippies. They were very aware of the religiousness of their "movement." One Hippie wrote,

> Drugs could be related to the search for authentic human existence where a person is not alienated from himself and estranged from his fellow man and where God is more than a dead word. It is just possible that the drug scene is more closely tied to the province of existential questions of

---

[105.] Again, a considerable literature describing the history, philosophy and lifestyles of the Hippies exists. This discussion is not meant to inform the reader substantially about that history. Instead, it is focused on the religious-like aspects of the "Hippie movement."

philosophy and theology than those of law or even medicine (quoted in Miller 1991: 29 - 30).

The Hippies even had "dope churches" such as Timothy Leary's *League for Spiritual Discovery*. During an interview, Leary described the League by saying, "we're not a religion in the sense of the Methodist Church seeking new adherents. We're a religion in the basic primeval sense of a tribe living together and centered around shared spiritual goals." In a different interview, Leary ([1966] 2001: 120) said, "my work is basically religious because it has as its goal the systematic expansion of consciousness and the discovery of energies within, which men call divine." The religiosity of the Hippie movement is captured in numerous songs. For example, Gitlin (1987: 204) notes that the Jefferson Airplane's song "Let's Get Together[106]" "brought religious yearning into Sixties pop." Similarly, the mysticism of the movement is evident in what can be considered their quintessential statement written about their quintessential mass gathering, Woodstock. In the song by that title, Joni Mitchell glorifies nature ("gotta get back to the land and try to set my soul free"), a spiritual search and "finding one-self" through experience ("I don't know who I am, but life is for learning"), the Hippie collective or "movement" ("by the time I got to Woodstock, we were half a million strong, and everywhere was a song, hope and celebration") and their pursuit of peace and harmony ("I dreamed I saw the bomber jet planes, flying shotgun in the sky, turning into butterflies above our nation"). Most importantly, however, she glorifies the individual and their need to find spiritual peace in the song's chorus by singing,

> We are stardust, we're just billion year old carbons.
> We are golden.
> We just got caught up in some devil's bargain

---

[106]. The song "Let's Get Together" was penned by Dino Valenti and first released nationally in 1966 by *Jefferson Airplane*; however, it was popularized, reaching top 10 status on the pop charts, by *The Youngbloods* when it was re-released in 1969.

and we got to get ourselves back to the garden.
To some resemblance of a garden

The religious reference is not accidental.

The Subculture as a Search for Re-Enchantment

To the growing mass of discontented youth in the 1950s, Beat ideology, this "new religion," presented itself as an attractive alternative to the conformist dominant culture. The Hippies had even a larger discontented group of young followers to attract. White middle-class suburban youth became fascinated with Beat and Hippie culture because it condemned the mindless consumption of consumer capitalism of which they thought their parents guilty. It attracted them because it justified their own sense of angst that adolescents so often feel. It glorified much that was "off limits" and "taboo," thereby evoking a sense of excitement. It explained why they should rebel against the system. It held the promise that the world was not mundane. Life was magical, if only one could get "beat enough" to see it. The subculture was a source of spiritual meaning in a world that appeared increasingly devoid of it. It offered a path to re-enchantment in the disenchanted world of modern America.

While drugs were a part of the bohemian world of the Beats and Hippies, they were but a part only. The drug subculture is more than simply a group of people sharing drugs. It is a worldview, a "pseudo religion." Like more traditional religions, the drug subculture has its rituals that unite the group. The passing of a joint; the cutting of lines of cocaine; the elaborate preparations for injecting drugs; the infamous "rap sessions" of the Hippies; all of these behaviors are ritualistically performed by experienced drug users (see, for example, Partridge 1973). Like more traditional religions, the subculture has its "sacred" objects and symbols. Images of the marijuana leaf are found on tee shirts, posters, pipes, license plates and other paraphernalia decorating the shrines of the devoted druggie. Like more traditional religions, the drug subculture has its own specialized language, its argot, which is truly meaningful only to the initiated.

Like traditional religions, the drug subculture has its "prophets" who disseminate the "word." For the Beats, the prophets were the likes of Kerouak, Ginsberg, Burroughs and Holmes. Jazz, poetry and literature comprised the gospel. For the Hippies, Timothy Leary and Abbey Hoffman were obvious spokespersons, but they also produced a highly active underground press where numerous commentators discussed Hippie values (see Spates and Levin 1972; Brake 1980). Mostly, however, Hippies found the "holy word" in the lyrics of the Beatles, Bob Dylan, Joan Baez, Jefferson Airplane and scores of other folk, rock and acid-rock musicians (see, for example Gitlin1987).

Most importantly, however, like traditional religions, the subculture has its community of followers, its congregation, if you will. And, it is this congregation, sanctified in Beat and Hippie ideology, that offers a sense of enchantment. Discussing the "search for meaning" among Hippies, Karr and Dent (1970: 191) state, "in the individual's flailing search for meaning, he is drawn toward the prominently visible 'hippie' group." Once the individual is accepted as a member, he or she "feels that meaning is inherent in the group, and that group membership will pour substance into his vacuousness" (Karr and Dent 1970: 192). Similarly, Partridge (1973: 45) argues, "the evening brings the balm of companionship, shared experiences, and rededication to the spirit of how life ought to be." The subculture becomes "a powerful fraternity . . . to many it is more attractive even than the effects of the drugs themselves" (Pope 1971: 77; also see Partridge 1973).

Does all this not ring of Durkheim's ([1915] 1968) analysis of religion? The power gained through a sense of membership in a religious community is, according to Durkheim, the basis of religion. It is what attracts us to our religious groups, makes us create totems and sacred objects, and transform our groups -- or the idolized image of them -- into gods. Just as extravagant "Cathedrals of Consumption" are attempts to re-enchant the dominant culture in modernized societies (see Ritzer 1999), the drug subculture offers the illicit drug user hope for meaning, magic, enchantment. It is a congregation of like-minded souls participating in highly ritualized behaviors that reaffirm their beliefs. In it they find "an illusion of personal acceptance and a subculture with which to identify; a united front in rebellion against parent and society; and an illusion of meaning in the group" (Karr and Dent 1970: 192).

Thus, the importance of the subculture in promoting use cannot be overstated. First, having drug-using peers, or involvement in the drug subculture, is consistently the best predictor of drug use among adolescents. Next, there is a strong inverse relationship between an adolescent's level of religiosity and his or her involvement in the drug subculture and with drug use. Understanding the drug subculture can help explain this correlation. By definition, those who are strongly religious are the ones who have avoided the secularizing effects of the modernization and rationalization processes. If Weber, Habermas and others are correct, these individuals would be the least disenchanted. Their world, filled with religious explanations and a religious understanding of the events and circumstances of everyday life, would not be devoid of magic. There is magic in God and God's workings. Thus, there would be little need to search for an alternative source of enchantment. Conversely, those individuals who have accepted the "modern" and "rational" worldview of science and rejected the "magical" worldview of religion would be the ones who experienced the sense of disenchantment Weber discusses. Therefore, these individuals would be the most likely to search for a source of re-enchantment. While some would undoubtedly attempt to find re-enchantment in consumerism (see Ritzer 1999) or through some other means, others would seek re-enchantment in the drug subculture. Once they entered the subculture, their "new religion" would encourage drug use. Indeed, the more one accepted the "doctrines" of the "religion," the more likely they would use drugs and use them heavily. From this perspective then, the correlation between drug use and religiosity is not negative; it is strong and positive. The direction of the relationship depends on the "religion" to which one belongs.

Consequently, understanding the drug subculture is vital to understanding drug use. It provides members access to drugs. It provides a group of people who can teach the user how to prepare and use the drugs. It provides a network of friends who teach the user how to define the drug's effects as pleasurable. It provides a group with whom to associate and recreate in activities that avoid detection and social control. It provides protection from social condemnation and legal sanctions. Users need the subculture for these rather mundane aspects of use. Yet, they also need the subculture for much more. We must realize that the attraction of the subculture is often not the allure of drugs per se; instead, the subculture provides an enchanting worldview. Even if this enchantment is fleeting, even if it is grossly misguided, it is nevertheless an attempt to find enchantment in a disenchanted world.

The Evolution of the Subculture: The Rationalization of Religion

The subculture has changed rather dramatically since the Beats and Hippies dominated it. However, even today, there remains a religious air to the drug subculture. While arguing that the drug subculture is a functional equivalent to a religion is not particularly new, arguing that the emergence of this "religion" is a result of the rationalization process is more novel. Yet, if the subculture was a result of the rationalization process, that process should have continued influencing the subculture. That is, has the evolution of the drug subculture occurred in a manner consistent with the rationalization process in general? I would argue that it has. The subculture's evolution since the 1940s has followed the general processes outlined throughout this book. Like in the dominant culture, modernization and rationalization secularized the subculture.

Recalling the discussion of how society was secularized strengthens the argument that the subculture is a pseudo-religion. With respect to societies, in early pre-modern societies, where the worlds of nature and the supernatural are undifferentiated, almost everything is sacred. Moreover, structurally, there is little differentiation in social positions. That is, these early pre-modern societies were relatively egalitarian (see Lenski 1966). In such societies, magic is directed toward manipulating this world, including both its natural and social phenomena.

To quote Weber ([1922] 1964: 1) again, "the most elementary forms of behavior motivated by religious or magical factors are oriented to *this* world" (emphasis in the original). As modernization advances, there is an eventual move away from naturalism toward an abstract, symbolic understanding of the world. As this process occurs, the gods become increasingly personified and connected to a community. A worldview develops that explains the world in a unified, orderly fashion based on "universal truths." A professional priesthood emerges and society becomes more differentiated in general and stratification increases. With further rationalization, however, secular values become increasingly acceptable and religious views become less dominant. The realm of the sacred diminishes in importance and a scientific worldview, complete with its emphasis on instrumentally rational action, dominates. This results in the principles of efficiency and calculation being applied to an ever-increasing range of behaviors and social contexts. Stratification begins to decrease, although it is still more prevalent than in pre-modern societies.

The subculture has taken a parallel path of evolution. First, as mentioned above, the Beat's spirituality was based on "the sacralization of everyday life and the sacramentalization of human relationships" (Prothero 1991: 214). To the Beats, every moment was sacred. Thus, like in "elementary forms" of religion, there was little differentiation between the sacred and the profane. Moreover, like in pre-modern societies, the Beat subculture was relatively egalitarian. In fact, the elites of the movement tried their best to be like the most downtrodden Beat members. Poverty was glorified and the "fallanheen" were noble. Plus, like in "elementary forms" of religion, the Beats used magic -- in this case drugs -- for personal enlightenment and to enhance creativity. That is, their drug use, or magic, was for manipulating this world and their place in it. As a movement, the Beats were largely unconcerned with social injustices or how to change them (Brake 1980). The Beat subculture, therefore, shared numerous similarities with early pre-modern societies. Indeed, pre-modern societies were superior to modern society and they sought to regain a connection to the "primitive."

As the Beats evolved into the Hippies, the subculture changed. First, the reasons for using drugs changed. While use was still a means of "self actualization," it became part of a "universalistic worldview" based on love, togetherness, passive resistance and other expressive values (see, for example, Spates 1976). Drug use was a pivotal part of this cosmology. Thus, the subculture became more "other worldly," in a sense, by focusing on the larger society beyond their group. As Spates (1976: 871) says,

> Hippies emphasized a concern for the welfare of others and affiliation as the sole determinants of interpersonal relations. From the former emphasis, the whole love ethic evolved -- hippies believed in loving all other human beings unconditionally and in showing complete tolerance for their life choices no matter how bizarre they might appear.

While the Beats rejected the dominant society and wanted nothing to do with it, the Hippies rejected it but hoped to change it by deliberating attacking it (see Brake 1980; Hoffman 1980). While their solutions were idealistic at best and naive and whimsical at worst, they nevertheless developed a culture that "was a subversive force cutting away at society's roots, a lived-out critique of the materialism and philistinism of contemporary industrial society" (Brake 1980: 96). Their concerns with society -- as opposed to a pure rejection of it -- were expressed in song lyrics and the underground press (see Peterson and Berger 1972; Gleason 1972; Spates and Levin 1972; Gitlin 1987). Their efforts to implement this ideology of change led to the development of new institutions. For example, the "diggers" efforts to provide free food to the Haight/Ashbury scene and their opening of "free stores" offered an alternative to more traditional social services (see Grogan 2001; also see Cavan 1972). Even if these tasks were accomplished with stolen food, they were symbolic attacks against the dominant economic system of the larger society. They were organized attempts to reorganize the basic social system. The Hippies ideology also influenced the dominant culture. As Brake (1980: 97) states,

A concern with ecology led to the development of pure food shops, preventive medicine, pollution politics and organic farming. The necessity to develop new alternative legal and social services led to a new interest in community politics. Techniques of consciousness raising and 'rap' groups developed a new consciousness of oppression outside of traditional class lines, which became essential in the development of feminism and gay politics in their struggle against patriarchy.

Thus, as in pre-modern or advanced pre-modern societies where religion becomes more "other worldly," Hippie "religion" became more "other worldly" in the sense of being focused beyond itself.

Next, as in pre-modern and advanced pre-modern societies, the gods become associated with a specific community. With the Hippies, their "god," if you will, became associated with their community. While drugs were certainly associated with the Beats, they were also associated with the lower classes, ethnic minorities, and street-people such as prostitutes who had little to do with the Beat movement (see, for example, Morgan 1981; Musto 1987). The Beats, partially by choice, were simply associated with these groups. The Hippies, however, were a much more identifiable group than the Beats. Moreover, they were primarily middle-class youth. They were far more visible and drew far more media attention than the Beats had. The media told America that Hippies were drug users and drug users were Hippies. Despite the inaccuracy, the association was firmly rooted in the populace's mind. Thus, drug use became intricately associated with Hippies and their community.

Third, as in pre-modern and advanced pre-modern societies, society in general, including religious institutions, differentiates and becomes stratified. Religion becomes independent from the state and less involved in economic affairs. Moreover, it becomes stratified when a professional priesthood emerges. This process also occurred in the subculture. As Brake (1980) points out, the Hippie subculture differentiated and became increasingly stratified. First, the Hippies divided into the "militants" and "mystics." The militants focused on the social critique and confrontational dimensions of the subculture. They were highly involved in larger social movements such as the Civil Rights Movement, the anti-war movement and the environmental movement (see Hoffman 1980; Grogan 2001; Kallen 2001). The mystics focused on developing alternative forms of consciousness and subjectivity as a means of self-attainment. Thus, like the modernizing society of the advanced pre-modern era that begins to differentiate between the profane and sacred worlds, the subculture began to divide between the "profane" militants and the "sacred" mystics. Second, the Hippie culture was more stratified than the Beat's. Brake, while discussing the findings of several ethnographic studies of the Hippie movement, (1980: 97) states that,

> There was . . . a blurred yet distinct social system. The top elite was the 'aristopopcracy' of high status and wealthy groups such as super stars. . . Next was the 'alternative bourgeoise' who had specialist knowledge (such as electronics, or production) or else were bohemians symbiotic to the underground. The 'lower-middle-class-drop-out' lacked the skills of the above, but was employed in a minor capacity by them. Finally there were the 'lumpenhippies,' the 'street people,' working-class and vagrant, living rough and 'street wise.'

Once again, the subculture's development paralleled that of a modernizing society, albeit over a much shorter period.

Thus, throughout the 1950s and 1960s, the rationalization process was affecting the subculture. This rationalization of the subculture continued in the 1970s. By the early 1970s, the Hippie movement had ended and most who had been involved pursued other lifestyles (see Spates 1976; Gitlin 1987; Russell 1993). Yet, the subculture obviously outlived the Hippies. As in society, however, a more "rational" perspective was replacing the subculture's "religion." By the 1970s, drug users largely dropped the pretense that their use was part of a spiritual quest. While users undoubtedly offered familiar rationalizations such as use promoted creativity or provided significant personal insight, there was little attempt to hide the recreational aspect of drug use. Use was mainly an expression of a rebellious and hedonistic lifestyle that revolved around "sex, drugs and rock-n-roll." The subculture's ideology became less coherent and far less "spiritual." For example, the music of the subculture, which had acted as a unifying voice and a source for subcultural "gospel" for the Hippies, no longer expressed the socially critical ideology of the subculture per se. Instead, rock music had largely reverted to its roots by singing about personal triumphs and failures, sexual liaisons and conquests, lost and found love and (of course) drugs and alcohol. While in the 1960s, Gil Scott Heron sang about how "the revolution will not be televised," Bob Dylan told us about how the "times they are a changin'" and Graham Nash claimed "we can change the world," by the mid-1970s we were reminded to "rock-n-roll all night, and party everyday" and Bruce Springsteen simply noted, "baby, we were born to run." By the mid-1970s, music and protest were no longer intricately linked.[107] The major artists of the 1970s (Led Zeppelin, The Eagles, Pink Floyd, Springsteen, Bad Company, Peter Frampton, Fleetwood Mac, to name a few of the biggest record sellers) were not known as protest musicians, despite their popularity in the drug subculture. Similarly, the subculture lost its voice in the underground press. While countless underground papers and magazines filled college campuses and city streets in the 1960s, by the mid-1970s, the presence of the underground press was noticeably lacking.

With the waning of the larger social movements that gave credence to their "cause," the ideological unity of the subculture weakened. Although still more liberal, more anti-materialistic, more individualistic, less traditionally religious and more eco-conscious than their non-drug-using peers, drug users in the 1970s were far less solidified than the Beats or the Hippies. Despite being the largest subculture in terms of numbers of members, the 1970s subculture mostly disappeared from the public's mind. There is much evidence to support this claim. First, these users, despite being the most numerous group of drug users in U.S. history, did not have a specific name such as the "Beats" or "Hippies." Even the

---

[107.] For example, if one searches *Sociofile*, the words "music" and "protest" do not appear simultaneously from 1969 until the late 1980s when rap music became a popular protest genre.

media, which popularized the term "hippie" and referred to the group as such more often than the group itself did, failed to think of a name for the 1970s druggies. Second, try to find a book or journal article dedicated to the drug subculture of the 1970s. While hundreds of books, academic articles, newspaper stories and magazine stories have been written about the Beats and Hippies, there is precious little written about the large number of drug users who roamed America's streets, concert halls and bars in the 1970s. Despite outnumbering the Beats and Hippies by millions, scholars and journalists largely ignored these youth.[108]

One reason for the "disappearance" of the subculture in the 1970s was they became less confrontational, at least openly. One reason they became less confrontational is that the larger culture and subculture were beginning to merge. For example, the underground press that did remain began to resemble the mainstream press (see Spates 1976). In the late 1960s, 73% of mainstream articles expressed instrumental values while only 20% of underground articles did. By the early 1970s, however, over 28% of underground articles expressed instrumental values. Moreover, the number of underground articles that conveyed an expressive value orientation decreased from 46.7% in the 1967 - 1969 period to 22.9% in the 1970 - 1972 period (Spates 1976: 875). With this drastic reduction in expressive values, the underground press was nearly identical to the mainstream press (see Spates 1976). As often happens with rebellious movements, the mainstream commercialized the Hippies by appealing to "Hippie" customs, styles and manners as they targeted the large-and-still-growing youth market. For example, television shows such as *Laugh In, The Monkeys* and *The Mod Squad* popularized "Hippie fashion." Similarly, "Hippie" style found its way into popular fashion magazines, movies, record stores and other commercial outlets. As Gitlin (1987: 205) notes when discussing the dissemination of counterculture's music and ideology, "thanks to modern mass media, and to drugs, notions which had been the currency of tiny

---

[108.] For example, a search on *Sociofile* for the term "drug subculture" or "drug users" returns only 69 cites between 1970 and 1980. By comparison, such a search conducted for the years 1980 to 1990 returns171 citations. This is despite the fact that overall rates of drug use declined substantially between 1980 and 1990.

groups were percolating through the vast demographics of the baby boom." As happens with pop culture groups (see Melly 1971; also see Spates 1976), the commercializing and popularizing of the subculture's style attracted waves of young people who wanted to join the movement, or, at least look like they had joined the movement. As the style of the subculture became increasingly popular, those attracted to the style, but unfamiliar with the group's values, began to swell the ranks of the subculture. However, these "inauthentic" subculture members, visually indistinguishable from "true" and "devout" members, were familiar with the style of the group only. The values and cultural capital of the group were not purchasable. As Melly notes, the infusion of these "wannabes" and "phonies" often alienates the "truly devout" from their own group. Moreover, the "phonies" trivialize the group's ideology. To use the religious analogy again, performing the group's rituals without knowing the meaning behind them detracts from their sacredness. Just as the Biblical teaching claims that faith without action is a hollow faith, action without faith is similarly insincere. As Shakespeare's Hamlet said, "My words fly up, my thoughts remain below: Words without thoughts never to heaven go" (*Hamlet* III, iii, 103). This process of dilution occurred to both the Beats (see, for example, Matza 1961; Maynard 1991; van Elteren 1999) and the Hippies (see, Spates and Levin 1972; Gitlin 1987). In both cases, commercialization popularized the subculture's "style" but ignored, or at least failed to promote, the group's ideology. This process sterilized the subculture and made it tolerable to the dominant society (in general, see Melly 1971).

The commercialization of the subculture had the same effect on it as commercialization has on the dominant culture: it rationalized it. Through this increased rationalization of the subculture, the subculture and dominant culture started to resemble each other, at least in style. The dividing line between the drug subculture and youth culture in general became almost non-existent. The subculture's argot became widely incorporated into "youth speak" and was even widely accepted by the youths' parents. Terms such as "cool," "bad," "rapping" and "munchies" that once meant something different to a drug user and a non-drug user no longer separated these groups. Even the language of drug use, such as "rushing," "tripping" and "high," became common-place terms to explain drug-like, but not necessarily drug-induced, states. The dress and sense of fashion -- or, lack thereof by some standards -- of the subculture also became common. Long hair, tie-dye and bell-bottoms no longer automatically identified a drug user. By the mid-1970s, one could purchase *new* clothes that were faded, patched or torn and tie-dye dresses were sold via mass-mailed catalogs. The dominant culture, or at least its youth, began performing other subculture rituals also. The subculture's music (soon to be, but not yet, labeled "classic rock") was no longer truly alternative. In fact, it was now a huge business with multi-millionaire stars. Similarly, I recall dressing in cut-offs and tie-dye, walking around barefoot, picking up litter and talking about "Mother Earth" and "peace" on the first Earth Day in 1970. I did these "hippie things" with approximately 125 other first, second and third graders. Dressing, talking and, to some extent, acting like Hippies was even encouraged by our (probably) Hippie-teacher. While we did these "hippie things," and had obviously heard of "hippies," we, as six and seven year-olds, were far from being hippies. While I cannot say with absolute certainty, I am confident that none of us "playing hippie" that day had tried any illicit drugs at that point in our lives. Weekly trips to traditional religious institutions, not through LSD-assisted rap sessions, fulfilled our spiritual needs. We were not Hippies; we just played as if we were. Thus, even by 1970, "Hippie culture" had pervaded youth culture. By the mid-1970s, the subculture and youth culture were almost indistinguishable.

The growing similarity between the mainstream and underground presses, the commercialization of subculture style, the assimilation of some of the subculture's values and argot into mainstream society, the increasingly blurry distinction between the subculture and the dominant culture and the near-total fusion of youth culture and the subculture all illustrate the further rationalization of the subculture. And, this greater rationalization further helped the subculture become nearly invisible. The "religion" of the subculture had been rationalized and demystified in a similar manner as the great religions had. Drug use was once a means to "spiritual enlightenment." By the 1970s, use, and the hedonism it represents, became an end itself.

By the 1980s, through the 1990s and into the new millennium, members of the drug subculture were hardly distinguishable from members of the dominant culture. While vestiges of the old bohemian culture could still be found in "Deadheads" and later "Phishheads" (see Cooper 2000), most illicit drug users' style and manners differed little from their non-drug-using peers.[109]   In fact, the similarities in style, language, mannerisms and everyday behaviors between today's users and non-users are so strong that users have found a new strategy for deviating. They blend! The rituals of the subculture have so thoroughly disseminated throughout the dominant culture that drug users no longer stand out unless they truly want to. This "blending" affords them the ability to deviate without detection. They can play both sides of the fence. They can dabble in the religion of the subculture without totally rejecting their old religion, their old society, their old friends and family. They can pursue their pastime taking a path of least resistance.

To be sure, the drug subculture is still a pseudo-religion complete with rituals, basic beliefs and a community of followers. However, like the dominant culture in which this subculture exists, the subculture's "religious" aspect has largely been replaced with more secular, less otherworldly, more practical concerns. While drug users still occasionally refer to the old "spiritual" reasons for use, most are likely to admit to the sheer hedonism of their use. For example, a former student told me, "When things get too weird, I like to smoke (marijuana). It provides clarity and helps me make sense of what people are doing. It helps me make sense of the world . . . makes it seem like there is a purpose to it all." While these utterances are reminiscent of the old Beat or Hippie justifications for use and

---

[109.]   Deadheads are fans of the band Grateful Dead and Phishheads are fans of the band Phish. These groups and their ideologies can be directly connected to the Beat and Hippie subcultures that came before them (Cooper 2000).

reflect an attempt to experience life, achieve self-actualization and a sort of spirituality, the student also said, "plus, I just like it. It's a lot of fun." Similarly, as part of her thesis project, a former student interviewed 181 Phishheads before a concert in 1999. These youth are today's equivalent of the Beats or Hippies. Unlike most of today's youthful drug users, many Phishheads openly embrace a style and fashion that highlights their subculture (see Cooper 2000). Thus, compared to most drug users today, these fans would be the most "Beat-like" and most likely to express a more spiritual justification for their drug use. Nevertheless, over one-quarter of the young (average age of 20) fans mentioned the pursuit of fun and excitement when asked to describe Phish fans (Cooper 2000). According to these fans, their following of the band and use of drugs was "all about feeling good" and "fun and energy" (quoted in Cooper 2000: 31). Thus, even among the subculture's most devout, drug use was for practical purposes such as "enhancing the concert experience" or "having fun" rather than for the pursuit of some new vision or new spirituality (quoted in Cooper 2000: 33).

Next, rationalization has also influenced drug use itself. Technologically sophisticated growing techniques produced hybrid marijuana plants that were more potent and had no troublesome seeds. Crack use replaced cocaine snorting so the user could achieve a quicker and more intense "buzz." "Designer drugs," made only with relatively advanced knowledge of chemistry and pharmacological principles, became popular. The knowledge of science has been applied to recreational drug use. As in the dominant society, technical expediency has replaced spirituality.

Similarly, while today's subculture clings to many of the values of the Beats and Hippies such as hyper-individualism and anti-materialism, few adopt the lifestyle of Beats or Hippies. The student and Phishheads mentioned before were not pursuing "true" Beat or Hippie "cosmology." The student, for example, graduated with two majors from a major state university with a perfect 4.0 grade point average. He scored a near-perfect score on the MCATs and is now attending one of the nation's most prestigious medical schools. He drove a used BMW he bought by saving money he made working as a web-page designer. He, as he said, "appreciates the finer things in life." He wants to be wealthy, healthy and wise. Rarely do today's youthful users totally reject their society. Rarely do they purposefully pursue a path of poverty. Many actually embrace capitalism, albeit to a lesser extent than most Americans do. For example, a Phishead said,

> A typical Phish fan is a generation X product which (*sic*) finds more fulfillment in experiences which they can achieve, rather than material things which they can earn. They are still very capitalistic though on a much more minimal scale (quoted in Cooper 2000: 30).

Similarly, among the 181 Phishheads interviewed, nearly two-thirds were in college. Nearly eighty percent of these young "Bohemians" worked at least part time. Like many of their contemporaries, these youths "turned on" and may have "tuned in." They certainly have not "dropped out." Most of today's users are analogous to the millions of "moderately devout" Christians, Jews, Muslims, Hindus or Buddhists who evoke their religion when needed, but do not devote their entire life to the religion's pursuit.

Thus, over time, the same forces of rationalization that made the dominant culture disenchanted altered the nature of the subculture. The subculture and the justifications for use that it provides have been rationalized. Ironically, if this argument is correct, the "religion" of the subculture was a response to the disenchantment of the dominant culture; however, over time, the same rationalization process that demystified the dominant culture brought the subculture and dominant culture closer in style. Eventually, they became similar enough to borrow from each other, shape each other, and melt together into a hybrid society that embraces both Bohemian and "traditional" values. Thus, the drug subculture and the dominant culture are dialectically related.

The "conservative," hyper-materialistic, "sterile" dominant culture that was devoid of magic gave rise to the subculture. The subculture was an attempt to find enchantment in a disenchanted world. It offered a "new vision" that was filled with the "magic" of drugs. It was the antithesis of the dominant culture in that it rejected material well-being and functional rationality. It rejected the technology that helped achieve functional rationality. It sought a substantive rationality that the dominant culture seemed to lose sight of as it pursued faster cars, super computers, space travel and the ability to destroy the planet with one bomb. For all its failings, the subculture sought, at least initially, what its members saw as a better way of life. Yet the forces of rationalization and modernization are too powerful. Eventually, the very forces its founders despised and that gave birth to their "movement" preyed on the subculture itself. Eventually, the subculture itself was rationalized. Because of the rationalization of the subculture, it became increasingly similar to the dominant culture. As its values, lifestyles, beliefs, attitudes and behaviors became increasingly common and decreasingly threatening, the dominant culture engulfed the subculture. The dominant culture's consumer capitalism marketed the culture's style, and the dominant culture spent billions of dollars buying it.

Yet, as in any dialectic relationship, by assimilating the antithesis, the thesis changed. As Brake (1980: 97) notes, the subculture's ideology influenced the dominant culture by promoting ecology, preventive and alternative medicine, and alternative lifestyles. The subculture's fashion is still evident today in any major city or on any major college campus. The subculture promoted tolerance and, as a society, we are more tolerant than before the subculture emerged. As Melly (1971: 240), writing in the early 1970s, noted, "doing your own thing, living openly with someone without marrying him, having a child out of wedlock and so on, lead to no general public outcry as would have been the case not so long ago." This statement is even truer today than it was when Melly wrote it. We are more accepting of different ethnic groups and sexual orientations today than we were in the 1950s. Alternative family relations, such as single-parent families or gay couples with children, are tolerated more today than prior to the subculture's "glory days." Divorce no longer raises suspicions and cranks the wheels of the rumor mill as it did during the Eisenhower post-war boom. Unlike in the 1950s, mothers of young children can now work full-time without anyone thinking less of her. Cohabitation is rather common. Oral sex is not "sex" by middle school standards and not defined as "deviant" at all by many college students.[110] Promiscuity in general is more acceptable for both genders. And, recreational drug use is "normal" (see Parker et al 1998; Parker et al 2002). While the subculture cannot take complete credit or blame for these changes by any stretch of the imagination, it nevertheless played a vital role in promoting these changes. If for no other reason, the subculture provided the corporate capitalists the style they so ably sold to America's youth.

So, today, the subculture and dominant culture have blended in style like never before. Both the dominant culture and the subculture have been rationalized. Each has lost some of its enchantment. This blending has permitted a new style of drug user to emerge and deviate without notice. The "mainstreamer" is able to deviate because he or she conforms. This blending can also possibly explain the decrease in overall rate of drug use witnessed since the end of the 1970s. Since the subculture lost some of its enchanting aspects, the subculture is less able to use disenchantment as a powerful recruiting force. Assuming this argument is accurate, the subculture would still attract those looking to rebel, but it would be less attractive to those looking for some "spiritual meaning" in their rebellion.

Yet, like the dominant culture, the religion of the subculture can coexist with the secularized, rationalized world. Just as many practitioners of traditional religious are more likely to explain a tornado as the convergence of favorable meteorological conditions that result in an intensified low-pressure system instead of the wrath of God, many of today's drug users acknowledge their use is for fun

---

[110]. In a class of 118 students, I asked them to rank 60 behaviors in terms of how deviant they considered each to be on a scale of 1 (not deviant at all) to 10 (extremely deviant). An adult performing oral sex on a consenting adult of an opposite received a mean rating of 1.61. Oral sex was deemed "less deviant" than all of the other behaviors, including a 19 year-old female and a 19 year-old male engaging in sexual intercourse (a mean of 1.91), masturbation (2.01), watching a pornographic movie (2.50), getting multiple tattoos (2.58), and a 35 year old female drinking enough alcohol to become intoxicated (2.85). For reference, an adult smoking marijuana for recreation was scored as only "moderately deviant" by my class with a mean ranking of 4.59.

and excitement instead of some pursuit of a "new vision." Moreover, the drug subculture, even today, is not totally devoid of spirituality. Just as extremely devout Christians, Jews, Muslims, Buddhist and Hindus hold to their religious beliefs in spite of the secularizing forces they confront, so do many devotees of the drug subculture. There remain "true believers." More importantly, those introduced to the subculture often find some "connection to something larger than themselves" and "meaning" in the group they form. As some Phish fans commented when describing their drug-using group, "everybody is willing to help out with anything you need," "you just feel like your part of a big family or something," and "(we're) a close knit community" (cited in Cooper 2000: 48 – 49). This "community" in-and-of-itself can be enchanting. The camaraderie this community offers is "magical," especially to the disaffiliated, the disillusioned, the angst-ridden. The subculture may no longer hold promise of a "new vision," but youth are still attracted to the subculture simply by the magical element of companionship. This "magic" is why associating with drug-using peers is so highly correlated with drug use.

The enchanting aspects of the drug subculture can also explain why illicit drug use is more pronounced among the younger members of our societies than among the older. For whom is the modern world most disenchanting? Youth! Society is not set-up to benefit youth. Social institutions are controlled by, and reward most handsomely, adults. Families and the manners in which we, as societies, have decided to arrange them make most sense to parents, not children. At least traditionally and in most families, parents have the power. Similarly, most societies, at least advanced capitalistic societies, lock children out of the dominant economic system. While we exploit their labor in fast food stores, supermarkets, restaurants and other service establishments, children cannot legally own property until they are 18 years old. Few, if any, under the age of 18 can get a job that pays handsomely. In most modernized societies, children are not permitted to participate meaningfully in political institutions. They cannot vote nor hold office until they are adults. Even in the educational system, a system designed primarily for children, children do not have the power. It is little wonder, then, why youth, especially those in advanced capitalistic societies, tend to be disillusioned and sense the disenchantment of our rationalized societies most intensely. It is little wonder, then, that youth are the ones who supply the critical mass to the drug subculture. They are the ones most likely to seek a "new vision," a new sense of enchantment, by joining the "religion" of drugs. Even Weber, who was unconcerned about the use of drugs, noted the tendency of youth to respond to the disenchantment of a rationalization society by creating religious-like groups. He wrote ([1919] 1946: 155),

> It is, however, no humbug but rather something very sincere and genuine if some of the youth groups who during recent years have quietly grown together give their human community the interpretation of a religious, cosmic, or mystical relation.

Even then, youth responded to disenchantment by creating a "new religion." The contemporary drug subculture is no different.

SUMMARY

The theory proposed in this manuscript is a macro-level theory. It argues that the rationalization and modernization processes have altered social definitions, social relations and social structures in such a way to promote the use of drugs. However, I believe the theory can account for many of the same facts about the use of drugs by individuals as more micro-level theories explain. Most of the personality characteristics of drug users are highly related to -- indeed, logical outcomes of -- a scientific worldview and individualism. These perspectives and values are products of the rationalization and modernization processes. Similarly, this macro-level theory can also account for the correlation between behavioral variables such as associating with drug-using peers and drug use. If understood as a pseudo-religion, the drug subculture offers a means of re-enchantment. The disenchantment associated with rationalization created a society that some found intolerable. They searched for a "new vision" and their attempts created the Beat subculture. This group, and the ideology its members professed, offered a source of enchantment to disillusioned youth who had found their "straight" world disenchanting. While the processes described in this book have also rationalized the subculture, the sense of community and belonging the subculture offers remains a powerful, and to some extent magical, attraction to millions of youth.

POSTSCRIPT
CONSIDERING POLICY

I have tried to maintain a "value-free" perspective throughout the manuscript. Such a perspective requires personal values to be suspended. In so doing, we cannot discuss what we *should do* about drugs or the "drug problem." Discussing such issues assumes there is a "problem," and believing that depends on one's values. There are perspectives that can morally justify drug use, even recreational drug use. While few may agree with such perspectives, no one can *empirically falsify* them. Simply put, you cannot scientifically determine policy since policy, by its very nature, is value oriented. However, this does not mean that policy cannot be informed by science. Indeed, most effective policy is. So, assuming the typical policy objective that we do want to "do something" about drug use, we can ask "what are the policy implications of all this?" Assuming the arguments I have made are accurate, what can we do about drug use?

Well, the honest answer is "not much." It is unlikely we can reverse, or even slow, the forces of modernization and rationalization. Chances are we do not even want to reverse them. Modernization and rationalization have served western societies well. We are far more comfortable because of modernization. We have far more material goods available to us because of modernization. Even the poorest person in the most modernized societies are better off materially than most of humanity was through most of history. Yes, modernization has its good points. Personally, I love living in a modernized society. I do not want to walk over the hill to haul water that may or may not be potable. However, just as increased crime is an unfortunate consequence of a free society; increased rates of drug use may be a consequence of a modernized society. Yet, for those who are disposed toward "doing something about drug use," this work does have some implications.

First, we must understand that the use of intoxicating drugs is as much about conforming as it is about deviating. Indeed, cross-culturally and historically speaking, drug use is often not deviant. The use of peyote during the ritual celebrations of the Native American Church is not deviant. Nor is the use of hallucinogens by dozens of shamanistic societies in remote areas of the world. The use of wine by Christians celebrating communion is not deviant. Of course, these uses of drugs are not what people have in mind when they talk about "doing something about drugs." Scholars, parents, religious leaders and politicians typically worry about the recreational use of illicit drugs, especially by youth. Yet, even the recreational use of intoxicants can be "normal," or non-deviant, behavior. The use of alcohol in westernized societies is not deviant, for example. Moreover, even illegal drug use, although "deviant" by definition, is about conformity. The use of marijuana by youth worldwide, for example, is often as much of a result of *conforming* to the subculture as it is about *deviating* from the dominant culture.

Failing to understand the social context of recreational drug use, failing to understand that much illicit drug use is about conforming to group norms, directs us down the wrong path. Focusing on individual "pathology" blinds us to the social reasons people use drugs. We need to remember that most illicit drug users are not that different from most non-users. They are looking for a connection to something greater, something communal. They are looking for social companionship with like-minded souls. They are looking for meaning in a disenchanted world. Many in westernized nations turn to consumerism and the Cathedrals of Consumption (see Ritzer 1999). Others turn to traditional religion. Some turn to sports. Still others turn to the drug subculture. What we should realize is that the *form* of the process is similar, even if the *content* differs. Thus, if we want to keep our youth from joining the drug subculture, we need to find a way to direct their search away from the subculture toward some other, more socially acceptable, group. We need to find a way to offer an alternative to the enchantment they find in the collective of the subculture.

To do this, we should recall Talcott Parson's basic insight that action is directed and governed by norms (see Parsons and Shils 1951). Therefore, understanding where the subculture's norms originated and how they have evolved and continue to evolve is critical to understanding the allure the subculture holds for youth. Offering programs that do not align with these norms are likely to fail. The alienated youth will not be attracted to the anti-drug program any more than he or she was to the offerings of the traditional society. We must also understand the drug norms of our dominant cultures. Our youth are directed toward the subculture by what our rationalized culture creates and promotes. Understanding how western values promote drug use can better equip us for channeling the individualism and scientific rationalism that, if left undirected or are misdirected, often result in the glorification of hyper-individualism and hedonism the subculture encourages. We need to find a way to promote the individualism we cherish while also instilling a sense of social responsibility to, and meaningful participation in, the collective. If we are to deter adolescents from using drugs, we must manipulate the leisure activities of our youth. We must offer more visible and instrumental activities (see Hawdon 1996; Hawdon 2004). By structuring these activities in ways that make them more attractive to drug-using youth, and by possibly fostering friendship networks with non-delinquent peers, we could offer membership in a collective that glorifies discretion instead of hedonism and sanctifies the group instead of the individual. While achieving this would be difficult, this strategy would undoubtedly be easier to accomplish than trying to undue years of poor socialization. Simply put, it is easier to manipulate behavior than to change an individual's psyche.

With respect to rehabilitating those who already use drugs, this research implies that religious-like programs such as A.A. or N.A. (see Trevino 1992) would likely be most successful, at least for those that remain spiritually oriented. If it is true that drug users turned to the subculture because of its enchanting "religious" nature, substituting an abstinence-promoting religion for the use-promoting one would offer a similar type of connectedness and, therefore, appeal. Users could find similar social networks in this new group and, therefore, remain a "part of something." Of course, traditional religious communities could possibly serve the same function; however, these are less likely to appeal to reforming drug users. These options were available to the user long before he or she began using drugs. For some reason, the individual rejected the traditional options when they were younger. Is it probable they will embrace these options now they are older? Most likely, they will be more attracted to a "new" religion such as those found in 12-step programs. For those who have rejected "spirituality" all together or have adopted a more rational or "scientific" worldview, a group-oriented program could still be successful. The program, however, would need to be devoid of a religious tone and present itself in a more secular manner. While such a program would retain functional elements of religion such as a community of "believers," it would have to be presented to its adherents in a way that promoted "rationality" instead of "spirituality" (see Atkins (2000; 2003) for a discussion of such programs and a review of evaluations comparing spiritual and secular approaches to addiction recovery). As Atkins (2003: 7) argues,

> The recognition that 'one size does not fit all' is also very promising for the future of substance abuse treatment. We are currently working to better understand the diverse needs of those recovering from substance abuse problems. Matching treatment clients to the right recovery support group is an important factor that should help to improve treatment for everyone, the spiritually oriented and the secularly oriented alike.

With respect to our international policies regarding illicit drugs, we should stop treating illicit drugs as "magical substances." We need to realize these commodities are simply that: commodities. They, like all commodities, follow laws of supply and demand. Thus, if we truly want to stop the flood of illicit drugs, especially plant-based illicit drugs, we need to (1) reduce the demand for them in modernized nations and (2) offer viable alternative sources of economic activity to those nations that are addicted to narco-dollars. To achieve the first, we need to understand the allure of the subculture as I just described. To accomplish the second, we must promote economic development. Tying development funds to drug eradication efforts is counter-productive. Eradication simply does not work when the best path to a feeding one's family is by harvesting coca or poppy. Of course, to help the lesser-developed nations develop strong economies will be expensive and require cooperation and coordinated efforts among the more developed nations of the world. Then again, it would probably be cheaper and take less coordination than trying to penetrate the drug distribution systems and intercept their products.

In any case, whether we are trying to deter our youth from starting to use drugs, trying to rehabilitate those who already do or attempting to disrupt the international drug distribution system, the road is a difficult one. While it is obviously possible to succeed, we are fighting forces that are old and powerful. We are fighting deeply ingrained forces that permeate our culture and grow stronger each day. We are fighting forces that are as relentless as the rising sea. Like the sea, the forces of modernization and rationalization are too powerful to stop. Yet, like the sea, we may be able to harness their power and redirect it. With respect to drug consumption, this may be too optimistic. The sea may yet swallow us. I hope, however, I have at least made us more aware of the forces we confront.

REFERENCES

Able, Ernest L. 1982. *Marihuana: The First Twelve Thousand Years*. New York: McGraw Hill.

Adlaf, Edward M. and Angelia Paglia. 2003. *Drug Use Among Ontario Students, 2003*. (CAMH Research Document, Number 14). Toronto: Centre for Addiction and Mental Health.

Adler, Patricia A. and Peter Adler. 1978. "Tinydopers: A Case Study of Deviant Socialization," *Symbolic Interaction* 1: 90 - 105.

Adler, Patricia A. 1985. *Wheeling and Dealing: An Ethnography of an Upper Level Drug-Dealing and Smuggling Community*. New York: Columbia University Press.

Affinnih, Yahya H. 1999. "A Review of Literature on Drug Use in Sub-Saharan Arica Countries and its Economic and Social Implications." *Substance Use and Misuse*, 34: 443 - 454.

Africa News Service. 2003. "Experts Believe School Drug Abuse Growing." April 11, u 8.

Africa News Service. 2003. "Bisho 'High' School Drug Alarm." April 14, u 9.

Agar, Michael H. 1977. "Into that Whole Ritual Thing: Ritualistic Drug Use among Urban American Heroin Addicts" Pp. 127 - 148 in *Drugs, Rituals and Altered States of Consciousness*, edited by Brain M. Du Toit. Rotterdam, The Netherlands: A.A. Balkema.

Akers, Ronald L. 1977. *Deviant Behavior: A Social Learning Perspective*. Belmont, Ca.: Wadsworth.

Akers, Ronald. L. 1992. *Drugs, Alcohol and Society*. Belmont, Ca.: Wadsworth.

Akers, Ronald L., Marvin D. Krohn, Lonn Lanza-Kaduce and Marcia Radosevich. 1979. "Social Learning and Deviant Behavior: a Specific Test of a General Theory." *American Sociological Review*, 44, 636- 655.

Albers, Patricia, and Seymour Parker. 1971. "The Plains Vision Experience: A Study of Power and Privilege." *Southwestern Journal of Anthropology*, 27 (3): 203 - 233.

Alford, B. W. E. 1996. *Britain in the World Economy Since 1880*. London: Longman.

Allen, Catherine J. 1988. *The Hold Life Has: Coca and Cultural Identity in an Andean Community*. Washington, D.C.: Smithsonian Institution Press.

Allen, G. C. 1981. *A Short Economic History of Modern Japan*. New York: St. Martin's Press.

Allen, G. C. 1958. *Japan's Economic Recovery*. London: Oxford University Press.

Anslinger, Harry J., and Courtney Ryley Cooper. 1937. "Marijuana: Assassin of Youth." *American Magazine*, July, Pp. 18 - 19 , 150 - 153.

366

Arneklev, Bruce J., Harold G. Grasmick and Robert J. Bursik, Jr. 1999. "Evaluating the Dimensionality and Invariance of 'Low Self-control'." *Journal of Quantitative Criminology* 15, 307 - 331.

Ary, Dennis, Elizabeth Tildesley, Hyman Hops and Judy Andrews. 1993. "The Influence of Parent, Sibling, Peer Modeling and Attitudes on Adolescent Use of Alcohol." *International Journal of the Addictions*, 28, 853 - 880.

Aseltine, Robert H. .1995. "A Reconsideration of Parental and Peer Influences on Adolescent Deviance." *Journal of Health and Social Behavior*, 36, 103 - 121.

Ashley, Richard. 1975. *Cocaine: Its History, Uses and Effects*. New York: St. Martin's.

Atkins, Randolph, Jr. 2000. *No Outside Enterprise: Rational Recovery's Countermovement Challege to the Institutionalization of the Twelve-Step Movement in American Addiction Care.* Unpublished dissertation. University of Virginia. Charlottesville, Virginia.

Atkins, Randolph. 2003. "The Efficacy of Spirtitual vs. Secular Approaches to Addition Recovery." Lecture to the South Place Ethical Society, 12 October 2003.

Aust, Rebbecca, Clare Sharp and Chris Goulden. 2002. *Prevalence of Drug Use: Key findings from the 2001/02 British Crime Survey*. Home Office Research Findings. London: Home Office.

Australian Institute of Health and Welfare 2002. 2001 *National Drug Strategy Household Survey: First results*. AIHW cat. no. PHE 35. Canberra: AIHW (Drug Statistics SeriesNo. 9).

Australian Institute of Health and Welfare 1999. *Drug Statistics Series*. AIHW cat. no. PHE 15. Canberra: AIHW.

Avineri, Shlomo and Avner De-Shalit. 1992. *Communitariansim and Individualism*. New York: Oxford University Press.

Baer Gerhard. 1992. "The One Intoxicated by Tobacco: Matsigenka Shamanism." Pp. 79- 100 in *Portals of Power: Shamanism in South America*, edited by E. Jean Matteson Langdon and Gerhard Baer. Albuquerque: University of New Mexico Press.

Bailey, Susan L. and Robert L. Hubbard. 1991. "Developmental Changes in Peer Factors and the Influence on Marijuana Initiation among Secondary School Students." *Journal of Youth and Adolescents*, 20, 339-360.

Barbour, Ian G. 1997. *Religion and Science. Historical and Contemporary Issues*. San Francisco: Harper.

Barrett, Leonard E. Sr. 1988. *The Rastafarians: Sounds of Cultural Dissonance*. Boston: Beacon Press.

Bartram, David. 2000. "Japan and Labor Migration: Theoretical and Methodological Implications of Negative Cases." *International Migration Review*, 34: 5 - 32.

Batchelder, Tim. 2001. "Drug Addictions, Hallucinogens and Shamanism: The View from Anthropology." *Townsend Letter for Doctors and Patients*, July, 2001. 74 - 78.

367

Bawer, Bruce. 2004. "The Other Sixties." *The Wilson Quarterly*, 28 (2): 64 - 84.

BDIS (Bahamas Drug Information System). 2002. *Annual National Report, 2001.* Nassau, Bahamas: Health Information and Research Division, Ministry of Health.

Bean, Lowell John and Katherine Siva Saubel. 1972. *Temalpakh (from the earth): Cahuilla Indian Knowledge and Usage of Plants.* Banning, Calif: Malki Museum Press.

Becker, Howard. 1953. "Becoming a Marihuana User," *American Journal of Sociology,* 59: 235 - 242.

Becker, Howard. 1963. *Outsiders: Studies in the Sociology of Deviance.* New York: Free Press.

Beckett, Katherine. 1994. "Setting the Public Agenda: 'Street Crime' and Drug Use in American Politics." *Social Problems* 41: 425-447.

Beckett, Katherine. 1995. "Media Depictions of Drug Abuse: The Impact of Official Sources." *Research in Political Sociology* 7: 161-182.

Beckett, Katherine and Theodore Sasson. 2000. "The War on Crime as Hegemonic Strategy: A Neo-Marxian Theory of the New Punitiveness in U.S. Criminal Justice Policy." Pp. 61 - 84 in *Of Crime and Criminality: The Use of Theory in Everyday Life,* edited by Sally S. Simpson. Thousand Oaks, CA.: Pine Forge.

Bédarird, François. 1979. *A Social History of England, 1851 - 1975.* London: Methuen.

Bellah, Robert N., Richard Madeson, William Sullivan, Ann Swidler and Steven Tipton. 1985. *Habits of the Heart: Individualism and Commitment in American Life.* New York: Harper and Row.

Bellenir, Karen (editor). 2000. *The Drug Abuse Sourcebook.* Detroit: Omnigraphics.

Benet, Sula. 1975. "Early Diffusion and Folk Uses of Hemp." Pp. 39 - 50 in *Cannabis and Culture,* edited by V. Rubin. Chicago: Aldine.

Benoist, Jean. 1975. "Réunion: Cannabis in a Pluricultural and Polyethnic Society." Pp. 227 - 234 *Cannabis and Culture,* edited by Vera Rubin. Chicago: Aldine.

Bent, Jaap van der. 1988. "How Low Can You Go: The Beat Generation and American Popular Culture." *European Contributions to American Studies,* 13: 145 - 156.

Ben-Yehuda, Nachman. 1986. "The Sociology of Moral Panics: Toward a New Synthesis." *The Sociological Quarterly,* 27: 495 - 513.

Berger, Peter, Brigitte Berger, and Hansfried Kellner. 1973. *The Homeless Mind.* New York: Vintage.

Bergman, Gerald R. 1996. "Religious Beliefs of Scientists: A Survey of the Research. *Free Inquiry,* 16(3): 41 - 47.

Black, Donald. 1976. *The Behavior of Law.* New York: Academic Press.

Black, Donald. 1998. *The Social Structure of Right and Wrong,* revised edition. New York: Academic Press.

Black, Donald. 2000. "The Purification of Sociology." *Contemporary Sociology*, 29: 704 - 709.

Blau, Peter, M. 1977. *Inequality and Heterogeneity: A Primitive Theory of Social Structure*. New York: Free Press.

Bogner, William C. and Howard Thomas. 1996. *Drugs to Market : Creating Value and Advantage in the Pharmaceutical Industry*. Tarrytown, NY: Pergamon

Boreham, Richard and Andrew Shaw (editors). 2001. *Smoking, Drinking and Drug Use Among Young People in England in 2000*. A Report to Department of Health. London: The Stationery Office.

Bourgois, Philippe. 1995. *In Search of Respect: Selling Crack in El Barrio*. New York: Cambridge University Press.

Brake, Mike. 1980. *The Sociology of Youth Culture and Youth Subcultures*. London: Routledge.

Brecher, Edward M. 1972. *Licit and Illicit Drugs*: *The Consumers Union Report on Narcotics, Stimulants, Depressants, Inhalants, Hallucinogens and Marijuana, including Caffeine, Nicotine and Alcohol*. Boston: Little, Brown.

Brook, J.S., D. W. Brook, H. S. Gordon, M. Whiteman and P. Cohen. .1990. "The Psychosocial Etiology of Adolescent Drug Use: A Family Interactional Approach." *Genetic, Social and General Psychology Monograph*, 116, 111- 267.

Brown, J.E. 1953. *The Sacred Pipe: Black Elk's Account of the Seven Rites of the Oglala Sioux*. Norman: University of Oklahoma Press.

Buchinsky, Moshe, and Jennifer Hunt. 1996. *Wage Mobility in the United States*. National Bureau of Economic Analysis Working Paper #5455.

Buchinsky, Moshe. 1994. "Changes in the U.S. Wage Structure 1963-1987: Application of Quantile Regression." *Econometrica*, 62: 405 - 458.

Buhner, Stephen Harrod. 2001. *Sacred Plant Medicine: Explorations in the Practice of Indigenous Herbalism*. Coeur d'Alene, Idaho: Raven Press.

Bursik, Robert J. and Harlod G. Grasmick. 1993. *Neighborhoods and Crime: The Dimensions of Effective Community Control*. New York: Lexington Books.

Bursik, Robert J. and Harlod G. Grasmick. 1995. "Neighborhood-based Networks and the Control of Crime and Delinquency." Pp. 103 - 130 in *Crime and Public Policy: Putting Theory to Work*. Edited by H. Barlow . Boulder: Westview Press.

Calabrese, Joseph D. 1997. "Spiritual Healing and Human Development in the Native American Church: Toward a Cultural Psychiatry of Peyote." *Psychoanalytic Review*, 84: 237 - 255.

Cavan, Sherri. 1972. *Hippies of the Haight*. St. Louis: New Critics Press.

Central Intelligence Agency (CIA). 2001. *The World Fact Book, 2001*. Washington D.C.: US Government Printing Office.

Chase-Dunn, Christopher. 1998. *Global Formation: Structures of the World-Economy*, updated edition. New York: Rowman and Littlefield.

Cheng, Mariah Mantsun and Arne L. Kalleberg. 1997. "How Permanent Was Permanent Employment? Patterns of Organizational Mobility in Japan, 1916-1975." *Work and Occupations*, 24 (1): 12 - 23.

Cheng, Mariah Mantsun and Arne L. Kalleberg. 1996. "Labor Market Structures in Japan: An Analysis of Organizational and Occupational Mobility Patterns. *Social Forces*, 74: 1235 - 1260.

Cherlin, Andrew J. 1981. *Marriage, Divorce, Remarriage*. Cambridge, MA.: Harvard University Press.

Cleckner, Patricia J. 1977. "Cognitive and Ritual Aspects of Drug Use Among Young Black Urban Males." Pp. 149 - 168 in *Drugs, Rituals and Altered States of Consciousness*, edited by Brain M. Du Toit. Rotterdam, The Netherlands: A.A. Balkema.

Codere, Helen. 1975. "The Social and Cultural Context of Cannabis Use in Rwanda." Pp. 217 - 226 in *Cannabis and Culture*, edited by Vera Rubin. Chicago: Aldine.

Coe, Sophie D. and Michael D. Coe. 1996. *The True History of Chocolate*. London: Thames and Hudson.

Cohen, Erik. 1989. "Citizenship, Nationality and Religion in Israel and Thailand." Pp. 66 - 92 in *The Israeli State and Society: Boundaries and Frontiers*, edited by Baruch Kimmerling. Albany: State University of New York Press.

Cohen, Mark Nathan. 1989. *Health and the Rise of Civilization*. New Haven: Yale University Press.

Cohen, Stanley. 1980. *Folk Devils and Moral Panics: The Creation of the Mods and Rockers*. London: MacGibbon and Kee.

Colson, Elizabeth and Thayer Scudder. 1982. *For Prayer and Profit. The Ritual, Economic and Social Importance of Beer in Gwembe District, Zambia, 1950-1982*. Stanford, CA: Stanford University Press.

Comitas, Lambros. 1975. "The Social Nexus of Ganja in Jamaica." Pp. 119 - 132 in *Cannabis and Culture*, edited by Vera Rubin. Chicago: Aldine.

Cooper, Jessica. 2000. *Can't Live While I'm Young: The Evolution of the Bohemian Subculture*. Unpublished Master's Thesis. Clemson University. Clemson, South Carolina.

Cortese, Carol Ann. 1999. "Drug Services and Cultural Adaptation." *Drugs: Education, Prevention and Policy*, 6: 361 - 366.

Cosgrove, Joanna. 2000. "Coffee and Tea." *Beverage Industry*, 91: 14 - 21.

Courtwright, David T. 2001. *Forces of Habit: Drugs and the Making of the Modern World*. Cambridge, Mass.: Harvard University Press.

Crow Dog, Leonard. 1984. "How Grandfather Peyote Came to the Indian People." Pp. 65 - 69 in *American Indian Myths and Legends*, edited by Richard Erdoes and Alfonso Ortiz. New York: Pantheon.

Daly, John W., and Bertil B. Fredholm. 1998. "Caffeine: An Atypical Drug of Dependence." *Drug and Alcohol Dependence*, 51: 199 - 206.

Davey, Jeremy, Tamzyn Davey and Patricia Obst. 2002. "Alcohol Consumption and Drug Use in a Sample of Australian University Students." *Youth Studies Australia*, 21 (3): 25 - 32.

Davies, P. 2003. "Malawi: Addicted to the Leaf." *Tobacco Control*, 12: 91 - 94.

Davis, Fred. 1967. "Why All of Us May Be Hippies Someday." *Trans-Action*, 5: 10 - 18.

Davis, Fred and Laura Munoz. 1968. "Heads and Freaks: Patterns and Meanings of Drug Use Among Hippies," *Journal of Health and Social Behavior*, 9: 156 - 164.

Deardorff, Kevin E. and Patricia Montgomery. 2003. *National Population Trends*. U.S. Census Bureau. http://www.census.gov/ population/www/ pop-profile/ nattrend.html

De Pinho, Alvaro Rubim. 1975. "Social and Medical Aspects of the Use of Cannabis in Brazil." Pp. 294 - 302 in *Cannabis and Culture*, edited by Vera Rubin. Chicago: Aldine.

Del-Boca, Frances K., and Jane A. Noll. 2000. "Truth or Consequences: The Validity of Self-Report Data in Health Services Research on Addictions." *Addiction*, 95, supplement 3, Nov, 347-360.

Dell, Colleen Anne and Karen Garabedian. 2003. *2002 National Report Drug Trends and the CCENDU network*. Canadian Community Epidemiology Network on Drug Use Annual Report. Ottawa: Canadian Centre on Substance Abuse.

DiFranza, Joseph, and Joe Tye. 1995. "Who Profits from Tobacco Sales to Children?" Pp. 57 - 62 in *The American Drug Scene: An Anthology*, edited by James Inciardi and Kenneth McElrath. Los Angeles: Roxbury.

Dobkin de Rios, Marlene. 1975. "Man, Culture and Hallucinogens: An Overview." Pp.401 - 416 in *Cannabis and Culture*, edited by V. Rubin. Chicago: Aldine.

Dobkin de Rios, Marlene. 1984. *Hallucinogens: Cross-Cultural Perspectives*. Albuquerque: University of New Mexico Press.

Donovan, John E. 1996. "Problem-Behavior Theory and the Explanation of Adolescent Marijuana Use." *Journal of Drug Issues*, 26 (2): 379 -404.

Donovan, John E., Richard Jessor and Frances M. Costa. 1991. "Adolescent Health Behavior and Conventionality-Unconventionality: an Extension of Problem-behavior Theory." *Health Psychology* 10 (1), 52 - 61.

Donovan, John E., Richard Jessor and Frances M. Costa. 1993. "Structure of Health-Enhancing Behavior in Adolescence: a Latent-variable Approach." *Journal of Health and Social Behavior*, 34: 346-362.

Donovan, John E., Richard Jessor, and Frances M. Costa. 1999. "Adolescent Problem Drinking: Stability of Psychosocial and Behavioral Correlates across a Generation." *Journal of Studies on Alcohol*, 60:352 - 361.

Doogan, Kevin. 1996. "Labour mobility and the changing housing market."*Urban Studies*, 33: 199 - 222.

Dreher, Melanie Creagan. 1983. "Marijuana and Work: Cannabis Smoking on a Jamaican Sugar Estate." *Human Organization*, 42 (1): 1 - 8.

Dreher, Melanie Creagan. 1982. *Working Men and Ganja: Marihuana Use in Rural Jamaica*. Philadelphia: Institute for the Study of Human Issues.

Drug Policy Alliance. 2002. Drug Policies around the World. http://www.drugpolicy.org/global/drugpolicyby/asia/.

Drug Scope. 2000. *Annual report on the State of the Drugs Problem in the European Union and Norway.* Brussels: European Monitoring Centre for Drugs and Drug Addiction.

Drug Scope. 2002. *Annual report on the State of the Drugs Problem in the European Union and Norway.* Brussels: European Monitoring Centre for Drugs and Drug Addiction.

Drug Scope. 2004. *Betel Nut.* http://www.drugscope.org.uk/druginfo/drugsearch/ds_results.asp?file=%5Cwip%5C11%5C1%5C1%5Cbetelnut.htm

Dudley, Jill, Gill Turner and Simon Woollacott. 2003. *Control of Immigration: Statistics United Kingdom 2002.* London: The Home Office.

Durkheim, Emile. [1915] 1966. *The Division of Labor in Society.* New York: The Free Press.

Durkheim, Emile. [1915] 1968. *The Elementary Forms of the Religious Life.* New York: The Free Press.

Durkheim, Emile. [1899] 1975. "Concerning the Definition of Religious Phenomena." Pp. 74 - 99 in *Durkheim on Religion: A Selection of Readings,* edited by W.S.F. Pickering. London: Routledge.

Du Toit, Brian M. 1975. "Dagga: The History and Ethnographic Setting of Cannabis Sativa in Southern Africa." Pp. 81 - 116 in *Cannabis and Culture,* edited by Vera Rubin. Chicago: Aldine..

Du Toit, Brian M. 1976. "Continuity and Change in Cannabis Use by Africans in South Africa." *Journal of Asian and African Studies,* 11: 203 - 208.

Duster, Troy. [1970] 1989. "The Legislation of Morality: Creating Drug Laws." Pp. 29 - 39 in *Deviant Behavior: A Text-Reader in the Sociology of Deviance,* 3rd edition, edited by Delos H. Kelly. New York: St. Martin's Press.

Earle, Charles W. 1880. "The Opium Habit: A Statistical and Clinical Lecture." *Chicago Medical Review,* 2: 442 -446.

Echols, Alice. 2002. *Shaky Ground: The Sixties and its Aftershocks.* New York: Columbia University Press.

The Economist. "Economic Indicators." August 15th, 1998, v. 348, n. 8081: 84.

Ehman, Mark A. 1983. "The Pure Land S_tras." Pp. 118 - 126 in *Buddhism: A Modern Perspective,* edited by Charles Prebish. State College: Pennsylvania State University Press.

Eide, Arne H., Stanley Wilson Acuda and Espen Roysamb. 1998. "Cultural Orientation and Alcohol-type Preferences Among Adolescents in Four Sociocultural Subgroups in Zimbabwe. *Journal of Cross-Cultural Psychology,* 29: 343 - 358.

Eide, Arne H. and Stanley Wilson Acuda. 1997. "Cultural Orientation and Use of Cannabis and Inhalants among Secondary School Children in Zimbabwe." *Social Science and Medicine* 45: 1241 - 1249.

Eide, Arne H. and Stanley Wilson Acuda, 1996. Cultural Orientation and

Adolescents' Alcohol Use in Zimbabwe. *Addiction*, 91: 807 - 814.

EIS (European Information Service). 2002. "Drugs: Consumption Patterns in Europe are Converging. *European Report*, Oct 5, 2002: 471.

Elekesa, Zsuzsanna and László Kovács. 2002. "Old and New Drug Consumption Habits in Hungary, Romania and Moldova." *European Addiction Research*, 8:166-169.

Elliott, Delbert S., David Huizinga and Suzanne Ageton. 1985. *Explaining Delinquency and Drug Use*. Beverly Hills: Sage.

Elwood, William N. 1994. *Rhetoric in the War on Drugs: The Triumphs and Tragedies of Public Relations*. Westport, Ct.: Praeger.

EMCDDA (European Monitoring Centre for Drugs and Drug Addiction). 1998. *Annual Report on the State of the Drugs Problem in the European Union*. Luxembourg: Office for Official Publications of the European Communities.

EMCDDA (European Monitoring Centre for Drugs and Drug Addiction). 2002. *Annual Report on the State of the Drugs Problem in the European Union*. Luxembourg: Office for Official Publications of the European Communities.

EMCDDA (European Monitoring Centre for Drugs and Drug Addiction). 2003. *Annual Report on the State of the Drugs Problem in the European Union*. Luxembourg: Office for Official Publications of the European Communities.

Emdad-ul Haq, M. 2000. *Drugs in South Asia: From the Opium Trade to the Present Day*. London: MacMillan Press.

EORG (European Opinion Research Group). 2002. "Attitudes and Opinions of Young People in the European Union on Drugs." *Eurobarometer 57.2*.

Erickson, Patricia G, Edward M. Adlaf, Glenn F. Murray, and Reginald G. Smart. 1987. *The Steel Drug: Cocaine in Perspective*. Lexington, MA.: Lexington Books.

ESPAD (European School Survey Project on Alcohol and Drugs). 1997. *Alcohol and Other Drug Use Among Students in 26 European Countries*. Stockholm: Swedish Council on Alcohol and Other Drugs.

ESPAD (European School Survey Project on Alcohol and Drugs). 2001. *Alcohol and Other Drug Use Among Students in 30 European Countries*. Stockholm: Swedish Council on Alcohol and Other Drugs.

Estes, Richard. 1988. *Trends in World Social Development: The Social Progress of Nations*. New York: Praeger.

European Commission. 2002. "The State of Health in the European Community 2000. Luxemburg: Office for Official Publication of the European Communities.

Fazey, Cindy. 2002. "Estimating the World Illicit Drug Situation: Realty and the Seven Deadly Political Sins." *Drugs: Education, Prevention and Policy*, 9 (1): 95 - 103.

Feather, David L. and Robert M. Hauser. 1978. *Opportunity and Change*. New York: Academic.

Firth, Raymond W. 1936. *We, The Tikopia: A Social Study of Kinship in Primitive Polynesia*. London: Allen and Unwin.

Fisher, James. 1975. "Cannabis in Nepal: An Overview." Pp. 257 - 255 in *Cannabis and Culture*, edited by Vera Rubin. Chicago: Aldine.

Flaker, Vito. 2002. "Heroin Use in Slovenia: A Consequence or a Vehicle of Social Changes"*European Addiction Research*, 8:170-176.

Flood-Page, Claire, Siobhan Campbell, Victoria Harrington, and Joel Miller. 2000. *Youth Crime: Findings from the 1998/99 Youth Lifestyles Survey*. London: Home Office Research Study 209.

Fort, Joel. 1973. *Alcohol: Our Biggest Drug Problem*. New York: McGraw Hill.

Frank, John N. 1995. "And the 1994 per cap Winners Are . . ." *Beverage Industry*, 86 (4): 10 - 12.

Fraser, F. 2002. *Drug Misuse in Scotland: Findings From the 2000 Scottish Crime Survey*. Edinburgh: Scottish Executive Central Research Unit.

French, Laurence Armand. 2000. *Addictions and Native Americans*. Westport, Ct.: Praeger.

Fuess, Scott M. Jr. 2003. "Immigration Policy and Highly Skilled Workers: The Case of Japan." *Contemporary Economic Policy*, 21: 243 - 257.

Furst, Peter T. 1976. *Hallucinogens and Culture*. San Francisco: Chandler and Sharp.

Furst, Peter. (editor) 1990. *Flesh of the Gods*. Prospect Heights, Illinois: Waveland Press.

Gately, Iain. 2002. *Tobacco: A Cultural History of how an Exotic Plant Seduced Civilization*. New York: Grove Press.

Gimbutas, Marija.1982. *The Goddesses and Gods Europe, 6500-3500 B.C.: Myths and Cult Images*. Berkeley: University of California Press.

Ginsberg, Allen. 1980. "The New Consciousness." Pp. 63 - 93 in *Composed on the Tongue*, edited by Allen Ginsberg. Bolinas, Ca.: Grey Fox.

Ginsberg, Allen. 1984. *Collected Poems, 1947 - 1980*. New York: Harper and Row.

Gitlin, Todd. 1987. *The Sixties: Tears of Hope, Days of Rage*. New York: Bantam.

Gleason. Ralph J. 1972. "A Cultural Revolution." Pp. 127 – 136 in R. Serge Denisoff and Richard A Peterson (eds). *The Sounds of Social Change: Studies in Popular Culture*. Chicago: Rand McNally.

Glenn, Curtis E. 1996. *Russia: A Country Study*. Washington, D.C.: Federal Research Division, Library of Congress.

Goode, Erich. 1972. *Drugs in American Society*. New York: Knopf.

Goode, Erich. 1990. "The American Drug Panic of the 1980s: Social Construction or Objective Threat?" *The International Journal of the Addictions*, 25: 1083-1098.

Goode, Erich. 1997. *Between Politics and Reason: The Drug Legalization Debate*. New York: Worth Publishers.

Goode, Erich. 1999. *Drugs in American Society*, 5th edition. Boston: McGraw-Hill.

Goode, Erich. 2003. "Drug Use and Self-Control: A Crack in the Invariant Hypothesis?" Paper presented at the annual meeting of the *Southern*

374

Sociological Society, New Orleans, Louisiana.

Goode, Erich, and Nachman Ben-Yehuda. 1994. *Moral Panics: The Social Construction of Deviance*. Cambridge, MA: Blackwell.

Goode, Erich and Nachman Ben-Yehuda. 1994b. "Moral Panics-Culture, Politics, and Social Construction," *Annual Review of Sociology* 20: 149 - 171.

Goodman, Jordan. 1993. *Tobacco in History: The Culture of Dependence*. London: Routledge.

Goss, Richard J. 1994. "The Riddle of the Religious Scientist." *The American Rationalist*, May-June, 39(1):105-107.

Gottfredson, Michael. R. and Travis Hirschi. 1990. *A General Theory of Crime*. Stanford, Ca.: Stanford University Press.

Graham, John, and Benjamin Bowling. 1995. *Young People and Crime*. London: Home Office Research and Statistics Department.

Greberman, Sharyn Bowman, Kiyoshi Wada.1994. "Social and Legal Factors Related to Drug Abuse in the United States and Japan." *Public Health Reports*, 109: 731 - 738.

Greener, Mark. 2001. *A Healthy Business: A Guide to the Global Pharmaceutical Industry*. London: Urch Publishing.

Greenwell, Lisa, and Mary-Lynn Brecht. 2003. "Self-reported Health Status among Treated Methamphetamine Users." *American Journal of Drug and Alcohol Abuse*, 29: 75 - 105.

Griffith, Ivelaw L. 1998. "The Geography of Drug Trafficking in the Caribbean." Pp. 97 - 120 in *From Pirates to Drug Lords: The Post-Cold War Caribbean Security Environment*, edited by Michael C. Desch, Jorge I. Domínguez and Andrés Serbin. Albany: State University of New York Press.

Grogan, Emmett. 2001. "The Diggers Feed the People." Pp. 124 - 131 in Stuart Kallen (editor) *Sixties Counterculture*. San Diego, CA.: Greenhaven Press.

Gureje, Oye and And'Dapo Olley. 1992. "Alcohol and Drug Abuse in Nigeria: A Review of the Literature." *Contemporary Drug Problems*, 19: 491 - 504.

Gusfield, Joseph R. 1963. *Symbolic Crusade: Status Politics and the American Temperance Movement*. Urbana: University of Illinois Press.

Gusfield, Joseph R. 1981. *The Culture of Public Problems: Drinking - Driving and the Symbolic Order*. Chicago: University of Chicago Press.

Gusfield, Joseph R. 1996. *Contested Meanings: The Construction of Alcohol Problems*. Madison: University of Wisconsin Press.

Halsey, A.H. 1988. "Introduction Statistics and Social Trends in Britain." Pp. 1 - 35 in *British Social Trends since 1900: A Guide to the Changing Social Structure of Britain*, edited by A. H. Halsey. London: MacMillan Press.

Halsey, A. H. 1988b. "Higher Education." Pp. 268 - 296 in *British Social Trends since 1900: A Guide to the Changing Social Structure of Britain*, edited by A. H. Halsey. London: MacMillan Press.

Habermas. Jurgen 1970. *Toward a Rational Society: Student Protest, Science and Politics*. Boston: Beacon.

Habermas, Jurgen. 1979. *Communication and the Evolution of Society*. Boston:

Beacon.

Habermas, Jurgen. 1984. *The Theory of Communicative Action, vol 1, Reason and the Rationalization of Society.* Boston: Beacon.

Habermas, Jurgen. 1984b. *The Theory of Communicative Action, vol 2, Lifeworld and System: A Critique of Functionalist Reason.* Boston: Beacon.

Halsey, A. H. 1988. "Introduction: Statistics and Social Trends in Britain." Pp. 1 - 35 in *British Social Trends since 1900*, edited by A. H. Halsey, London: MacMillan.

Halsey, A. G. 1988b. "Population Trends." Pp. 267 - 293 in *British Social Trends since 1900*, edited by A. H. Halsey, London: MacMillan.

Hanson, Glen R., Peter J. Venturelli, and Annette E. Fleckenstein. 2002. *Drugs and Society*, 7th edition. Sudbury, MA.: Jones and Bartlett.

Harney, Steve. 1991. "Ethnos and the Beat Poets." *Journal of American Studies*, 25: 363 - 380.

Hartnoll, R., U. Avico, F. R. Ingold, K. Lange, L Lenke, A. O'Hare, and A. de Roij-Motshagen. 1989. *A Multi-city Study of Drug Misuse in Europe.* Washington, D.C.: United Nations Office on Drugs and Crime.

Hasan, Khwaja A. 1975. "Social Aspects of the Use of Cannabis in India." Pp. 235 - 246 in *Cannabis and Culture*, edited by Vera Rubin. Chicago: Aldine.

Hattox, Ralph S. 1985. *Coffee and Coffeehouses: The Origins of a Social Beverage in the Medieval Near East.* Seattle: University of Washington Press.

Hawdon, James E. 1996a. "Deviant Lifestyles: The Social Control of Daily Routines." *Youth and Society,* 28: 162 - 188.

Hawdon, James E. 1996b. "Cycles of Deviance: Social Mobility, Moral Boundaries and Drug Use, 1880 - 1990. *Sociological Spectrum* 16: 183 - 207.

Hawdon, James E. 1996c. *Emerging Organizational Forms: The Proliferation of Regional Intergovernmental Organizations in the Modern World System.* Westport, Ct.: Greenwood Press.

Hawdon, James E. 1999. "Daily Routines and Crime: Using Routine Activities as Measures of Hirschi's Involvement." *Youth and Society,* 30: 395 - 415.

Hawdon, James E. 2001. "The Role of Presidential Rhetoric in the Creation of a Moral Panic: Reagan, Bush, and the War on Drugs." *Deviant Behavior*, 22: 419 - 446.

Hawdon, James E. 2003. "Adolescent Drug Use: Is it Self Control or Social Control?" Paper presented at the annual meeting of the *Southern Sociological Society*, New Orleans, Louisiana.

Hawdon, James E. 2004. "Drug Use in Middle School: Assessing Attitudinal and Behavioral Predictors." *Free Inquiry in Creative Sociology*, 32: 17 - 41.

Hawdon, James E. and John Ryan. 2003. "Social Networks, Social Capital, Collective Efficacy and Crime: Extending Social Disorganization Theory." A paper presented at the annual conference of the *National Funding Collaborative on Violence Prevention.* Washington: D.C.

Hasan, Khwaja A. 1975. "Social Aspects of the Use of Cannabis in India." Pp.

235 - 246 in *Cannabis and Culture*, edited by V. Rubin. Chicago: Aldine.

Headley, Sue. 2001. "Illicit Drug Use in Regional Australia, 1988 - 1998. *Youth Studies Australia*, 20 (3): 58 - 69.

Health Canada, 1995. *Horizons Three Young Canadians' Alcohol and Other Drug Use: Increasing Our Understanding*. Edited by David Hewitt, Garry Vinje and Patricia MacNeil. Ottawa: Publications Unit Health Canada.

Hebdige, Dick. 2002. *Subculture: the Meaning of Style*. London: Routledge.

Hechter, Michael. 1987. *Principles of Group Solidarity*. Berkeley: University of California Press.

Helmer, John. 1975. *Drugs and Minority Oppression*. New York: Seabury Press.

Hirschi, Travis. 1969. *Causes of Delinquency*. Berkeley: University of California Press.

Hirschfelder, Arlene, and Paulette Molin. 1992. *The Encyclopedia of Native American Religions*. New York: Facts on File.

Hobsbawm, Eric. 1968. *Industry and Empire*. Baltimore: Penguin.

Hoffman, Abbey. 1980. *Soon to be a Major Motion Picture*. New York: Putnam

Hoffman, Joseph H., Grace M. Barnes, John W. Welte, and Barbara A. Dintcheff. 2000. "Trends in Combinational Use of Alcohol and Illicit Drugs Among Minority Adolescents, 1983-1994." *American Journal of Drug and Alcohol Abuse*, 26: 311 - 325.

Holmes, John Clellon. 1952. "This is the Beat Generation." *New York Times Magazine*, November 16th: SM 10 - 14.

Holmes, John Clellon. [1958] 1988. "The Philosophy of the Beat Generation." Pp. 65 - 77 in *Passionate Opinions: The Cultural Essays*, edited by John Clellon Holmes. Fayetteville: University of Arkansas Press.

Home Office. 1982. *Results from the British Crime Survey, 1982*. London: Home Office Research, Development and Statistics Directorate.

Home Office. 1988. *Results from the British Crime Survey, 1988*. London: Home Office Research, Development and Statistics Directorate.

Hope, Trina L. and Kelly R. Damphousse. 2002. "Applying Self-control Theory to Gang Membership in a Non-Urban Setting." *Journal of Gang Research* 9, (2): 41 - 61.

Hopkins, Thomas J. 1971. *The Hindu Religious Tradition*. Belmont, Ca.: Wadsworth.

Hout, Michael. 1988. "More Universalism, Less Structural Mobility: The American Occupational Structure in the 1980s." *American Journal of Sociology*, 93: 1358 - 1400.

Hout, Michael. 1984. "Status, Autonomy, and Training in Occupational Mobility." *American Journal of Sociology*, 89: 1379 - 1409.

Htdocs. 1991. "Caffeine." Http:// 193.51.164.11/htdocs/monograph/vol51.

Hutchinson, Harry William. 1975. "Patterns of Marihuana Use in Brazil." Pp. 173 - 183 in *Cannabis and Culture*, edited by V. Rubin. Chicago: Aldine.

Hultkrantz, Ake. 1992. *Shamanic Healing and Ritual Drama: Health and Medicine in Native North American Religious Traditions*. New York: Crossroads Publishing.

IMS Health 2000. "Improved Outlook for Latin America, 1999 - 2004. WWW. IMS-global.com.

IMS Health 2001. *World Review, 2000.* WWW.IMS-global.com.

IMS Health 2001b. "Strong Growth Forecast for Cental and East Europe." WWW.IMS-global.com.

IMS Health 2001c. "Asian Market to Reach 30 Billion by 2005." WWW.IMS-global.com.

IMS Health 2003. *World Review, 2003.* WWW.IMS-global.com.

Inciardi, James A. 1992. *The War on Drugs II: The Continuing Epic of Heroin, Cocaine, Crack, Crime, AIDS, and Public Policy.* Mountain View, Ca.: Mayfield.

Inciardi, James A., Ruth Horowitz and Anne E. Pottieger. 1993. *Street Kids, Street Drugs, Street Crime: An Examination of Drug Use and Serious Delinquency in Miami.* Belmont, Ca.: Wadsworth.

International Coca Organization (ICCO). 2000. *Quarterly Bulletin of Cocoa Statistics*, 26 (4), 1999/2000.

International Labour Organisation (ILO). 1996. *Country Report: Russia, 2d Quarter 1996.* Economist Intelligence Unit: London.

Inyengar, Shanto. 1991. *Is Anyone Responsible? How Television Frames Political Issues.* Chicago: University of Chicago Press.

ISDD (Institute for the Study of Drug Dependence). 1993. *Drug Misuse in Britain: Annual Audit of Drug Statistics.* London: Institute for the Study of Drug Dependence.

Iyotani, Toshio. 1998. "Globalization and Immigrant Workers in Japan." *National Institute for Research Advancement Review*, Winter 1998.

Jackson, Keith and Andy Bennett. 1998. *Control of Immigration: Statistics United Kingdom, Second Half and Year 1997.* London: The Home Office.

Jackson, Ralph. 1988. *Doctors and Diseases in the Roman Empire.* Norman, OK: University of Oklahoma Press.

Jacobson, Peter D., Jeffrey Wasserman, and John R. Anderson. 1997. "Historical Overview of Tobacco Legislation." *Journal of Social Issues*, 53: 75 - 96.

Jamieson, Ross W. 2001. "The Essence of Commodification: Caffeine Dependencies in the Early Modern World." *Journal of Social History*, 35: 269 - 294.

Jang, Sung Joon and Terence P. Thornberry. 1998. "Self-Esteem, Delinquent Peers, and Delinquency: A Test of the Self-Enhancement Thesis." *American Sociological Review*, 63: 586-598.

Japanese Ministry of Health. 2004. "Trends in Vital Statistics by Prefecture in Japan, 1899-1998." http://www.mhlw.go.jp/english/.

Javetz, Rachel, and Judith T. Shuval. 1990. "Normative Contexts of Drug Use among Israeli Youth." Pp. 427 - 459 in *The Socio-cultural Matrix of Alcohol and Drug Use: A Sourcebook of Patterns and Factors*, edited by Brenda Forster and Jeffrey Colman Salloway. Lewiston, NY: Edwin Mellen.

Javetz, Rachel, and Judith T. Shuval. 1982. "Vulnerability to Drugs among Israeli

Adolescents." *Israel Journal of Psychiatry* 19: 97 - 119.

Javetz, Rachel, and Judith T. Shuval 1984. "Drug Use among High School Students in Israel: A Syndrome of Social Vulnerability." *Youth and Society.* 16: 171 - 194.

Jeffery, Debra, Axel Klein and Les King. 2002. *UK Drug Report on Trends in 2001: Report from the United Kingdom Focal Point to EMCDDA.* Lisbon, Portugal: European Monitoring Centre for Drugs and Drug Addiction (EMCDDA).

Jensen, Eric L., Jurg Gerber, and Ginna M. Babcock. 1991. "The New War on Drugs: Grass Roots Movement or Political Construction?" *Journal of Drug Issues* 21: 651-667.

Jeri, F. Ral. 1984. "Coca-paste Smoking in Some Latin American Countries: A Severe and Unabated Form of Addiction." *Bulletin on Narcotics*, 1984, issue 2: 15 - 31.

Jessor, Richard. 1987. "Problem-Behavior Theory, Psyosocial Development and Adolescent Problem Drinking." *British Journal of Addiction* 82: 331 - 342.

Jessor, Richard and Shirley L. Jessor. 1977. *Problem Behavior and Psychosocial Development: a Longitudinal Study of Youth.* New York: Academic Press.

Jessor, Richard and Shirley L. Jessor. 1980. "A Social-psychological Framework for Studying Drug Use." Pp. 54 - 71 In *Theories on Drug Abuse*, edited by Dan J. Lettieri. Rockville, MD.: National Institute on Drug Abuse.

Jessor, Richard., James A. Chase, and John E. Donovan. 1980. "Psychosocial Correlates of Marijuana Use and Problem Drinking in a National Sample of Adolescents." *American Journal of Public Health*, 70: 604-613.

JICA (Japan International Cooperation Agency). 1992. *Anti-drug Activities in Japan.* Tokyo: National Police Agency of Japan.

Johnson, Bruce D. 1973. *Marihuana Users and Drug Subcultures.* New York: Wiley-Interscience.

Johnston, Lloyd D., Patrick M. O'Malley, Jerald G. Bachman, and John E. Schulenberg. 2004. Monitoring the Future National Results on Adolescent Drug Use: Overview of Key Findings, 2003. (NIH Publication No. 04-5506). Bethesda, MD: National Institute on Drug Abuse.

Johnston, Lloyd. D, Patrick M. O'Malley, and Jerald G. Bachman. 2003. *Monitoring the Future national survey results on drug use, 1975-2002. Volume I: Secondary school students* (NIH Publication No. 03-5375). Bethesda, MD: National Institute on Drug Abuse.

Jonnes, Jill. 1996. *Hep-cats, Narc, and Pipe Dreams: A History of America's Romance with Illegal Drugs.* New York: Scibner.

Joseph, Roger. 1975. "The Economic Significance of Cannabis Sativa in the Moroccan Rif." Pp. 184 - 193 in *Cannabis and Culture*, edited by Vera Rubin. Chicago: Aldine.

Juhn, Chinhui, Kevin M. Murphy, and Brooks Pierce. 1993. "Wage Inequality and the Rise in Returns to Skill." *Journal of Political Economy*, 101:. 410 - 442.

Julien, Robert M. 2001. *A Primer of Drug Action*, 9th edition. New York: Worth.

Jutkowitz, Joel M. and Hongsook Eu. 1994. "Drug Prevalence in Latin American and Caribbean Countries: A Cross-National Analysis." *Drugs: Education, Prevention and Policy*, 1 (3): 199 - 252.

Kallen, Stuart A. (editor) 2001. *Sixties Counterculture*. San Diego, CA.: Greenhaven Press.

Kandel, Denise B. 1980. "Developmental Stages in Adolescent Drug Involvement." Pp. 120 - 127 in *Theories on Drug Abuse*, edited by Dan J. Lettieri. Rockville, MD.: National Institute on Drug Abuse.

Kandel, Denise B. (editor). 2002. *Stages and Pathways of Drug Involvement: Examining the Gateway Hypothesis*. Cambridge, U.K.: Cambridge University Press.

Kandel, Denise B., and KazuoYamaguchi. 1993. "From Beer to Crack: Developmental Patterns of Involvement in Drugs." *American Journal of Public Health*, 83: 851 - 855.

Kandel, Denise B., and Kazuo Yamaguchi. 1999. "Developmental Stages of Involvement in Substance Use." Pp, 50 -74 in *Sourcebook on Substance Abuse: Etiology, Assessment and Treatment*, edited by R. E. Tarter, R. J. Ammerman and P. J. Ott. New York: Allyn and Bacon.

Kandel, Denise B. and Kazuo Yamaguchi. 2002. "Stages of Drug Involvement in the U.S. Population." Pp. 65 - 89 in *Stages and Pathways of Drug Involvement: Examining the Gateway Hypothesis*, edited by Denise B. Kandel. New York: Cambridge University Press.

Kaplan, Howard B. 1975. *Self-Attitudes and Deviant Behavior*. Pacific Palisades Ca.: Goodyear.

Kaplan, Howard B., and Hiroshi Fukurai. 1992. "Negative Social Sanctions, Self Rejection, and Drug Use." *Youth and Society*, 23: 275 - 298.

Kaplan, Howard B. and Robert J. Johnson. 1991. "Negative Social Sanctions and Juvenile Delinquency: Effects of Labeling in a Model of Deviant Behavior." *Social Science Quarterly*, 72: 98 - 122.

Kaplan, Howard B., Robert J. Johnson and C.A. Bailey. 1986. "Self-rejection and the Explanation of Deviance: Refinement and Elaboration of a Latent Structure." *Social Psychology Quarterly*, 49: 110 - 128.

Kaplan, Howard B., Robert J. Johnson and C.A. Bailey. 1987. "Deviant Peers and Deviant Behavior: Further Elaboration of a Model." *Social Psychology Quarterly*, 50: 277 - 284.

Kanomata, Nobuo. 1997. "Dynamic Changes of Social Mobility in Japan, 1955 - 1995." Keio University. http://web.iss.u-tokyo.ac.jp/rc28/PKanomata.pdf.

Kanovsky, Eliyahu. 1970. *The Economic Impact of the Six-Day War: Israel, Occupied Territories, Egypt and Jordan*. New York: Praeger.

Karr, Stephen D. and Oran B. Dent. 1970. "In Search of Meaning: The Generalized Rebellion of the Hippie." *Adolescence*, 18 (5): 187 - 196.

Kato, Masaaki. 1969. "An Epidemiological Analysis of the Fluctuation of Drug Dependence in Japan", *The International Journal of the Addictions*, 4: 591 - 621.

Kato, Masaaki. 1983. "A Birds Eye View of the Present State of Drug Abuse in

380

Japan." *Drug and Alcohol Dependance,* 11: 55-56.

Katz, Charles M., Vincent J. Webb, Patrick R. Gartin, and Chris E. Marshall. 1997. "The Validity of Self-reported Marijuana and Cocaine Use." *Journal of Criminal Justice,* 25 (1): 31 - 42.

Kennedy, Paul. 1987. *The Rise and Fall of the Great Powers: Economic Change and Military Conflict from 1500 - 2000.* New York: Random House.

Kerimi, Nina. 2000. "Opium Use in Turkmenistan: A Historical Perspective." *Addiction,* 95: 1319 - 1333.

Khalifa, Ahmad M. 1975. "Traditional Patterns of Hashish Use in Egypt." Pp. 195 - 205 in *Cannabis and Culture,* edited by Vera Rubin. Chicago: Aldine.

Khan, Munir A., Assad Abbas, and Knud Jensen. 1975. "Cannabis Usage in Pakistan: A Pilot Study of Long Term Effects on Social Status and Physical Health. " Pp. 345 - 354 *Cannabis and Culture,* edited by Vera Rubin. Chicago: Aldine.

Kimmerling, Baruch. 1989. "Between "Alexandria-On-The-Hudson" and Zion." Pp. 237 - 265 in *The Israeli State and Society: Boundaries and Frontiers.* Edited by Baruch Kimmerling. Albany: State University of New York Press.

Kolb, Lawerence, and A. G. Dumez. 1924. "The Prevalence and Trend of Drug Addiction in the United States and Factors Influencing It." *Public Health Reports,* 39: 1179 - 1204.

Kosai, Yutaka. 1997. "The Postwar Japanese Economy, 1945 - 1973." Chapter Four in *The Economic Emergence of Modern Japan,* edited by Kozo Yamamura. Cambridge: Cambridge University Press.

Kozlova, Marina. 2002. "Central Asia Links Rising HIV to Drug Use" *United Press International,* June 28.

Krohn, Marvin D., Allan J. Lizotte, Terence P. Thonrberry, Carolyn Smith, and David McDowall. 1996. "Reciprocal Causal Relationships among Drug Use, Peers, and Beliefs." *Journal of Drug Issues,* 26: 405-428.

Kumagai, Fumie. 1995. "Families in Japan: Beliefs and Realities." *Journal of Comparative Family Studies,* 26 (1): 135 - 164.

Kurlantzick, Joshua. 2002. "China's Drug Problem and Looming HIV Epidemic" *World Policy Journal* 19: 70 - 76.

La Barre, Weston. 1975. *The Peyote Cult,* 4th edition. Hamden, Ct.: Archon Books.

Ladislav Csémya, Ludk Kubikaa and Alojz Nociarb. 2002. "Drug Scene in the Czech Republic and Slovakia during the Period of Transformation." *European Addiction Research,* 8:159-165.

Lagerspetza, Mikko and Jacek Moskalewicz. 2002. "Drugs in the Postsocialist Transitions of Estonia, Latvia, Lithuania and Poland." *European Addiction Research,* 8:177-183.

LaGrange, Randy L. and Helene Raskin White. 1985. "Age Differences in Delinquency: A Test of Theory," *Criminology,* 23, 19 - 43.

Langdon, E. Jean Matteson. 1992. "Introduction: Shamanism and Anthropology." Pp. 1 - 21 in *Portals of Power: Shamanism in South America,* edited by E. Jean Matteson Langdon and Gerhard Baer. Albuquerque: University of

New Mexico Press.

Langdon. E. Jean Matteson. 1992b. "Dau: Shamanic Power in Siona Religion and Medicine." Pp. 41 - 61 in *Portals of Power: Shamanism in South America,* edited by E. Jean Matteson Langdon and Gerhard Baer. Albuquerque: University of New Mexico Press.

Langdon, E. Jean Matteson and Gerhard Baer (editors). 1992. *Portals of Power: Shamanism in South America.* Albuquerque: University of New Mexico Press.

Laidler, Karen A. Joe, David Hodson and Harold Traver. 2000. *The Hong Kong Drug Market.* A Report for the UNDCP Global Study in IllicIt Drug Markets, Center for Criminology, University of Hong Kong, November 2000.

Laumann. Edward O. 1973. *Bonds of Pluralism: The Form and Substance of Urban Social Networks.* New York: Wiley.

Leary, Timothy. [1966] 2001. "Spreading the Psychedelic Message." Pp. 115 - 123 in *Sixties Counterculture,* edited by Stuart Kallen. San Diego: Greenhaven Press.

Lee, A. Robert. 1996. "Introduction." Pp. 1 - 9 in *The Beat Generation Writers,* edited by A. Robert Lee. London: Pluto.

Lee, A. Robert. 1996b. "Black Beats: The Signifying Poetry of LeRoi Jones/Amiri Baraka, Ted Joans and Bob Kaufman." Pp 158 - 177 in *The Beat Generation Writers,* edited by A. Robert Lee. London: Pluto.

Leeman, Robert F. and Seymour Wapner. 2001. "Some Factors Involved in Alcohol Consumption of First-Year Undergraduates." *Journal of Drug Education,* 31, 249 - 262.

Lenski, Gerhard, E. 1966. *Power and Privilege: A Theory of Social Stratification.* New York: McGraw-Hill.

Levine, Harry G., and Craig Reinarman. 1988. "The Politics of America's Latest Drug Scare." Pp. 89-103 in *Freedom and Risk: Secrecy, Censorship, and Repression in the 1980s,* edited by Richard Curry. Philadelphia: Temple University Press.

Levinthal, Charles F. 2002. *Drugs, Behavior and Modern Society,* third edition. Boston: Allyn and Bacon.

Li, Hui-Lin. 1975. "The Origin and Use of Cannabis in Eastern Asia: Their Linguistic-Cultural Implications." Pp. 51 - 62 in *Cannabis and Culture,* edited by Vera Rubin. Chicago: Aldine.

Lickey, Marvin E., and Barbara Gordon. 1991. *Medicine and Mental Illness.* New York: Freeman.

Lipstiz, George. 1990. *Time Passages.* Minneapolis: University of Minnesota Press.

Lipsmeyer, Christine S. 2003. "Welfare and the Discriminating Public: Evaluating Entitlement Attitudes in Post-Communist Europe." *Policy Studies Journal,* 31: 545 - 565.

Liu, Jian, Bob Jones, Cary Grobe, Christofer Balram and Christiane Poulin. 2003. *New Brunswick Student Drug Use Survey 2002 Highlights Report.* New Brunswick: Department of Health and Wellness.

382

Lloyd, T. O. 2002. *Empire, Welfare State, Europe: History of the United Kingdom, 1906 - 2001*, fifth edition. Oxford: Oxford University Press.

Long, Jason. 2001. Urbanization, Internal Migration, and Occupational Mobility in Victorian Britain. http://faculty-web.at.northwestern.edu/economics/long/long.pdf.

Luhmann, Niklas. 1982. *The Differentiation of Society*. New York: Columbia University Press.

Lyman, Michael D., and Garry W. Potter. 1998. *Drugs in Society: Causes, Concepts and Controls*, 2nd edition. Cincinnati, Oh.: Anderson.

Lyon, William S. 1998. *Encyclopedia of Native American Shamanism: Sacred Ceremonies of North America*. Santa Barbara, Ca.: ABC-Clio.

Macdonald, Dave. 1996. "Drugs in Southern Africa: An Overview." *Drugs: Education, Prevention and Policy*, 3: 127 - 144.

Macdonald, Kenneth, and John Ridge. 1988. "Social Mobility." Pp. 202 - 226 in *British Social Trends since 1900: A Guide to the Changing Social Structure of Britain*, edited by A. H. Halsey. London: MacMillan Press.

Machin, Stephen. 1996. "Wage Inequality in the U.K.." *Oxford Review of Economic Policy*, 12: 47-64.

Mack, Douglas R. A., Douglas R Ewart and Sekou Sankara Tafari. 1999. *From Babylon to Rastafari : Origin and History of the Rastafarian Movement*. Chicago: Research Associates, School Times Publications.

Maddison, Angus. 1995. *Monitoring the World Economy, 1820 - 1992*. Paris: Organisation for Economic Co-Operation and Development.

Mann, Michael. 1986. *The Sources of Social Power: A History of Power from the Beginning to A.D. 1760*. Cambridge: Cambridge University Press.

Marcos, Anastasios C., Stephen J. Bahr and Richard E. Johnson. 1986. "Test of a Bonding/Association Theory of Adolescent Drug Use." *Social Forces*, 65, 135 - 161.

Marks, Robert. 1990. "A Freer Market for Heroin in Australia: Alternatives to Subsidizing Organized Crime. Part 1." *The Journal of Drug Issues*, 20: 131-176.

Martin, Christopher S., Amelia M. Arria, Ada C. Mezzich, and Oscar G. Bukstein. 1993. "Patterns of Polydrug Use in Adolescent Alcohol Abusers." *American Journal of Drug and Alcohol Abuse*, 1993 19: 511 - 522.

Martin, Dominique. 1999. "Modernization in the Crisis: From Talcott Parsons to Jurgen Habermas. Pp. 23 - 44 in *System Change and Modernization: East-West in Comparative Perspective*, edited by W. Wladyslaw, Jan Buncak, Pavel Machonin, and Dominique Martin. Warsaw, Poland: Ifis.

Martin, Marie Alexandrine. 1975. "Ethnobotanical Aspects of Cannabis in Southeast Asia." Pp. 63 - 75 in *Cannabis and Culture*, edited by V. Rubin. Chicago: Aldine.

Mason, W. Alex and Michael Windle. 2002. "Gender, Self-control, and Informal Social Control in Adolescence: a Test of Three Models of the Continuity of Delinquent Behavior." *Youth and Society*, 33: 479 - 514.

Mattison, J. B. 1883. "Opium Addiction Among Medical Men." *Medical Record*, 23: 621 - 623.

Matza, David. 1961. "Subterranean Traditions of Youth." *The Annals of the American Academy of Political and Social Sciences*, 338: 102 - 118.

Maynard, John Arthur. 1991. *Venice West: The Beat Generation in Southern California*. New Brunswick: Rutgers University Press.

Mayock, Paula. 2002. "Drug Pathways, Transitions and Decisions: the Experiences of Young People in an Inner-city Dublin Community." *Contemporary Drug Problems*,29: 117 - 158.

McBroom, James R. 1994. "Correlates of Alcohol and Marijuana Use among Junior High School Students: Family, Peers, School Problems, and Psychosocial Concerns." *Youth and Society*, 26: 54 - 69.

McKenna, Terrence. 1992. *Food of the Gods: The Search for the Original Tree of Knowledge, a Radical History of Plants, Drugs, and Human Evolution*. New York: Bantam Books.

McLaughlin, G. T. 1973. "Cocaine: The History and Regulation of a Dangerous Drug." *Cornell Law Review*, 58: 537 - 572.

McNeil, Helen. 1996. "The Archeology of Gender in the Beat Movement." Pp. 178 - 199 in *The Beat Generation Writers*, edited by A. Robert Lee. London: Pluto.

Measham, Fiona. 2002. "Doing Gender–Doing Drugs: Conceptualizing the Gendering of Drugs Cultures." *Contemporary Drug Problems*, 29: 335 - 373.

Measham, Fiona, Russell Newcombe and Howard Parker. 1994. "The Normalization of Recreational Drug Use amongst Young People in North-West England." *The British Journal of Sociology*, 45: 287-312.

Measham, Fiona, Judith Aldridge and Howard Parker. 2001. "Unstoppable? Dance Drug Use in the U.K. Club Scence." Pp. 80 - 97 in *UK Drugs Unlimited*, edited by H. Parker, J. Aldridge and R. Egginton. New York: Palgrave.

Mehr, Farhang. 1991. *The Zoroastrian Tradition*. New York: Element Books.

Melly, George. 1971. *Revolt into Style: The Pop Arts*. Garden City, NY: Anchor Books.

Merriam, John. 1989. "National Media Coverage of Drug Issues, 1983-1987." Pp. 21-28 in *Communication Campaigns About Drugs: Government, Media and the Public*, edited by Pamela J. Shoemaker. Hillsdale, N.J.: Lawrence Erlbaum Associates.

Merrill, Frederick Thayer. 1942. *Japan and the Opium Menace*. New York: International Secretariat Institute of Pacific Relations and the Foreign Policy Association.

Miller, Michelle. A., Jess K. Alberts, Michael L. Hecht, Melanie R. Trost and Robert L. Krizek. 2000. *Adolescent Relationships and Drug Use*. Mahwah, New Jersey: Lawrence Erlbaum Associates.

Miller, Timothy. 1991. *The Hippies and American Values*. Knoxville: University of Tennessee Press.

Millman, Robert B. and Ann Bordwine Beeder. 1998. "The New Psychedelic Culture: LSD, Ecstasy, 'Rave,' and the Grateful Dead." Pp. 194 - 198 in *The American Drug Scene: An Anthology*, second edition, edited by James Inciardi and Kenneth McElrath. Los Angeles: Roxbury..

Milner, Murray, Jr. 1987. "Theories of Inequality: An Overview and Strategy for Synthesis." *Social Forces* 65: 1053 - 1089.

Mingay. G. E. 1986. *The Transformation of Britain, 1830 - 1939*. London: Routledge and Kegan Paul.

Model, Suzanne, and David Ladipo. 1996. "Context and Opportunity: Minorities in London and New York." *Social Forces*, 75: 485 - 510.

Moen, Darrell Gene. 1999. "The Postwar Japanese Agricultural Debacle." *Hitotsubashi Journal of Social Studies*, 31 (1): 29 - 52.

Mooney, James. 1891. "The Sacred Formulas of the Cherokees," *Seventh Annual report of the Bureau of Ethnology to the Secretary of the Smithsonian Institute 1885-1886*. Washington, D.C." Government Printing Office.

Morgan, H. Wayne. 1974. *Yesterday's Addicts. American Society and Drug Abuse, 1865 - 1920.* Norman, Ok.: University of Oklahoma Press.

Morgan, H. Wayne. 1981. *Drugs in America: A Social History, 1800 - 1980.* Syracuse, NY: Syracuse University Press.

Moscarini, Giuseppe, and Francis Vella. 2002. "Aggregate Worker Reallocation and Occupational Mobility in the United States: 1971 - 2000. *The Institute for Fiscal Studies*, working paper 02/18. http://www.ifs.org.uk/ workingpapers/ wp0218.pdf.

Mott, J. 1989. "Self Reported Cannabis Use in Great Britain in 1981." *British Journal of Addiction*, 80: 37 - 43.

Musto, David F. 1987. *The American Disease: Origins of Narcotic Control.* Expanded edition. New York: Oxford University Press.

Musto, David F. 2002. *Drugs in American History: A Documentary History.* New York: New York University Press.

National Academy of Sciences. 1999. *Science and Creationism.* 2d. ed. Washington DC: National Academy Press.

NCMDA (The National Commission on Marihuana and Drug Abuse). 1972. *Marihuana, A Signal of Misunderstanding.* Washington D.C.: U.S. Government Printing Office.

Nietzsche, Friedrich W. [1887] 1964. *Complete Works.* Translated by Oscar Levy. New York: Russell and Russell.

Noble, Trevor. 2000. "The Mobility Transition: Social Mobility Trends in the First Half of the Twenty-First Century." *Sociology*, 34: 35 - 51.

Noonan, Peggy. 1995. "The Healthiest Foods: Are You Eating Enough?" *Longevity*, February: 53 - 61.

Obot, Isidore. 1990. "Substance Abuse, Health and Social Welfare in Africa: An Analysis of the Nigerian Experience." *Social Science and Medicine*, 31: 699 - 705.

OECD (Organisation for Economic Co-operation and Development). 1995. *OECD Economic Surveys: The Russian Federation 1995.* Paris: OECD.

Office of National Statistics. 1999. Young Adults Who Used Drugs: by Age, 1994 and 1996: Social Trends 29 Dataset. www.statistics.gov.uk/statbase.

Office of National Statistics. 1999b. People Aged 16 to 59 in Scotland Who Used Drugs, 1993 and 1996: Social Trends 29 Dataset www.statistics.gov. uk/statbase.

Office of National Statistics. 2003. "Pupils in primary and secondary schools, 1946/47 to 1998/99." Social Trends 30. www.statistics.gov.uk/statbase.

Office of National Statistics. 2003b. "Population Change, 1901-2021." Social Trends 33. www.statistics.gov.uk/statbase.

OGD (Observatoire Géopolitique des Drogues or Geopolitical Drug Watch). 1996. *The Geopolitics of Drugs.* Translated by Laurent Laniel, Deke Dusinberre, and Charles Hoots. Boston: Northeastern University Press.

OGD (Observatoire Géopolitique des Drogues or Geopolitical Drug Watch). 1997. *Annual Report, 1995/1996.* Boston: Northeastern University Press.

Orcutt, James D. and J. Blake Turner. 1993. "Shocking Numbers and Graphic Accounts: Quantified Images of Drug Problems in the Print Media." *Social Problems,* 40: 190-206.

Osgood, Wayne D., Janet K. Wilson, Patrick M. O'Malley, Jerald G. Bachman and Lloyd D. Johnston. 1996. "Routine Activities and Individual Deviant Behavior." *American Sociological Review,* 61: 635 - 655.

Palacios, Wilson R., and Melissa E. Fenwick. 2003. "'E' is for Ecstasy: A Participation Observation Study of Ecstasy Use." Pp. 277 - 283 in *In Their*

Own Words: Criminals on Crime, 3rd edition, edited by Paul Cromwell. Los Angeles: Roxbury.

Palgi, Phyllis. 1975. "The Traditional Role and Symbolism of Hashish among Moroccan Jews in Israel and the Effect of Acculturation." Pp. 207 - 216 in *Cannabis and Culture*, edited by Vera Rubin. Chicago: Aldine.

Parker, Howard. 2001. "Unbelievable? The UK's Drugs Present." Pp. 1 - 13 in *UK Drugs Unlimited*, edited by H. Parker, J. Aldridge and R. Egginton. New York: Palgrave.

Parker, Howard, Judith Aldridge and Fiona Measham. 1998. *Illegal Leisure: The Normalization of Adolescent Recreational Drug Use*. London: Routledge.

Parker, Howard, Lisa Williams and Judith Aldridge. 2002. "The Normalization of 'Sensible' Recreational Drug Use: Further Evidence from the North West England Longitudinal Study." *Sociology*, 36: 941 - 965.

Parker, Richard B. 1996. *The Six-Day War: A Retrospective*. Gainesville, Fl.: University Press of Florida.

Parliment, Thomas H., Chi-Tang Ho, and Peter Schieberle (editors). 2000. *Caffeinated Beverages: Health Benefits, Physiological Effects, and Chemistry*. Washington, D.C.: American Chemical Society.

Parrinder, Geoffrey (editor). 1983. *World Religions: From Ancient History to the Present*. New York: Facts on File Publications.

Parsons, Talcott. 1951. *The Social System*. Free Press.

Parsons, Talcott. [1963] 1967. "On the Concept of Political Power." Pp. 297-354 in *Sociological Theory and Modern Society*. New York: Free Press.

Parsons, Talcott. [1964] 1967. "Evolutionary Universals in Society." Pp. 490-520 in *Sociological Theory and Modern Society*. New York: Free Press.

Parsons, Talcott. 1977. *Social Systems and the Evolution of Action Theory*. New York: Free Press.

Parsons, Talcott and Edward A. Shils. 1951. *Toward a General Theory of Action*. Cambridge: Harvard University Press.

Parssinen, Terry M. 1983. *Secret Passions, Secret Remedies: Narcotic Drugs in British Society, 1820 - 1930*. Philadelphia: Institute for the Study of Human Issues.

Partridge, William L. 1973. *The Hippie Ghetto: The Natural History of a Subculture*. New York: Holt, Rinehart, and Winston.

Passarides, Christopher A., and Jonathan Wadsworth. 1989. "Unemployment and the Inter-Regional Mobility of Labour." *The Economic Journal*, 99: 739 - 755.

Pego, Christona M., Robert F. Hill, Glenn W. Solomon, Robert M. Chisholm, and Suzanne E. Ivey. 1995. "Tobacco, Culture, and Health among American Indians: A Historical Review." *American Indian Culture and Research Journal*, 19: 143 - 165.

Perrin, Michel. 1992. "The Body of the Guajiro Shaman: Symptoms of Symbols? Pp. 103 - 125 in *Portals of Power: Shamanism in South America*, edited by E. Jean Matteson Langdon and Gerhard Baer. Albuquerque: University of New Mexico Press.

Peterson, Richard A. and David G. Berger. 1972. "Three Eras in the Manufacturing of Popular Music Lyrics. Pp. 282 – 303 in *The Sounds of Social Change: Studies in Popular Culture*, edited by R. Serge Denisoff and Richard A Peterson. Chicago: Rand McNally.

Phillipes, Joël L. And Ronald D. Wynne. 1980. *Cocaine: The Mystique and the Reality*. New York: Avon.

Plant, Martin, and Patrick Miller. 2000. Drug Use Has Declined in United

Kingdom. *British Medical Journal*, 320: 1536 - 1537.

Plowman, Timothy. 1984. "The Ethnobotany of Coca." *Advances in Economic Botany* 1: 62-111.

Poggi, Gianfranco. 1978. *The Development of the Modern State: A Sociological Introduction*. Stanford: Stanford University Press.

Pollak, Charles P. and David Bright. 2003. "Caffeine Consumption and Weekly Sleep Patterns in US Seventh-, Eight-, and Ninth-graders. *Pediatrics*, 111: 42 - 47.

Pollock, Donald. 1992. "Culina Shamanism: Gender, Power, and Knowledge." Pp. 25 - 40 in *Portals of Power: Shamanism in South America*, edited by E. Jean Matteson Langdon and Gerhard Baer. Albuquerque: University of New Mexico Press.

Polsky, Ned. 1967. Hustlers, Beats and Others. Chicago: Aldine.

Ponte, Stefano. 2002. "The 'Latte Revolution'? Regulation, Markets and Consumption in the Global Coffee Chain. *World Development*, 30: 1099 - 1123.

Pope, Harrison, Jr. 1971. *Voices from the Drug Culture*. Boston: Beacon.

Poznyaka, Vladimir B., Vadim E. Pelipas, Anatoliy N. Vievskic, L. Miroshnichenkob. 2002. "Illicit Drug Use and Its Health Consequences in Belarus, Russian Federation and Ukraine: Impact of Transition." *European Addiction Research*, 8:184-189

Prasad. Eswar S. 2000. "Wage inequality in the United Kingdom, 1975-99." *IMF Staff Papers*, 49: 339 - 364.

Prebish, Charles S. (editor). 1983. *Buddhism: A Modern Perspective*. State College: Pennsylvania State University Press.

Prothero, Stephen. 1991. "On the Holy Road: The Beat Movement as Spiritual Protest." Harvard Theological Review, 84: 205 - 222.

Quindlen, Anna. 2000. "The Drug That Pretends it Isn't." *Newsweek*, April 10, 2000: 88.

Quindlen, Anna. 2000b. "America's Most Pervasive Drug Problem is the Drug that Pretends It Isn't." *Salt Lake Tribune*, April 20: A - 11.

Rawls, John. 1971. *A Theory of Justice*. Cambridge: Cambridge University Press.

Reasons, Charles. 1974. "The Politics of Drugs: An Inquiry in the Sociology of Social Problems." *The Sociological Quarterly*, 15: 381-404.

Reid, Gary and the Asian Harm Reduction Network (AHRN). 1998. *The Hidden Epidemic: A Situation Assessment of Drug Use in South East and East Asia in the Context of HIV Vulnerability*. Chiang Mai, Thailand: Asian Harm Reduction Network.

Reinarman, Craig. 1996. "The Social Construction of Drug Scares." Pp. 77 - 86 in *Deviance: The Interactionist Perspective*, edited by Earl Rubington and Martin Weinberg. Needham Heights, Ma.: Allyn and Bacon.

Reinarman, Craig, and Harry G. Levine. 1989. "The Crack Attack: Politics and Media in America's Latest Drug Scare." Pp. 115 -137 in *Images of Issues; Typifying Contemporary Social Problems,* edited by Joel Best. New York: Aldine de Gruyter.

Ritzer, George. 1999. *Enchanting a Disenchanted World: Revolutionizing the Means of Consumption*. Thousand Oaks, Ca.: Pine Forge.

Rocha-Silva, Lee. 1998. *The Nature and Extent of Drug Use among South African Young People: A Review of Research Conducted Between the Mid-1970s and the Mid-1990s* Pretoria, South Africa: CADRE (Centre for Alcohol/Drug-Related Research).

Rosenbaum, Marsha, Patricia Morgan, and Jerome E. Beck. 1998. "Ethnographic

Notes on Ecstasy Use Among Professionals." Pp. 199 - 204 in *The American Drug Scence: An Anthology*, 2nd edition, edited by James A. Inciardi and Karen McElrath. Los Angeles: Roxbury.

Rosenfeld, Richard, Steven F. Messner, and Eric P. Baumer. 2001. "Social Capital and Homicide." *Social Forces*, 2001, 80 (1): 283-309.

Rousseau, Jean-Jacques. [1762] 1979. *Emile*. Allan Bloom Translator. New York: Basic Books.

Rubin, Jerry. 1970. *Do It: Scenarios of the Revolution*. New York: Simon and Schuster.

Rubin, Vera (editor). 1975. *Cannabis and Culture*. Chicago: Aldine.

Rubin, Vera. 1975. "The 'Ganja Vision' in Jamaica." Pp. 257 - 266 in *Cannabis and Culture*, edited by V. Rubin. Chicago: Aldine.

Rublowsky, John. 1974. *The Stoned Age: A History of Drugs in America*. New York: Putnam.

Ruggiero, Vincenzo, and Nigel South. 1995. *Eurodrugs: Drug Use, Markets and Trafficking in Europe*. London: UCL Press.

Russell, Cheryl. 1993. *The Master Trend. How the Baby Boom is Remaking America*. New York: Plenum Press.

SAMHSA (Substance Abuse and Mental Health Data Archive). 2003. *Findings from the 2001 National Household Survey on Drug Abuse*. http: //www.samhsa.gov.

SAMHSA (Substance Abuse and Mental Health Services Administration). 1999. *Summary of Findings from the 1998 National Household Survey on Drug Abuse*. Rockville, MD.: Office of Applied Statistics.

SAMHSA (Substance Abuse and Mental Health Services Administration). 2002. *Results from the 2001 National Household Survey on Drug Abuse: Volume I. Summary of National Findings* (Office of Applied Studies, NHSDA Series H-17, DHHS Publication No. SMA 02-3758). Rockville, MD.

Sampson, Robert J. 2001. "Crime and Public Safety: Insights from Community-Level Perspectives on Social Capital." Pp. 89 - 114 in *Social Capital and Poor Communities*, edited by Susan Saegert, J. Phillip Thompson, and Mark R. Warren. New York: Russell Sage Foundation.

Sampson, Robert J., Stephen W. Raudenbush, and Felton Earls. 1997. "Neighborhoods and Violent Crime: A Multilevel Study of Collective Efficacy." *Science*, 277: 918 - 924.

Sargent, Margaret. 1990. "Drinking and the Perpetuation fo Social Inequality in Australia." Pp. 499 - 519 in *The Socio-cultural Matrix of Alcohol and Drug Use: A Sourcebook of Patterns and Factors*, edited by Brenda Forster and Jeffrey Colman Salloway. Lewiston, NY: Edwin Mellen.

Savelsberg, Joachim J. 1994. "Knowledge, Domination, and Criminal Punishment." *American Journal of Sociology*, 99: 911-943.

Schaefer, Stacy B. 2000. "The Crossing of the Souls: Peyote, Perception and Meaning among the Huichol Indians." Pp. 138 - 168 in *People of the Peyote: Huichol Indian History, Religion and Survival*, edited by Stacy Schaefer and Peter Furst. Albuquerque: University of New Mexico Press.

Schaefer, Stacy B., and Peter T. Furst. 2000. "Introduction." Pp. 1 - 25 in *People of the Peyote: Huichol Indian History, Religion and Survival*, edited by Stacy Schaefer and Peter Furst. Albuquerque: University of New Mexico Press.

Schultes, Richard Evans, and Albert Hofman. 1979. *Plants of the Gods: Origins of Hallucinogenic Use*. New York: McGraw-Hill.

Scoditti, Giancarolo M. G. 1989. *Kitawa: A Linguistic and Aesthetic Analysis of Visual Art in Melanesia*. New York: Mouton de Gruyter.

Seligson F. H, D. A. Krummel, and J. L. Apgar. 1994. "Patterns of Chocolate Consumption" *American Journal of Clinical Nutrient*, 60 (6 supplement): 1060S-1064S.

Shalev, Michael. 1989. "Jewish Organized Labor and the Palestinians: A Study of State/Society Relations in Israel" Pp. 93 - 133 in *The Israeli State and Society: Boundaries and Frontiers*, edited by Baruch Kimmerling. Albany: State University of New York Press

Shaw, Clifford R. and Henry D. McKay. [1942] 1972. *Juvenile Delinquency and Urban Areas*, 3rd edition. Chicago: Chicago University Press.

Shaw, John. 2001. *The Old Pubs of Lichfield*. Walsall, England: WM Print.

Shenon, Philip. 1998. "Asia's Having One Huge Nicotine Fit." Pp. 57 - 60 in *The American Drug Scene: An Anthology*, second edition, edited by James Inciardi and Kenneth McElrath. Los Angeles: Roxbury..

Shoemaker, Pamela J., Wayne Wanta, and Dawn Leggett. 1989. "Drug Coverage and Public Opinion, 1972-1986." Pp. 67-80 in *Communication Campaigns About Drugs: Government, Media and the Public,* edited by Pamela Shoemaker. Hillsdale: Lawrence Erlbaum Associates.

Siegel, Ronald K. 1995. "Cocaine: Recreational Use and Intoxication." Pp. 162 - 171 in *The American Drug Scene: An Anthology*, edited by James A. Inciardi and Karen McElrath. Los Angeles: Roxbury.

Simmel, Georg. 1955. *Conflict and the Web of Group Affiliations*. Glencoe, Il.: Free Press.

Simmel, Georg. [1908] 1971. "Freedom and the Individual." Pp. 217 - 226 in *On Individuality and Social Forms*, edited by Donald N. Levine. Chicago: University of Chicago Press.

Simmel, Georg. [1908] 1971b. "Group Expansion and the Development of Individuality." Pp. 251 - 293 in *On Individuality and Social Forms*, edited by Donald N. Levine. Chicago: University of Chicago Press.

Siringi, Samuel. 2001. "More Drug Addicts in Africa Are Injecting Drugs." *The Lancet*, July 21, 358, 9277: 218.

Smith, David and Douglas White. 1988. "Structure and Dynamics of the Global Economy: Network Analysis of International Trade: 1965 - 1980." Paper presented at the 83rd annual meetings of the American Sociological Association, Atlanta Georgia.

Snell, Clete. 2001. *Neighborhood Structure, Crime, and Fear of Crime: Testing Bursik and Grasmick's Neighborhood Control Theory*. New York: LFB Scholarly Publishing.

Snyder, David and Edward L. Kick. 1979. "Structural Position in the World System and Economic Growth, 1955 - 1970: A Multiple Network Analysis of Transnational Interactions." *American Journal of Sociology*, 84: 1096 - 1125.

Snyder, Solomon. 1971. *Uses of Marijuana*. New York: Oxford University Press.

Spates, James L. 1976. "Counterculture and Dominant Culture Values: A Cross-National Analysis of the Underground Press and Dominant Culture Magazines" *American Sociological Review*, 41: 868-883.

Spates, James L., and Jack Levin. 1972. "Beats, Hippies the Hip Generation and the American Middle Class: An Analysis of Values." *International Social Science Journal*, 24: 326 - 353.

Spillane, Joseph. 2000. *Cocaine : From Medical Marvel to Modern Menace in the United States, 1884-1920*. Baltimore: Johns Hopkins University Press.

Stares, Paul B. 1996. *Global Habit: The Drug Problem in a Borderless World*. Washington, D.C.: Brookings Institution.

Stephens, Richard C. 1991. *The Street Addict Role: A Theory of Heroin Addiction.* Albany, N.Y.: State University of New York Press.

Strang, John, Paul Griffiths, and Michael Gossop. 1997. "Heroin Smoking by 'Chasing the Dragon:' Origins and History." *Addiction*, 92: 673 - 683.

Sutherland, Edwin H. 1939. *Principles of Criminology*. Philadelphia: Lippincott.

Suwanwela, Charas, and Poshyachinda, V. 1986. "Drug Abuse in Asia." *Bulletin on Narcotics*, 38/1/2: 41-53.

Szasz, Thomas. 1974. *Ceremonial Chemistry: The Ritual Persecution of Drugs, Addicts, and Pushers*. Garden City: Anchor Press.

Tamura, M. 1989. *Japan: Stimulant Epidemics Past and Present*. Tokyo: National Research Institute of Police Science.

Tedlock, Barbara. 2001. "Divination as a Way of Knowing: Embodiment, Visualisation, Narrative, and Interpretation." *Folklore*, 112: 189 - 207.

Ter Bogt, Tom, Rutger Engels, Belinda Hibbel, Stijn Verhagen, and Frits Van Wel. 2002. "Dancestasy: Dance and MDMA Use in Dutch Youth Culture." *Contemporary Drug Problems*, 29: 157 - 183.

Terry, Charles E. and Mildred Pellens. (1928) 1970. *The Opium Problem*. New York: Bureau of Social Hygiene.

Thompson, J. Eric S. 1970. *Maya History and Religion*. Norman: University of Oklahoma Press.

Thompson, John M. 2004. *Russia and the Soviet Union: An Historical Introduction from the Kievan State to the Present*, 5th edition. Boulder, Co.: Westview.

Thornberry, Terence P. 1987. "Toward an Interactional Theory of Delinquency." *Criminology*, 25: 863 - 891.

Thornberry, Terence P. 1996. "Empirical Support for Interactional Theory: A Review of the Literature." Pp. 198 - 235 in *Delinquency and Crime: Current Theories*, edited by J.D. Hawkins. New York: Cambridge University Press.

Thornberry, Terence P., Allen Lizotte, Marvin Krohn, M. Farnworth, and Sung Joon Jang. 1991. "Testing Interactional Theory: an Examination of Reciprocal Casual Relationships among Family, School, and Delinquency." *Journal of Criminal Law and Criminology*, 82, 3 - 35.

Thornberry, Terence P., Allen Lizotte, Marvin Krohn, M. Farnworth, and Sung Joon Jang. 1994. "Delinquent Peers, Beliefs, and Delinquent Behavior: A Longitudinal Test of Interactional Theory." *Criminology*, 32:47 - 83.

Toch, Evelyn. 1990. "Rapidly Developing Alcoholism in Oriental Jews: A Pilot Study of Enzymes and Attitudes." Pp. 486 - 497 in *The Socio-cultural Matrix of Alcohol and Drug Use: A Sourcebook of Patterns and Factors*, edited by Brenda Forster and Jeffrey Colman Salloway. Lewiston, NY: Edwin Mellen.

Trevino, A. Javier. 1992. "Alcoholics Anonymous as Durkheimian Relgion." *Research in the Social Scientific Study of Religion*, 4: 183 – 208.

Tucker, David. 1984. *The World Health Market: The Future of the Pharmaceutical Industry*. New York: Facts on File.

Turner, Michael and Alex R. Piquero. 2002. "The Stability of Self-Control." *Journal of Criminal Justice*, 30: 457 - 472.

UN (United Nations). 1948. *Statistical Yearbook*. New York: United Nations Publications.

UN (United Nations). 1952. *Statistical Yearbook*. New York: United Nations Publications.

UN (United Nations). 1958. *Statistical Yearbook*. New York: United Nations Publications.

UN (United Nations). 1963. *Statistical Yearbook*. New York: United Nations

Publications.

UN (United Nations). 1965. *Statistical Yearbook*. New York: United Nations Publications.

UN (United Nations). 1968. *Statistical Yearbook*. New York: United Nations Publications.

UN (United Nations). 1970. *Statistical Yearbook*. New York: United Nations Publications.

UN (United Nations). 1972. *Statistical Yearbook*. New York: United Nations Publications.

UN (United Nations). 1974. *Statistical Yearbook*. New York: United Nations Publications.

UN (United Nations). 1981. *Statistical Yearbook*. New York: United Nations Publications.

UN (United Nations). 1985. *Statistical Yearbook*. New York: United Nations Publications.

UN (United Nations). 1988. *Statistical Yearbook*. New York: United Nations Publications.

UN (United Nations). 1991. *Statistical Yearbook*. New York: United Nations Publications.

UN (United Nations). 1994. *Statistical Yearbook*. New York: United Nations Publications.

UN (United Nations). 2000. *Statistical Yearbook*. New York: United Nations Publications.

UNODC (United Nations Office on Drugs and Crime). 2004. *World Drug Report 2004*. Oxford, N.Y.: Oxford University Press.

UNODC (United Nations Office on Drugs and Crime). 2003. *Global Illicit Drug Trends*. New York: United Nations Publications.

UNODC (United Nations Office on Drugs and Crime). 2003b. *Caribbean Drug Trends 2001 - 2002*. Bridgetown, Barbados: UNODC Caribbean Regional Office.

UNODC (United Nations Office on Drugs and Crime). 2002. *World Drug Report 2002*. Oxford, N.Y.: Oxford University Press.

UNODC (United Nations Office on Drugs and Crime). 2002b. *Global Illicit Drug Trends*. New York: United Nations Publications.

UNODC (United Nations Office on Drugs and Crime). 2000. *World Drug Report 2000*. Oxford, N.Y.: Oxford University Press.

UNODC (United Nations Office on Drugs and Crime). 1999. *Global Illicit Drug Trends*. New York: United Nations Publications.

UNODC (United Nations Office on Drugs and Crime). 1999b. *The Drug Nexus in Africa*. Vienna: United Nations Publications.

UNODC (United Nations Office of Drugs and Crime). 1987. Review of Drug Abuse and Measures to Reduce the Illicit demand for Drugs By Region. *Bulletin on Narcotics*. 39/1/1: 3 - 30.

USDEA (United States Drug Enforcement Administration) 2001. *Ecstasy Rolling Across Europe*. Alexandria, Va.: Strategic Intelligence Unit of the Office of Strategic Intelligence.

USDEA (United States Drug Enforcement Administration). 2003. *Drug Intelligence Brief. Mexico: Country Profile, 2003*. (Publication Number DEA-03047) Alexandra, Va.: Mexico/Central America Strategic Intelligence Unit of the Office of Strategic Intelligence.

USDEA 2003b. Drug Intelligence Brief. Guatemala: Country Profile. (Publication Number DEA-03002). Alexandra, Va.: Mexico/Central

America Strategic Intelligence Unit of the Office of Strategic Intelligence.

USDEA 2003c. Drug Intelligence Brief. Costa Rica: Country Profile. (Publication Number DEA-03045) Alexandra, Va.: Mexico/Central America Strategic Intelligence Unit of the Office of Strategic Intelligence.

United States Bureau of the Census. 1975. *Historical Abstract of the United States*. Washington, D.C.: Author.

United States Bureau of the Census. 2000. "Population Estimates Program, Population Division," http://eire.census.gov/popest/archives/pre1980/popclockest.txt

United States Bureau of the Census. 2003. "Poverty Status of People by Family Relationship, Race, and Hispanic Origin: 1959 to 2001" http://www.census.gov/hhes/poverty/histpov/hstpov2.html

United States Treasury Department, Bureau of Narcotics. 1939. *Traffic in Opium and Other Dangerous Drugs*. Washington, D.C.: U.S. Government Printing Office.

Unwin, Tim. 1991. *Wine and the Vine: An Historical Geography of Viticulture and the Wine Trade*. London: Routledge.

van de Wijngaart, Govert F. 1990. "Heroin Addiction in the Netherlands." Pp. 521 - 551 in *The Socio-cultural Matrix of Alcohol and Drug Use: A Sourcebook of Patterns and Factors*, edited by Brenda Forster and Jeffrey Colman Salloway. Lewiston, NY: Edwin Mellen.

Van Eltern, Mel. 1999. "The Subculture of the Beats: A Sociological Revisit." *Journal of American Culture*, 22 (3): 71 - 99.

Valadez, Susana. 1992. *Huichol Indian Sacred Ritual*. Oakland, Ca.: Amber Lotus.

Vaughn, Michael S., Frank F.Y. Huang, and Christine Rose Ramirez. 1995. "Drug Abuse and Anti-drug Policy in Japan: Past History and Future Directions." *British Journal of Criminology*, 35: 491-524.

Vazsonyi, Alexandor T., Lloyd E. Pickering, Marianne Junger and Dick Hessing. 2001. "An Empirical Test of a General Theory of Crime: a Four-nation Comparative Study of Self-control and Prediction of Deviance." *Journal of Research in Crime and Delinquency*, 38: 91 - 131.

Vega, William A. and Andres G. Gil. 1998. *Drug Use and Ethnicity in Early Adolescence*. New York: Plenum.

Vega, William A., Eleni Apospori, Andres G. Gil, Rick S. Zimmerman and G. J. Warheit. 1996. "A Replication and Elaboration of the Esteem-Enhancement Model. *Psychiatry*, 59, 128 - 144.

Villatoro, Jorge A., Elena Medina-Mora, Francisco Juarez, Estela Rojas, Silvia Carreno and Shoshana Berenzon. 1998. "Drug Use Pathways among High School Students of Mexico."*Addiction*, 93: 1577 - 1588.

Vitebsky, Piers. 2001. *Shamanism*. Norman: University of Oklahoma Press.

Vogal, Karen. 2003. "Female Shamanism, Goddess Cultures, and Psychedelics." *ReVision*, 25: 18 – 29.

Wallerstein, Immanuel. 1974. *The Modern World-System: Capitalist Agriculture and the Origins of the European World-Economy in the Sixteenth Century*. New York: Academic Press.

Wang, Gabe T., Hengrui Qiao, Shaowei Hong and Jie Zhang. 2002. "Adolescent Social Bond, Self-Control and Deviant Behavior in China." *International Journal of Contemporary Sociology* 39 (1): 52 - 68.

Wang, Wen. 1999. "Illegal Drug Abuse and the Community Camp Strategy in China." *Journal of Drug Education*, 29: 97 - 114.

Warren, Chalres W., Leanne Riley, Samira Asthma, Michael P. Eriksen, Lawrence Green, Curtis Blanton, Cliff Loo, Scott Batchelor, and Derek Yach. 2000.

"Tobacco Use by Youth: A Surveillance Report from the Global Youth Tobacco Survey Project" *Bulletin of the World Health Organization*. Geneva, Switzerland: World Health Organization.

Watson, Steven. 1995. *The Birth of the Beat Generation: Visionaries, Rebels and Hipsters, 1944 - 1960*. New York: Pantheon.

Weber, Max. [1919] 1946. "Science as a Vocation." Pp. 129 -156 in *From Max Weber: Essays in Sociology*. New York: Oxford University Press.

Weber, Max. 1947. *The Theory of Social and Economic Organization*. Translated by A.M Henderson and Talcott Parsons. New York: Oxford University Press.

Weber, Max. [1904-1905] 1958. *The Protestant Ethic and the Spirt of Capitalism*. Translated by Talcott Parsons. New York: Scribners.

Weber, Max. [1922] 1964. *The Sociology of Religion*. Translated by Ephraim Fischoff. Boston: Beacon Press.

Weber, Max [1922] 1978. *Economy and Society*, volumes 1 and 2. Berkeley: University of California Press.

Weir, Shelagh. 1985. *Qat in Yemen: Consumption and Social Change*. London: The British Museum Publications Limited.

Williams-Garcia, Roberto. 1975. "The Ritual Use of Cannabis in Mexico." Pp. 133 - 145 in *Cannabis and Culture*, edited by Vera Rubin. Chicago: Aldine.

Winter, Joseph. 2000. "Traditional Uses of Tobacco by Native Americans." Pp. 9 - 58 in *Tobacco Use by Native Americans: Sacred Smoke and Silent Killer*, edited by Joseph Winter. Norman: University of Oklahoma Press.

Winter, Joseph. 2000b. "From Earth Mother to Snake Women: The Role of Tobacco in the Evolution of North American Religious Organization." Pp. 265 - 304 in *Tobacco Use by Native Americans: Sacred Smoke and Silent Killer*, edited by Joseph Winter. Norman: University of Oklahoma Press.

Wood, Bruce. 1988. "Urbanisation and Local Government." Pp. 322 - 356 in *British Social Trends since 1900: A Guide to the Changing Social Structure of Britain*, edited by A. H. Halsey. London: MacMillan Press.

World Bank. 1998. *World Development Indicators 1998*. Washington, D.C.: The World Bank.

WHO (World Health Organization). 2003. "Tobacco Use Among Young People." http://www.who.int/archives/ntday98/ad908e_4.htm.

WHO (World Health Organization). 2003b. "Adult Per Capita Alcohol Consumption."http://www.who.int/whosisalcohol/alcohol _apc_data&language=english

WHO (World Health Organization). 2002. "GYTS: Results in the Americas." *Epidemiological Bulletin*, 23 (2).

WHO (World Health Organization). 2002b. *The World Health Report 2002*. Geneva, Switzerland: World Health Organization.

WHO (World Health Organization). 2000. "Worldwide Trends in Tobacco Consumption and Mortality." Http://www.druglibrary.org/shaffer/tobacco /who-tobacco.htm.

WHO (World Health Organization). 1997. *Tobacco or Health: A Global Status Report*. Geneva, Switzerland: World Health Organization.

WHO (World Health Organization). 1999. *Confronting the Epidemic: A Global Agenda for Tobacco Control Research*. Geneva Switzerland: World Health Organization.

WHO Regional Office for Europe. 1997. "Smoking, Drinking and Drug Taking in the European Region." *WHO Regional Office for Europe Bulletin*. Copenhagen, Denmark: WHO Regional Office for Europe Health

Information Unit.

WHO Regional Office for Europe. 2001a. *Highlights on Health in Hungary.* Copenhagen, Denmark: WHO Regional Office for Europe Health Information Unit.

WHO Regional Office for Europe. 2001b. *Highlights on Health in Latvia.* Copenhagen, Denmark: WHO Regional Office for Europe Health Information Unit.

WHO Regional Office for Europe. 2001c. *Highlights on Health in Bulgaria.* Copenhagen, Denmark: WHO Regional Office for Europe Health Information Unit.

WHO Regional Office for Europe. 2001d. *Highlights on Health in the Czech Republic.* Copenhagen, Denmark: WHO Regional Office for Europe Health Information Unit.

WHO Regional Office for Europe. 2001e. *Highlights on Health in Poland.* Copenhagen, Denmark: WHO Regional Office for Europe Health Information Unit.

WHO Regional Office for Europe. 2001f. *Highlights on Health in Slovenia.* Copenhagen, Denmark: WHO Regional Office for Europe Health Information Unit.

Wurmser, Leon 1980. "Drug Use as a Protective System." Pp. 71 - 74 in *Theories of Drug Abuse*, edited by Dan J. Lettieri. Rockville, MD.: National Institute on Drug Abuse.

Yi-Mak, Kam, and Larry Harrison. 2001. "Globalisation, Cultural Change and the Modern Drug Epidemics: The Case of Hong Kong." *Health, Risk and Society,* 3: 39 - 57.

Yokoyama, Minoru. 1992. "Japan: Changing Drugs Laws: Keeping Pace with the Problem." *Criminal Justice International*, 8: 11-18.

Zickel, Raymond. 1989. *Soviet Union: A Country Study.* Washington, D.C.: Federal Research Division, Library of Congress.

Index

12-step programs, 362

Achieved statuses, v, 160, 238-239, 241, 243, 245-246

Advanced Pre-Modern Societies, 144-147

Africa, 20, 25-27, 31, 35, 37, 39, 42-43, 45, 47, 50-62, 67, 76-77, 79, 107, 111, 113, 115, 116, 132, 175, 192, 197, 208, 216-217, 233, 235, 281

Alcohol (also see beer, wine, distilled spirits), iii, 4-5, 8-9, 14, 21, 22, 27, 31, 34-39, 41, 46, 49, 63, 64, 73, 76, 86, 88, 115, 125, 126, 129, 167, 169, 171-172, 174, 176, 177, 179, 192, 197-200, 204, 208-213, 215-216, 218, 221-223, 226, 229, 230, 234-237, 245, 252-253, 262, 286, 315, 333, 347, 355, 360

American Civil War, 176, 219-220

American Revolutionary War, 176

Amphetamine (ATS; also see methamphetamine), iii, 50, 51-52, 56-57, 62, 73-76, 79-82, 86, 93-94, 100-101, 104, 106-111, 115-116, 129, 175-176, 205, 206-207, 209, 211, 213-217, 234, 252-253, 275-277, 287

Anti-war movement, 346

Ascribed statuses, xiv, 150, 159, 238, 246

Asia, 24-26, 29, 35, 37-38, 42-43, 45-46, 50-52, 62-63, 65-70, 72-77, 93, 107, 110-111, 113, 115-117, 125, 166, 179, 197, 216-217

Attitudinal Theories of Drug Use, 312-315

and individualism, 324-329
and science 320-324

Barbiturates, 86, 112, 204, 210, 252, 253, 288, 333

Beats, 330-337, 339, 342-345, 347, 349, 352

Beer (also see alcohol), 7, 34-38, 64, 122, 170, 172, 174, 188, 191, 204

Behavioral Theories of Drug Use, 315-318

Benzodiazepines, 42, 61, 75, 204, 209-210

Bhang (see marijuana), 50, 64, 192

Bible, v, 98, 167, 191, 198-200

Boggs Act, 258

Bouncing Bush festival, 122, 233

Buddhism, v, 145, 196, 198

Burroughs, William 330, 334-335, 340

Cacao (see chocolote), 25-27

Caffeine, iii, vii, 8, 14, 22-23, 25-27, 31, 41, 46, 49, 88, 115, 126, 174, 204, 212, 217, 230, 252

Cali cartel, ix

Cannabis (also see marijuana), 50-51, 53, 55-57, 61, 63-66, 76-80, 83, 86-87, 89-90, 97-98, 107, 111, 115, 123, 167, 169-170, 175-176, 179-180, 191, 205, 207, 209, 211, 232-233, 243, 275-276, 288, 297, 306-307

Caribbean, 37, 96-98, 100-101, 104-106, 116

Cassady, Niel 330

Catholic Church, 127

Charas (see marijuana), 50

Cherokee Black Drink, 190, 233

Cheyenne Buffalo Ceremony, 190
Chocolate, 25-26, 28
Christianity, v, 39, 98, 127, 133, 145, 167, 191-192, 196, 200, 220, 335, 353, 356, 360
Cigarette smoking (also see tobacco), 7, 171
Civil Rights Movement, 266, 346
Club Drugs, 75
Coca leaf chewing, 102
Coca paste, 102-103, 107
Coca-Cola, 27
Cocaine, iii, viii, ix, 7, 12-14, 16, 20, 22, 47, 50-52 56-58, 75-76, 82-84, 86, 89, 91-92, 94-96, 100-104, 106-110, 112-113, 115-117, 129-130, 176-177, 179, 192, 204-207, 209-210, 213-217, 219, 221-223, 227-228, 230, 234, 236-237, 241, 244-245, 250-253, 256, 265, 268-270, 276, 306, 323, 339, 352
Codeine-based cough syrups, 75, 209
Coffee, 23-24, 218
Confucianism., v, 198
Core nations (also see developed countries), 206-207, 209, 211, 216-217, 243, 246, 250
Crack, ix, 12, 91-92, 103, 107, 130, 209, 236-237, 245, 352
Dangerous Drugs Act, 228
Datura, 132, 167, 186, 190
Drug Enforcement Adminsitration (DEA), viii, 1
Deadheads, 334, 351
Declaration on the Guiding Principles of Drug Demand Reduction, 18
Developed countries (also see core nations), 6, 17, 24, 31, 33, 240
Developing countries (also see peripheral nations), 16, 26, 31-32, 35, 239
Deviance structure, v, 259-262, 290, 293, 296
Deviant parents, 4

Differential Association Theory, 316-317
Disenchantment, vi, x, xiii, 154, 329, 334, 336, 341, 353, 355, 357, 358
Distilled spirits (also see alcohol), 34-38
Defense of the Realm Act (DORA 40B), 228
Drugs defined, 7-10
Drug subculture, vi, 5, 126, 134, 173, 219, 329, 334, 339-342, 347-349, 351, 353, 356-357, 360
Drug Subculture as a Religion, 334-339
Drug-using peers, 4, 317, 319, 341, 347, 351, 356-357
Dutch East India Company, 24
Dylan, Bob 340, 347
Early Pre-Modern Societies, 137-142
Ecstasy (also see MDMA), 50, 57, 61, 73, 76, 79, 81-83, 86, 89, 92, 94, 96, 104, 107-112, 129, 172, 177, 179, 204-206, 209, 211, 213-217, 232, 236-237, 245, 269, 275-277
England (also see Great Britain), xvii, 23, 24, 173, 174, 178, 184, 222, 228, 272, 280, 281, 367, 368, 383, 385, 388
Europe, 20, 23-26, 29, 31, 34-37, 42-46, 50-52, 67-69, 76-88, 90, 93-94, 101, 107-108, 111, 113, 115-116, 118-119, 121, 166, 171-175, 177, 179, 205, 216-217, 221, 224, 256, 263, 290, 292, 296, 302, 304, 307
Fallanheen, 332-343
First Opium War, 71
Fly agaric mushroom, 167, 188
Ganja (also see marijuana), 50, 64, 97-98, 133, 169-172, 180
Gender gap in smoking, 33, 240
GHB, 94-95, 177
Ginsberg, Allen 330-331, 334-336, 340

Global Illicit Drug Trends, 19-21, 52, 113

Global Patterns of Illicit Drug Use, 50-52

Great Britain (drug use in; also see social mobility and drug use in Great Britain; England), vi, 82, 225, 227, 234, 241, 263, 283, 285, 289, 297, 303, 307, 384

Haight-Ashbury, 344

Hallucinogens, 11, 60-61, 87, 94, 106, 112, 166-167, 179, 187, 189, 204, 209-210, 236-237, 245, 306, 308, 333, 360

Harrison Act, 228

Hashish (also see marijuana), 50, 55, 64-65, 77, 170-171, 243, 29-300, 303

Hemp Monday, 63

Heroin, vii, ix, 8, 12, 14, 17, 20, 51, 56-57, 59, 67-72, 74-76, 81, 84-87, 93, 104, 107-110, 112, 115-118, 125, 129, 204-205, 208-211, 213-214, 222, 228-230, 236-237, 242-243, 245, 252-253, 265, 269, 271, 275, 288, 308, 333

Higher religions, iv, 145-145, 200-201, 321

Hinduism, 64, 145, 167, 191-192, 197, 200, 353, 356

Hippies, 177, 334, 337-340, 342, 344-348, 350, 352

Holy Communion, 125, 127, 133

Iboga, 60

Ice, 94, 177

Industrial revolution, 29, 218, 220-221, 224, 226

Inhalants, 60, 87, 94, 105, 112, 204-205, 210-211, 214, 232, 237, 245

Instrumentally rational action, 143, 148, 343

International Opium Convention, 227

Islam (also see Muslim), 10, 34, 38-39, 60, 63-64, 196, 199, 230, 325

Judaism /Jews / Jewish, 12, 39, 145, 167, 170-171, 181, 192, 196-198, 243, 280, 299, 302-303, 353, 356

Kerouak, Jack, 330, 334-335, 340

Ketamine, 75, 94

Khat (also see qat), 60

Kola nuts, 23, 26- 27

Law, v, ix, xiv, 8, 10, 34, 125, 133, 141, 145, 147, 151, 155-156, 158-159, 160, 163, 228, 230, 250, 255, 287, 309, 316, 338

League for Spiritual Discovery, 338,

Leary, Timothy, 338, 340

LSD, 11, 12, 57, 60-61, 88-89, 95, 107, 109, 129, 177, 204-205, 209, 230, 236-237, 245, 253, 265, 275-276, 306, 350

Marijuana (see cannabis), iii, 5, 7-8, 12, 14, 16, 18, 47, 50-56, 61-67, 72-73, 75-78, 82, 89-90, 92, 94, 96-99, 106-111, 125-126, 129, 130, 133, 161, 169-170, 172, 174-177, 179, 184, 187, 189, 191-192, 199-200, 204, 207-211, 213-217, 230, 231-233, 235-236, 244-245, 250-253, 255-257, 265, 268-269, 271, 275, 277, 303, 307-308, 315, 317, 333-334, 339, 351-352, 355, 360

MDMA (see ecstasy), 75, 95

Mechanical solidarity, 140-141

Medellin cartel, viii

Methamphetamine (amphetamine), 14, 73-75, 80-81, 94, 176, 210, 234, 236-237, 246, 265, 269, 308

Middle East, 35, 37, 39, 42, 47, 52, 62, 64, 67, 72, 75, 178, 200, 216, 234

Mitchell, Joni, 338

Modern Societies (see core nations), 147-158

Modernization, ii, iv-vi, xi-xvi, 2-3, 7, 121, 136-137, 144, 146, 148, 152, 155-158, 160, 163-165, 183, 203-204, 207-209, 211-218, 220-

222, 224-225, 229-232, 238-239, 241-242, 246-247, 249, 259, 262, 295, 308-309, 311, 319-320, 324, 328-329, 331, 337, 341-343, 354, 359, 363

Moral panics, 6, 18

Morphia (see morphine), 222, 228

Morphine, 8, 21, 87, 204, 221, 252, 256, 265, 270, 288, 299, 333

Muslim (also see Islam), 12, 38-39, 60, 133, 170-171, 216-217, 299, 353, 356

Narcotic Control Act, 258

National Household Survey on Drug Abuse, 92, 236, 243, 245

Native American Church, 124, 133, 192, 360

North Africa (and drug use), 47, 54, 57, 60-61, 233

North America, 35-37, 42-46, 50-51, 88-89, 91-94, 96, 101, 105, 107-108, 111-113, 115-116, 118, 175, 180, 186, 190, 204, 216-217, 382

Objective harm, 5, 14, 251-254

Oceania, 26, 35, 37-38, 50-52, 61, 67, 93, 101, 107-108, 110-113, 115, 216-217

Open Door Policy, 227

Opiates (also see heroin; also see morphine) 51, 59, 85, 104, 222

Opium, 64, 67-72, 84, 93, 115, 123, 176, 179, 199-200, 204, 208, 218, 220, 222, 224-225, 227-228, 234, 241, 255, 270-271, 286, 288, 299, 330

PCP, 95, 112, 204, 236-237, 265

Peripheral nations (also see developing nations), 205-207, 209, 216, 246

Peyote, 12, 123-125, 128, 133, 167-168, 179, 187, 189, 191-192, 204, 233, 333, 360

Pharmaceutical drugs, iii, ix, xiii, 8, 21-23, 40-46, 49, 96, 105, 176,

184, 204, 207, 213-215, 217, 287-288

Phishheads, 334, 351-353

Policy, i, vi, xi, xii, xv, xvi, 1, 5, 228, 306, 311, 359

Polydrug use, 129, 208-211

Polydrug users, 86, 95, 207, 210-211

Pre-Modern Societies, xii, 142-144, 150-151, 155, 158-162, 182, 184, 208, 212, 224, 233, 235, 238-239, 246, 249, 343, 345

Problem-behavior Proneness Theory, 314-315

Profane, (drugs as) iv, xii, xvi, 122-125, 127-137, 143, 147, 155, 159-161, 163, 165, 181, 184-185, 193, 198, 200, 201, 204, 311, 329, 343, 346

Psychoactive revolution, i

Qat (also see Khat), 60, 172, 174, 178, 200

Qu'ran, v, 39, 64, 199-200

Rationalization, iv, vi, x-xii, xvi, 3, 7, 121, 136-137, 142, 144-145, 147, 152, 154-158, 163-165, 181-183, 185, 193-195, 198, 200-201, 203, 213, 232, 235, 246-247, 249, 308, 311, 319, 323-324, 328-329, 331, 334, 341-343, 346, 349-350, 352-354, 357, 359, 363

Re-enchantment (drug use and), 335-336, 339, 341, 357

Regional Drug Consumption Patterns
Africa, 53-61
Asia and the Middle East, 62-76
Europe, 76-88
North America, 88-96
South America, Central America and the Caribbean, 96-107
Oceania 107-111

Rites, 122, 124, 135, 181

Ritlin, 94

Rohypnol, 94

Routine Activity Pattern Theory, 317-318

Russian Center for the Study of Drug Addiction, 306

Russian Federation, 31, 33, 80, 84, 85, 87, 209, 240, 305, 384, 386

Sacred (drugs as), iv, xii, xvi, 28, 121-125, 127-138, 143, 146-148, 155, 157, 158, 160-161, 163, 165-166, 181, 184-185, 188, 190, 197-198, 200, 204, 230, 311, 322, 329, 333, 335-336, 339-340, 342-343, 346

Sacred Pipe Ceremony, 128, 130

School Survey Project on Alcohol and Drugs, 80

Sedatives, 40, 87, 94, 112, 236-237, 246, 265, 269, 288

Self-control, ii, 3-4, 119-120, 164, 312-314, 318-319, 324, 326-327, 329

Self-control Theory, 313-314

Self-degradation Theory, 312-313

Self-esteem Theory, 312-313

Self-esteem, ii, 3, 11, 118-119, 164, 312-313, 315, 329

Shamanistic religions, iv, 66, 167, 186-187, 192-193, 197, 201, 233, 360

Shamans, 167, 186, 188-189, 233

Six-Day War, 65, 299-303

Social control, xii, xiii, xiv, xv, 7, 11-13, 120, 124, 127, 135, 141, 144-145, 156, 159-160, 174, 207, 212, 224, 230, 255, 262, 309, 318, 342

Social Learning Theory, 316-317

Social mobility, ii, v, xv, xvi, 5, 134, 140, 142, 144, 150, 152, 155, 159, 162, 238, 259-269, 271, 273-274, 277, 279-281, 283-285, 289-290, 293, 295, 297-300, 303-304, 307, 309, 311

Social Mobility and Rates of Drug Use, 259-263

Social Mobility and Drug Use in Britain, 269-285

Social Mobility and drug use in Isreal, 299-303

Social Mobility and Drug Use in Japan, 285-299

Social mobility and drug use in the Soviet Republics and Eastern Bloc Nations, 303-308

Social Mobility and Drug Use in the United States, 263-269

Solvents, 61, 75, 87, 94, 276, 288

South America, 25, 28, 35, 36, 37, 44-45, 50-52, 67, 91, 96-97, 99-106, 108, 111, 113, 115-116, 184, 188-190, 197, 208, 216-217, 302, 366, 380-381, 385-386

Southern Africa, 53-55, 57-58, 60-61

Steroids, 204

Stimulant Drug Control Law, 287

Subcultural Theories, 316-318

Subjective harm, 254-256, 309

Synthetic narcotics, 210

Tea, 24-25

The Religious Sub-functions of Drug Use, 185-192

achieve a higher state while praying or mediating, 191

divination, 189, 194

healing powers, 189

offerings to the gods, 190

possession, 98, 188, 192, 228, 286

prepare for other rituals, 191

purification rituals, 190

symbolically show devotion, 192

transcendental travel, 186, 195-197

transformation., 127, 186

visions and communication with spirits, 187

The Social Functions of Drug Use, 167-181

as Medicine, 179-181

as Recreation, 179

as Religious objects, 166-168

as Rite of Passage, 168-169

as Social Boundaries, 169-171

as Social Contract, 173-174

as Social Lubricant, 177-178

as Source of Magic, 168
to Promote Economic Production, 174-176
to Promote Efficient Warfare, 176
to Promote Solidarity, 171-173
Tobacco, iii, 1, 4-5, 8, 12, 14, 21-22, 28-33, 41, 46, 49, 54, 60, 70, 73, 88, 115, 121, 125, 128-131, 167-168, 171, 173, 180, 184, 186-192, 204, 208, 210, 212-213, 218, 230, 234-235, 239, 241-242, 252-253, 262
Tranquilizers, 40, 87, 94, 105, 112, 204-205, 236-237, 246, 265, 288
United Nations' Office of Drug Control and Prevention, 18, 19
United States (and drug use; also see social mobility and drug use in the United States), vi, ix, 4, 6, 8-9, 13-14, 17-18, 20, 22, 25-26, 29-30, 32-33, 37, 40-41, 43-45, 53, 69, 79, 88-96, 103, 107, 116, 120-121, 126, 169, 171-173, 176, 184, 204, 218-222, 224-225, 227-230, 236-237, 240-241, 244-245, 249, 255-256, 263-264, 266-270, 274, 277, 285-286, 289, 292, 296-297, 303, 307, 326, 329, 347, 367-368, 370, 374, 379-380, 384, 388, 390-391

US Office of National Drug Control Policy, 17
Valium, 42, 253
Vicodin, 94
Wars of German Unification, 82, 176
West Africa, 26, 54, 57-58
Wihgita festival, 179
Wine (also see alcohol), 7, 26-27, 34-37, 125, 127, 133, 161, 167, 179, 181, 191-192, 196, 198, 360
Woodstock, 338
World Drug Report, 19, 21, 52, 74
World Health Organization, 31, 53-54, 59
World War I, 82, 84, 176, 221, 224, 227, 274, 281
World War II, 27, 74, 175-176, 274, 281, 286, 289, 292, 299, 304
Yage, 167, 189, 233
Youth, xviii, 3-4, 20, 22, 31, 41, 46, 54-61, 64, 67, 74-75, 77-78, 80-81, 84-87, 94, 99, 102, 105-106, 109, 112, 119-120, 128, 134, 158, 169, 179, 204-205, 209-211, 213-215, 231, 236, 241, 250, 258, 265-266, 268, 275-277, 281, 284, 292, 296-298, 300, 303, 306, 308, 317, 319, 329, 333-334, 339, 345, 348-350, 352, 355-358, 360-361, 363
Zoroastrianism, 167, 191, 198

MELLEN STUDIES IN SOCIOLOGY

1. Sucheng Chan (ed.), **Social and Gender Boundaries in the United States**
2. Mathew J. Kanjirathinkal, **A Sociological Critique of Theories of Cognitive Development: The Limitations of Piaget and Kohlberg**
3. Sucheng Chan (ed.), **Income and Status Differences Between White and Minority Americans**
4. Eugene G. d'Aquili and Hans Mol, **The Regulation of Physical and Mental Systems: Systems Theory of the Philosophy of Science**
5. Frances Woods, **Value Retention Among Young Creoles: Attitudes and Commitment of Contemporary Youth**
6. Michael Stein, **The Ethnography of an Adult Bookstore: Private Scenes, Public Places**
7. Erika G. King, **Crowd Theory as a Psychology of the Leader and the Led**
8. Michael R. Ball, **Professional Wrestling as Ritual Drama in American Popular Culture**
9. Michel Peillon, **The Concept of Interest in Social Theory**
10. Pierre Hegy, **Myth as Foundation for Society and Values: A Sociological Analysis**
11. Rohit Barot (ed.), **The Racism Problematic: Contemporary Sociological Debates on Race and Ethnicity**
12. Steven Thiele, **Morality in Classical European Sociology: The Denial of Social Plurality**
13. Douglas McArthur, **Information, Its Forms and Functions: The Elements of Semiology**
14. Aysan Sev'er (ed.), **A Cross-Cultural Exploration of Wife Abuse: Problems and Prospects**
15. Xuanning Fu and Tim B. Heaton, **Interracial Marriage in Hawaii, 1983-1994**
16. Diana Joyce Fox, **An Ethnography of Four Non-Governmental Development Organizations: Oxfam America, Grassroots International, ACCION International, and Cultural Survival Inc.**
17. Mary Ann Romano, **Beatrice Webb (1858-1943) – The Socialist With a Sociological Imagination**
18. Robert Peter Siemens, **Introduction to Cultural Historical Sociology**
19. Thomas E. Jordan, **Victorian Child Savers and Their Culture: A Thematic Evaluation**
20. Jon Louis Winek, **A Qualitative Study of the Co-Construction of Therapeutic Reality: A Process and Outcome Model**
21. Madawi Al-Rasheed, **Iraqi Assyrian Christians in London: The Construction of Ethnicity**
22. Jerry Jensen and Larry Jensen, **Families-the Key to a Prosperous and Compassionate Society for the 21st Century**
23. Joseph D. Yenerall, **The Rural Elderly in America: The Old Folks at Home**

24. Herman J. Loether, **The Social Impacts of Infectious Disease in England, 1600 to 1900**

25. Frank D. Sisya, **The Political Life of a Public Employee Labor Union: Regional Union Democracy**

26. Frank W. Elwell, **A Commentary on Malthus' 1798 Essay on Population as Social Theory**

27. Lilli Perez Iyechad, **An Historical Perspective of Helping Practices Associated with Birth Marriage and Death Among Chamorros in Guam**

28. Robert A. Stebbins, **New Directions in the Theory and Research of Serious Leisure**

29. D.A. Lopez, **The Latino Experience in Omaha: A Visual Essay**

30. David Inglis, **A Sociological History of Excretory Experience: Defecatory Manners and Toiletry Technologies**

31. Bryan S. Green, **A Textual Analysis of American Government Reports on Aging**

32. Marios Christou Stephanides, **The History of the Greeks in Kentucky–1900-1950**

33. Bonnie Fisher, **Social Influences on the Writing of Marion Dane Bauer and Katherine Paterson: Writing as a Social Act**

34. Amitra A. Hodge, **The Perceptions and Experience of Undergraduate Males on a Predominantly Female Campus**

35. Max Farrar, **The Struggle for 'Community' in a British Multi-Ethnic Inner-City Area: Paradise in the Making**

36. Mary Ann Romano (ed.), **Lost Sociologists Rediscovered: Jane Addams, Walter Benjamin, W.E.B. Du Bois, Harriet Martineau, Pitirim A. Sorokin, Flora Tristan, George E. Vincent, and Beatrice Webb**

37. Ahmet F. Öncü, **A Sociological Inquiry into the History of the Union of Turkish Chambers of Engineers and Architects: Engineers and the State**

38. Stan C. Weeber, **Lee Harvey Oswald–A Socio-Behavioral Reconstruction of His Career**

39. Gila Hayim, **Instability and Cultural Configurations in the Complex Systems of Modernity–An Autopoetic Perspective**

40. Sarah N. Gatson and Amanda Zweerink, **Interpersonal Culture on the Internet–Television, the Internet, and the Making of a Community**

41. Meir Yaish, **Class Mobility Trends in Israeli Society, 1974-1991**

42. Andrew J. Weigert, **Religious and Secular Views on Endtime**

43. Tim B. Heaton, Stephen J. Bahr, and Cardell K. Jacobson, **A Statistical Profile of Mormons–Health, Wealth, and Social Life**

44. Don Swenson, **A Neo-Functionalist Synthesis of Theories in Family Sociology**

45. Fuad Baali, **The Science of Human Social Organization: Conflicting Views on Ibn Khaldun's (1332-1406) Ilm al-Umran**

46. Gregory C. Leavitt, **Incest and Inbreeding Avoidance: A Critique of Darwinian Social Science**

47. James E. Hawdon, **Drug and Alcohol Consumption as Functions of Social Structures: A Cross-Cultural Sociology**

48. Amy J. Fitzgerald, **Animal Abuse and Family Violence: Researching the Interrelationships of Abusive Power**

49. Koenraad Kortmulder and Yuri Robbers, **The Agonic and Hedonic Styles of Social Behaviour**